Physics Research and Technology

nova
science publishers
www.novapublishers.com

Physics Research and Technology

Four-Dimensional Atomic Structure and Law
Kunming Xu, PhD
2023. ISBN: 979-8-88697-989-3 (Hardcover)
2023. ISBN: 979-8-89113-062-3 (eBook)

Conjugate Construction of Quantum Optics: From Foundations to Applications
Jeffrey Zheng, PhD (Editor)
2023. ISBN: 979-8-88697-651-9 (Hardcover)
2023. ISBN: 979-8-88697-653-3 (eBook)

Ferrite Materials and Technologies
Ravi Panwar, PhD
Dharmendra Singh (Editors)
2023. ISBN: 979-8-89113-086-9 (Hardcover)

75 Years of the Pion
Igor Strakovsky, PhD
and Nicolai Popov, PhD (Editors)
2023. ISBN: 979-8-88697-917-6 (Hardcover)
2023. ISBN: 979-8-89113-164-4 (eBook)

Fields of Particles with Spin, Theory and Applications
Alina Ivashkevich, PhD, Nina Krylova, PhD, Elena Ovsiyuk, PhD,
Vasiiliy Kisel, PhD, Vladimir Balan, PhD and Viktor Red'kov, PhD
2023. ISBN: 979-8-89113-012-8 (Hardcover)
2023. ISBN: 979-8-89113-113-2 (eBook)

Introduction to Multidisciplinary Science in an Artificial-Intelligence Age: Properties of Matter: Elasticity, Permeability, Porosity, Viscosity, and Wettability
Luc Thomas Ikelle, PhD
2023. ISBN: 979-8-88697-522-2 (Hardcover)
2023. ISBN: 979-8-88697-578-9 (eBook)

More information about this series can be found at https://novapublishers.com/product-category/series/physics-research-and-technology/

Kunming Xu

Four-Dimensional Atomic Structure and Law

nova
science publishers
www.novapublishers.com

DOI: https://doi.org/10.52305/QHFH2255

NOTICE TO THE READER

The Publisher has taken reasonable care in the preparation of this book but makes no expressed or implied warranty of any kind and assumes no responsibility for any errors or omissions. No liability is assumed for incidental or consequential damages in connection with or arising out of information contained in this book. The Publisher shall not be liable for any special, consequential, or exemplary damages resulting, in whole or in part, from the readers' use of, or reliance upon, this material. Any parts of this book based on government reports are so indicated and copyright is claimed for those parts to the extent applicable to compilations of such works.

Independent verification should be sought for any data, advice or recommendations contained in this book. In addition, no responsibility is assumed by the Publisher for any injury and/or damage to persons or property arising from any methods, products, instructions, ideas or otherwise contained in this publication.

This publication is designed to provide accurate and authoritative information with regards to the subject matter covered herein. It is sold with the clear understanding that the Publisher is not engaged in rendering legal or any other professional services. If legal or any other expert assistance is required, the services of a competent person should be sought. FROM A DECLARATION OF PARTICIPANTS JOINTLY ADOPTED BY A COMMITTEE OF THE AMERICAN BAR ASSOCIATION AND A COMMITTEE OF PUBLISHERS.

Library of Congress Cataloging-in-Publication Data

Names: Xu, Kunming, author.
Title: Four-dimensional atomic structure and law / Kunming Xu.
Description: New York : Nova Science Publishers, [2023] | Series: Physics Research and Technology | Includes bibliographical references and index. |
Identifiers: LCCN 2023031742 (print) | LCCN 2023031743 (ebook) | ISBN 9798886979893 (hardcover) | ISBN 9798891130623 (adobe pdf)
Subjects: LCSH: Atomic structure. | Atomic structure--Mathematical models.
Classification: LCC QC173.4.A87 X84 2023 (print) | LCC QC173.4.A87 (ebook) | DDC 539/.14--dc23/eng/20230817
LC record available at https://lccn.loc.gov/2023031742
LC ebook record available at https://lccn.loc.gov/2023031743

Published by Nova Science Publishers, Inc. † New York

A unique perspective dedicated to the human civilization of the twenty-first century

Contents

Preface

Discovering the four-dimensional spherical structure with the law of motion is the theme of this book. Here I develop a quaternity theory to model the fundamental truth of nature. Quaternity space means four geometric elements in a sphere: two periphery points, two semicircular arcs, two hemispherical surfaces, and the inner and outer spheres, which form the basic shapes of four dimensions. The dimension stairs in the atomic sphere are the theoretical basis of electronic orbital quantization in quantum mechanics. The discovery of four space dimensions originated from the observation of molecular structures such as methane, ethane, and ethyne where anisotropic 2p orbitals within a carbon atom are more natural than orbital hybridization in explaining the orientations of their chemical bonds. However, we shall set forth the fundamental space theory first in this treatise and then apply it to the description of the organic molecules. The law of nature arose from the need to characterize the motion of an electron from one dimension to another that features general harmonic oscillation. Dynamic calculus is defined for this sinusoidal process in contrast to traditional calculus as an infinitesimal limit. Hence the theory of quaternity is founded on the secure foothold of the atomic structure with law. The math presented here is not intimidating, but still requires careful digestion due to its original conceptual abstraction. For example, we regard the imaginary number i as the gateway function of a spherical quantity, the angular velocity in general harmonic oscillation and the biological gonad of an organism.

We describe two-dimensional spacetime with the instantiation of a helium shell in rigorous mathematics and extend it to life phenomena. For examples, a husband and a wife are two dimensions of the family; and animals and plants are two kingdoms of the advanced lives. The interplay and transformation between both dimensions are the eternal theme of nature and can be abstracted as a mathematical rope. A DNA molecule, composed of space and time strands, is a rope; so is a human lineage with both sexes. We define four-dimensional space with the instantiation of a neon atom and establish its connection with eight trigrams in traditional Chinese medicine where the Taiji yin and yang fishes diagram is usually placed at the center of eight trigrams. Yin and yang fishes correspond to both 1s electrons in a helium shell; and eight trigrams correspond to the 2s2p electron octet in a neon shell. Electrons in both shells are spacetime continuous through dimension encapsulation mechanism outwards and calculus chain rules inwards. The atomic structure with law will vitalize the ancient medical philosophy.

We attribute the chirality of organic molecules to the electron configuration of the carbon center bonded by four different substituents. A carbon atom is potentially chiral because its four outer electrons $2p_x2p_y2p_z2s$ may take the various shapes of a point, a semicircular arc, a hemispherical surface, and a solid sphere, respectively. Besides optical activity, the anisotropic 2p orbitals may direct electron transfer in a selective pathway through the stereocenter in a macromolecule. This inspires me to identify a DNA molecule as a stepwise

LC oscillatory circuitry. The electronic structure of a carbon atom conforms to the five element theory in traditional Chinese medicine, and what's more, a carbon atom with six electrons is indeed a physical instance of the spherical quantity solutions to three-dimensional Schrödinger's equation. Hence we establish the connection between quaternity and quantum mechanics concerning about the atomic structure. The chiral carbon also sets a rotation model for the gene expression of a biological cell where five levels of interrelated biochemical processes are geared coherently by strict dynamic calculus of spherical quantities, the law of nature.

People are getting less patient with scientific research in this rapidly changing world, but I have to be patient in order to pin down the unfathomable atomic structure with law. Developing the theory of quaternity has become part of my life during the last decade and quaternity spacetime my perspective from which I try to reexamine the real entities in a fresh manner. This leading experience gives me confidence on the validity of quaternity theory for the atomic structure and natural pattern in general even when my research articles are repeatedly declined by established journals. From a fresh perspective, this research will modify the fundamental assumptions and perceptions of quantum mechanics. With dimensional gradients of anisotropic 2p orbitals (two periphery points, two semicircular arcs, and two hemispherical surfaces) in the X, Y, and Z directions replacing isotropic Laplacian operator, I tentatively fine tune three-dimensional Schrödinger's equation and provide alternative solutions to it. I have built the theory of quaternity with heart and soul and sincerely hope that the readers will not only find it meaningful and enlightening, but also share my joy and excitement in the exploration of the natural world.

Four-dimensional space in a sphere is a revolutionary idea. I have no doubt that this new perception will eventually change the modern spacetime worldview! Let's discuss it step by step in the text.

Kunming Xu
Xiamen University, China
May 31, 2023

About the Author

Kunming Xu
Xiamen University, China
Email: kunmingx@xmu.edu.cn.

Kunming Xu is an associate professor at Xiamen University. Born in 1965 in Xiamen, China, he received a BSc from Xiamen University in 1987 and MSc and PhD degrees from the University of Delaware at Lewes in 1997 and 2000 respectively. In addition to academic career, he had extensive industrial experience, including working as an associate engineer and a sales representative for the Xiamen Overseas Chinese Electronic Company from 1987 to 1994, and as a software developer for the U. S. Data Corporation in Richardson, Texas in 2000 and as a software engineer for eWings Technologies, Inc. in Plano, Texas in 2001.

Part I. The Atomic Structure

The world is an expression of natural laws. Scientific research in its various branches seeks not only to record particular occurrences in the world of our experience, but also to discover the general laws hidden behind the apparent phenomena. Although the laws are independent of the observers, people watching a specific event from various perspectives or in different contexts usually arrive at different depths of understanding and explanations. Some accounts are more fundamental than the others. Newton's perception of a falling apple under gravity was more fundamental than other people's casual observations. Archimedes' experience in his bathtub inspired in him a thorough understanding of the buoyancy principle. This treatise describes my thrilling discovery of dynamic calculus of spherical quantities, which is so wonderful and fascinating that I believe it reflects nature on the utmost fundamental level.

The theoretical basis of this research lies in the novel conception of time and space. In Euclidean space, it is difficult to understand a system with more than three dimensions of space. When describing the motion of an object, we tend to treat the object as a point particle and record its position in the X, Y, and Z directions at every instant of time. Such a three-dimensional curve with time is known as the world line of the object. This linear algebraic approach does solve a lot of problems in classical mechanics, but it assumes uniformity of time and hence is not feasible to characterize the oscillation of electrons where multiple levels of sinusoidal time and space are involved. Instead of adopting physical quantities in Euclidean space, here we introduce spherical quantities in quaternity spacetime. A spherical quantity assumes a fixed magnitude and a continuously varying dimension, complementary to a physical quantity. A spherical quantity characterizes electronic motion by dimension transformation instead of position movement. This nonlinear approach turns out to be extremely powerful in elucidating electronic orbitals within atoms. In this thesis, an atomic orbital refers to an unoccupied spacetime dimension in an atomic sphere. When occupied by an electron, it is called an electronic orbital. This usage is slightly different from the classical meaning of an atomic orbital.

The atomic structure is the dimensional composition of space and time within an atom and the law of nature refers to dynamic calculus that governs the transformation between two dimensions. The first part of this book explains the basic rationale of spherical quantities in dynamic calculus through the description of electronic orbitals within inert atoms. The first chapter introduces harmonic oscillation and space and time symmetry in the two-dimensional sphere of a helium atom. Time and space are dynamically represented by two electrons. The second chapter provides a geometrical model of four-dimensional spacetime with the electron octet in a neon shell as a material instance. The theory of quaternity regards space and time as inherent physical properties that cannot be divorced from physical objects themselves. In the atomic sphere, three orientations of X, Y, and Z do not mean three dimensions automatically, but rather each dimension is associated with a real electron. So the description of electrons naturally constitutes an illustration of typical spacetime properties. The third chapter characterizes spherical quantities for two-dimensional helium and four-dimensional neon in terms of vector calculus. It is helpful to understand the motion of electrons by dynamic calculus, especially their synchronization by the general Stokes' theorem. The fourth chapter unifies quantum mechanics and special relativity under the quaternity perspective, and hence demonstrates the great capacity of the new science. It seamlessly presents the elegant quaternity solution to Schrödinger's equations. By the principle of rotatory operations, the fifth chapter formulates a wave function for every electron in various inert atoms and arranges all chemical elements into a quaternity periodic table according to the property of their characteristic electrons. The table provides sharp insight into the complicated relationships between various electrons within a large atom. These five chapters establish quaternity spacetime step by step in rigorous mathematical logic. They outline the atomic structure with the pulsing rhythm of electrons in an elegant and consistent manner.

Modern physicists and chemists are too embarrassed to confess that they don't know the structure of a helium atom, the simplest molecule in existence, not to mention the organization of the world in general. Something fundamental must have been missing from traditional sciences, which prevents them from making a breakthrough in understanding the world order. Here we start by probing into the basis of the most successful theory in physics, namely classical mechanics, to find out what has been missing. Curiously, every concept has been developed at the expense of losing generality unconsciously. New knowledge of an alternative perception or complementary view will help us understand the natural pattern better. Open to us is the jungly frontier of fundamental science. So please fasten your seat belt and enjoy an exotic view as we drive through a vast, unbroken stretch of wild territory.

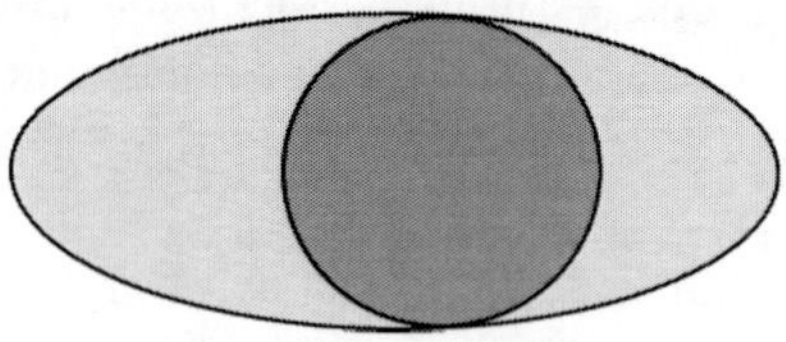

Motif 1. Circles, spheres, convolution, pupil, portal, and perception.

Chapter 1

Spherical Quantities in Two-Dimensional Spacetime

We begin by discussing some of the concepts fundamental to classical mechanics. The first concept on which to base all physics is physical quantity. Classical mechanics is the study of physical quantities and is framed by the conception of them. The second fundamental concept is three-dimensional linear space with one-dimensional uniform time. Linear space is normally modeled by a Cartesian coordinate system, and Cartesian space that satisfies the properties of scalar product operation on pairs of vectors is termed Euclidean space. Physical quantities and Euclidean space are the basic assumptions, explicit or implicit, of classical mechanics. They impart to it both strengths and weaknesses in describing the natural phenomena. The strengths are well known, but the weaknesses of the assumptions are hardly recognized. The purpose of this chapter is to dig out the roots of the weaknesses and introduce the concept of a spherical quantity in sinusoidal form to complement a physical quantity in Euclidean space. From a fresh perspective, we view both electrons within a helium atom as space and time dimensions of a two-dimensional system and characterize their harmonic oscillations by dynamic calculus. Denying what seems most obvious to anyone, we shall establish an atomic spacetime theory from scratch.

1.1. Introduction

A physical quantity has a measurable magnitude and a definite dimension, the magnitude being expressed by a certain number and the dimension by a certain unit. If we adopt one variable to account for the measurable number and another variable to account for the dimension unit, we soon discover that while the number variable could be either a continuous variable or a discrete one depending on the physical quantity of concern, the dimension variable is a discrete variable. A continuous variable can assume an infinite number of values between any two fixed points at least theoretically. For example, the water level in a dam may rise from 1.15m to 1.16m high continuously in an hour. A discrete variable has only certain fixed values, with no intermediate values possible in between. For example, the units of velocity and acceleration are m/s and m/s^2, respectively, but no physical quantity has a unit of $m/s^{1.16}$. For all physical quantities, the dimension variables take some discrete units composed of seven base units: meter, kilogram, second, ampere, kelvin, mole, and candela. For a specific physical quantity, the dimension variable is assumed to be unchangeable or static, in a fixed unit.

The basic assumption of a variable magnitude and an invariable dimension of a physical quantity, which is often taken for granted in physics, has shaped our mind in the direction of linear algebra that we can model any events by a certain combination of some physical quantities. However, physical quantities are discrete quantities in terms of their dimensions. Any combinations of finite discrete quantities cannot describe the evolution of an event in a continuous manner, inevitably leaving holes in spacetime without being covered properly. For example, a circular function cannot be expressed by a finite power series (cf. Review question 1.5). Based on the recognition of this limitation, we introduce a spherical quantity as an alternative measure to a physical quantity [1-3]. Switching the properties of both variables for a physical quantity the other way around, we get a spherical quantity – a quantity that assumes a fixed magnitude and a continuously varying dimension (Figure 1.1).

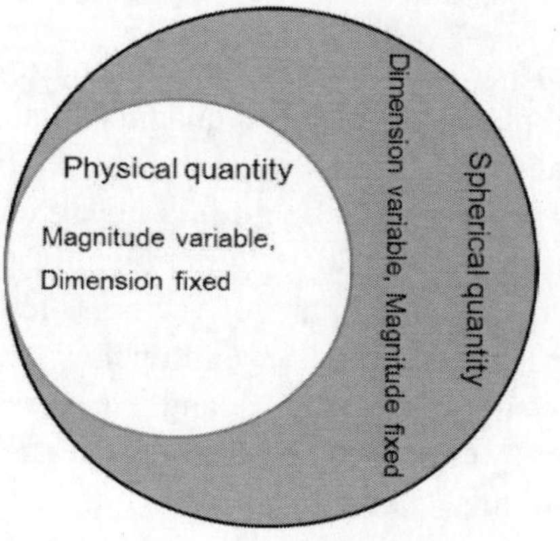

Figure 1.1. Complementary properties of spherical quantity and physical quantity.

The beginning of any science is to accept on faith something seemingly obvious. Sometimes the assumptions are explicitly declared, as is the case in geometry that is built upon several axioms; and sometimes the assumptions are taken for granted without being mentioned at all, as is the case in physics that describes physical phenomena based on physical quantities in Euclidean space. The latter case is the most dangerous adventure in science. One has to check the validity of the assumptions whenever the scope of physics exceeds that of classical mechanics. Lack of attention to this fact is the cause of the difficulties at present faced by electrodynamics and quantum mechanics. We will have to ask the readers to accept on faith the concept of spherical quantity at this stage.

The applicability of spherical quantities rests on atomic spacetime that will be explained shortly. Only when two physical quantities interconvert smoothly in certain situations is it rational to introduce a spherical quantity with a continuously varying dimension between the dimensions of both physical quantities. This is the case in electromagnetic waves where a changing electric field induces a magnetic field, and a changing magnetic field induces an electric field. The dimensions of electric and magnetic field strengths differ by a velocity factor, which may be ascribed to the interval between space and time dimensions. We shall define these two basic dimensions within a helium shell in rigorous mathematics. Before embarking on the journey, it is helpful for us to examine the conventional space and time concepts, their implicit assumptions, and limitations associated with those assumptions.

1.2. Euclidean Space and Newtonian Time

Since antiquity, human beings have believed that they know about space and time because of their direct experience, but the scientific conception of space and time turns out to be elusive. While their meanings seem intuitively clear, attempts to define them encounter remarkable difficulty. This is not because space and time are so complex but they are so fundamental that there is not any preceding rule for reference. Like the stage of an act, the theme of a song, or the context of a paragraph, space and time are the foundation and background of all sciences. Their importance can never be overestimated. How we define space and time determines our perspective and standpoint in making observations of the world around us. It goes without saying that any successful theories must handle the basic concepts properly. Significant examples are Newtonian mechanics and Einstein's relativity.

The ancient Egyptians learnt how to measure land and construct pyramids thousands of years ago. By Newton's time (1642-1727), the Cartesian coordinate system had been set up to account for three-dimensional space and people had been pretty good at elementary Euclidean geometry. But what is about time? Is it one-dimensional? Isaac Newton thought about this and defined that "Absolute, true and mathematical time, of itself, and from its own nature, flows equably, without relation to anything external." Given such a postulate, he developed three laws of motion that form the core of classical mechanics. Simple as it was, yet it has been the most practical and fruitful interpretation of time ever since.

Space and time had customarily been considered to be separate physical quantities until the 1900s when Minkowski (1864-1909) proposed a four-axis spacetime continuum, in which a time dimension is coupled together with three space dimensions through events. Minkowski's four-axis coordinates became the framework for Einstein's special relativity, which explains that space is relative, and time is relative, too. They are relative in the sense that when an object travels at a high speed, its time dilates while length contracts compared with those of rest objects. Such an unusual conclusion is demonstrated by Lorentz's transformation connecting an inertial reference frame and another frame moving at a constant velocity relative to it.

In the general theory of relativity, Albert Einstein (1879-1955), whose insight penetrated the sky, further proposed that spacetime curves in the vicinity of stars and other massive objects and came up with an equation relating the curvature tensor of the distance function to the distribution of matter and energy in spacetime.

In spite of the progress, the modern conception of space and time is still framed by Euclidean space and Newtonian time. Since Euclidean geometry does not incorporate time into its three-dimensional Euclidean space, time remains isolated from space. Einstein's innovation does not surpass this framework either. In both special and general relativities, timelike and spacelike dimensions are clearly distinguishable. Time is assumed to be one-dimensional and hence is not considered to be the counterpart of or in symmetry with space. While three-dimensional Euclidean space abstracted by a Cartesian coordinate system is rectilinear, one-dimensional Newtonian time recorded by the ticking of a mechanical clock is circular. Both are orthogonal in their measures.

Moreover, both Euclidean space and Newtonian time are dissociated from real entities. Euclidean space is modeled by Cartesian coordinates that are independent of any real objects. Because the earth is a solid sphere, positioning a rectangular coordinate system in it is cumbersome. A human body has so many organs and chambers that positioning a Cartesian coordinate system in the heterogeneous phases does not make any sense. On the other hand, Newtonian time is also dissociated from any circadian organisms, the natural subjects of uniform time sensing but sinusoidal time recording. The rotation and revolution of the earth are more important in determining four seasons than any accurate mechanical clock, which counts exactly 24 hours day in, day out. The detachment of space and time and their dissociation from reality must have limited their applications in the real world.

The applicability of classical mechanics based on Euclidean space and Newtonian time is a good tester of the validity of both space and time physical quantities. Classical mechanics characterizes the motion of objects on the ground to a certain precision, but it breaks down when applied to a large astronomical distance. For example, objects are no longer traveling in a straight line as predicted by Newton's first law of motion; instead geodesics represent the paths of freely falling objects in a given cosmic space. Relativity takes over in explaining the discrepancy. This indicates that the universe is curved or spherical that defies linear characterization.

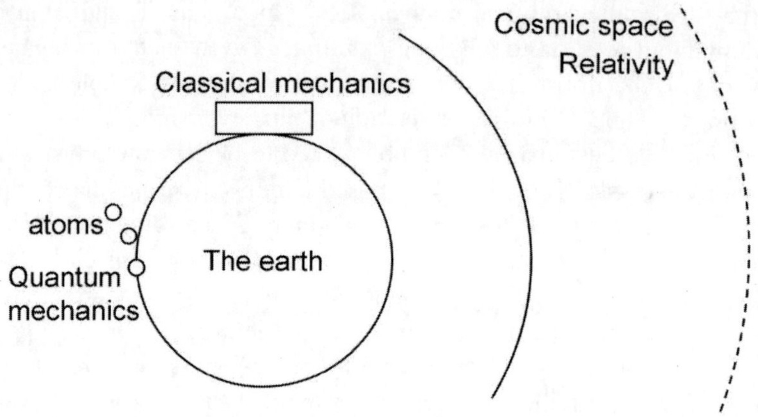

Figure 1.2. Between the scales of atoms and cosmic space, classical mechanics only survives in the narrow crevice.

Turning our focus to the microscopic world, the principle of uncertainty precludes the application of Newton's laws either. Quantum mechanics has to be called for to study the behavior of electrons. Three-dimensional Schrödinger's equation, the basis of quantum mechanics, could not be easily solved without transforming the wave function from Cartesian coordinates into spherical polar coordinates. Why? It is because a hydrogen atom is spherical in spacetime that defies linear characterization. If so, why don't we start with a wave equation formulated by spherical polar coordinates instead of one by Cartesian coordinates? This is exactly what we shall consider in this treatise.

From macrocosm to microcosm, the results of validity test indicate that classical mechanics barely survives in the narrow crevice between both ends (Figure 1.2). Using rectangles to measure land on the earth is only an approximation to the curved surface of the globe. The approach fails when the scale of the rectangles is larger than thousands of miles or smaller than a millimeter. In the same manner, physical quantities cannot describe the spherical nature of atoms and the universe as a whole, to which quantum mechanics and relativity should be applied, respectively.

The limitations of linear space and uniform time in the spherical world are conceivable. Unfortunately, modern physics does not relate curved cosmic spacetime to spherical atomic spacetime on different scales. While general relativity furnishes tensors to account for spacetime curvatures, quantum mechanics adopts statistical probability to describe electronic orbitals. Departing from classical mechanics in the contrary directions, both fields remain detached and to be unified. Since space and time are the most fundamental physical quantities, the limitations of classical mechanics and the lack of unification of relativity and quantum mechanics prompt us to search for new space and time concepts on a more profound level and in a more practical context.

1.3. Two-Dimensional Spacetime

If in the atomic microcosm or in the universe, space and time on certain scales are nonlinear, given that the validity test of classical mechanics has indicated so, then we need to be cautious on any presumptions that we have inadvertently introduced after Euclidean space or Newtonian time. Let's discard every antecedent belief and premise except saying that space and time are a pair of fundamental spherical quantities. Let's abandon every framework or stereotype and start to build conceptual models from scratch!

Our common sense is inclined to support two first-degree spherical quantities, which may refer to up and down, positive and negative, yin and yang, male and female, on and off, yes and no, etc. Many entities, such as a plant leaf, a cashew nut, a human body, and a planet, exhibit two symmetrical sides or two contrary poles. Technically, to describe a certain property, one must establish a reference first, e.g., up position is relative to down, north is contrary to south, and male is opposite to female. Hence at least two quantities are needed as fundamental spherical quantities to explain an event. Using one quantity as a reference, the other as a ruler, both spherical quantities can sufficiently measure everything in the world. We have seen that binary numbers can actually express everything that a computer or robot does, the capabilities of which appear unlimited. Given space and time quantities, any third quantity could be derived or computed from them through certain combinations or transformations. Hence it is reasonable to assume that there are exactly two fundamental or basic spherical quantities, namely space and time.

From the above reasoning, the definitions of space and time are relative, i.e., if the first one is designated as space, then the second one as time, or the other way around, even though we tend to associate space with a volume and time with a flow. The volume and the flow are interdependent and coupled intimately since reservoir volumes at different altitudes lead to a river between them and a flow naturally accumulates into a reservoir volume.

To expatiate on spherical quantities of space and time in a helium atom with exactly two electrons, we may associate space with one electron, and time with the other, or in a more general manner as will be introduced. A helium atom is such a conservative system that both electrons should best represent two basic dimensions. It is by this approach that we explore the property of space and time through the description of atomic structure as follows.

1.3.1. General Harmonic Oscillation

Consider the motion of electrons in a helium atom as harmonic oscillation. An electron living in two-dimensional spacetime possesses space and time components. As shown in Figure 1.3, an electron is confined to a spatial sphere as well as to a temporal sphere, both being inseparable. Space and time components are inherently coupled together making the electron.

They cannot be divorced from the electron just as two sides of a coin cannot be divorced from the coin itself. Because a helium atom is inert, both 1s electrons must be undergoing a certain kind of constant oscillation without damping; otherwise, the helium shell would not be a stable system. So the basic assumption of the atomic system is simple harmonic oscillation:

$$\frac{d^2\Psi}{dt^2} = -\omega^2\Psi\,, \tag{1.1}$$

where Ψ is the spherical quantity for the time component of 1s electrons and ω denotes the angular velocity. One is tempted to compare this equation with Schrödinger's equation, which is much more complicated in expression. But bear in mind that we are dealing with spherical quantities in which space and time have non-classical meanings, so do their related parameters. We borrow the term "harmonic oscillation" from classical mechanics, but will give it a fresh meaning here! Be aware of the big difference between spherical quantities and physical quantities in the simple harmonic oscillation equation! Throughout this monograph, we define "harmonic oscillation" or "simple harmonic oscillation" as the transformation process of a spherical quantity in sinusoidal form. We shall discuss Schrödinger's equation in section 1.3.4 and suffice it to say here that the complexity of Schrödinger's equation is unnecessary in spherical quantities. The simple harmonic oscillation equation provides much useful information on electronic orbitals with its typical solution as

$$\Psi = C_1(\cos\alpha - i\sin\alpha)\,, \tag{1.2}$$

where C_1 is a dimension constant and α is a radian angle related to the time component of the electrons, which satisfies the relation of

$$-\frac{d\alpha}{dt} = \omega\,, \tag{1.3}$$

where the minus sign implies contrary aligning direction of α displacement relative to the time dimension orientation. The solution can be verified by carrying out differential operations twice upon Ψ with respect to time t pending $\omega^2 = -1$. We have to caution you again that the solution to equation (1.1) in spherical quantities is slightly different from that in physical quantities without losing generality (cf. Review Question 1.12 in Appendix I).

Similarly, since space and time are relative and symmetric, the spherical quantity for the space component of the electrons satisfies:

$$\frac{d^2\psi}{dl^2} = -\frac{1}{r^2}\psi\,, \tag{1.4}$$

$$\psi = C_2(\cos\beta + j\sin\beta)\,, \tag{1.5}$$

$$\frac{1}{r} = \frac{d\beta}{dl}\,, \tag{1.6}$$

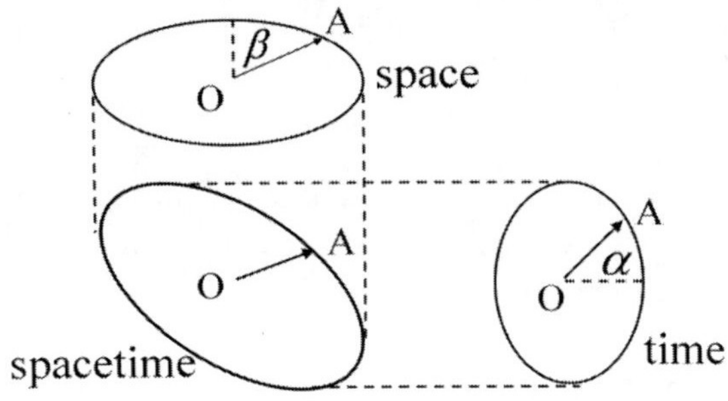

Figure 1.3. General harmonic oscillation of an electron within a helium atom where its time component is determined by the rotation of α angle and space component by the rotation of β angle. Both space and time variables are orthogonal.

$$v = \omega r\,, \tag{1.7}$$

where C_2 is a dimension constant, r is the orbital radius, v is the velocity, and j is an imaginary unit like i, but it marks an imaginary space component instead of an imaginary time component. One can verify the solution by carrying out integral operations twice upon ψ over space l pending $r = j$.

Since an electron is composed of time and space components that are orthogonal at any moment, we may express electronic orbitals by their coordination as

$$\Omega = \Psi \cdot \psi\,, \tag{1.8}$$

whence

$$\Omega = C_1 C_2 (\cos\alpha \cos\beta - i \sin\alpha \cos\beta - ij \sin\alpha \sin\beta + j \cos\alpha \sin\beta)\,. \tag{1.9}$$

Here the dot operation means coordination or association of two orthogonal spherical quantities making the electrons. However, to understand the significance of the spherical quantities, we must decipher the meanings of i and j notations in them. (In principle, a complex number is introduced when real numbers cannot express a two-dimensional vector. On the Cartesian XY plane, the unit imaginary number i is an operator casting a real number in the X-axis into an imaginary component in the Y-axis. The imaginary and the real parts of a complex number denote two perpendicular space orientations to express a vector on the plane.)

How to express various dimensions in a two-dimensional helium shell? If we start with a dimensionless spherical quantity, change it in the direction of reducing a time dimension and increasing a space dimension, and then reverse the dimension changes the other way around to complete a cycle, then we get four distinct dimensions. They are $(1, \omega, v, r)$ with SI units of (1, 1/s, m/s, m) respectively. These four dimensions represent a full set of possible combinations of one-dimensional space with one-dimensional time in a helium shell. Thus, the significance of complex notations in Ω is interpreted dimensionally as

$$\begin{pmatrix} 1 & i \\ ij & j \end{pmatrix} = \begin{pmatrix} 1 & \omega \\ v & r \end{pmatrix}. \tag{1.10}$$

How to interpret angular velocity ω as a unit imaginary number logically? First, because the angular velocity as a physical quantity has a magnitude with a dimension unit of 1/s, we may absorb its magnitude into its dimension unit in a spherical quantity, keeping its magnitude as 1. For example, if the physical quantity is 2×10^{16} Hz, then we assume 5×10^{-17} s as one spherical quantity dimension in time for the system. Second, the choice of a spherical quantity unit depends on the magnitude of ω in the system, independent of any predefined unit. Since the magnitude of ω is a characteristic of the electron, we may treat 5×10^{-17} s as a time dimension unit for the system. Thus, we may treat ω as a spherical quantity of unit magnitude with reciprocal time dimension. This is possible because we may adopt a proper dimension unit to absorb the magnitude. This unit perception of an angular velocity is important for it to serve as a unit imaginary number in mathematics consistently. Third, on the Cartesian plane with a scalar axis and a frequency axis, the number 1 at the abscissa axis multiplied by a unit dimension ω transforms into a number on the ordinate axis clockwise, and multiplied by ω twice returns the number to the abscissa axis but in the negative direction (see Figure Q3 in Appendix I). Since the unit spherical quantity ω may transform a factor into a dimension orthogonal to the factor, ω is a perfect imaginary unit i.

Since Ψ and ψ represent two orthogonal time and space components, the partial differentiation on Ω with respect to time or space dimension only affects its time or space component.

$$\frac{\partial^2 \Omega}{\partial t^2} = \psi \frac{d^2 \Psi}{dt^2}, \tag{1.11}$$

$$\frac{\partial^2 \Omega}{\partial l^2} = \Psi \frac{d^2 \psi}{dl^2}. \tag{1.12}$$

Combining equations (1.1), (1.4), (1.7), (1.11), and (1.12) yields an oscillation equation for electrons within the helium shell:

$$\frac{\partial^2 \Omega}{\partial t^2} = v^2 \frac{\partial^2 \Omega}{\partial l^2}, \tag{1.13}$$

which we shall call the duality equation. Conversely, upon separation of space and time variables, this partial differential equation would revert to two ordinary differential equations (1.1) and (1.4). There are four characteristic roots to the duality equation, representing four spherical quantities:

$$\begin{pmatrix} \Omega_0 \\ \Omega_1 \\ \Omega_2 \\ \Omega_3 \end{pmatrix} = C_1 C_2 \begin{pmatrix} \cos\alpha\cos\beta \\ -\omega\sin\alpha\cos\beta \\ -v\sin\alpha\sin\beta \\ r\cos\alpha\sin\beta \end{pmatrix}. \tag{1.14}$$

However, since there are only two electrons within a helium atom, each must comprise two adjacent roots. For example, one electron may take the form of ($\Omega_0 + \Omega_1$) while the other ($\Omega_2 + \Omega_3$). In this way, we have defined both electrons in a helium atom by four characteristic spherical quantities that observe the duality equation. Disregarding constant dimensions of C_1C_2, four spherical quantities are in space and/or time dimensions only.

1.3.2. Duality in a Helium Shell

Among four roots to the duality equation, there are strict differential or integral relationships:

$$-\frac{\partial\Omega_0}{\partial t} = \Omega_1, \tag{1.15}$$

$$\int \Omega_1 dl = \Omega_2, \tag{1.16}$$

$$-\int \Omega_2 dt = \Omega_3, \tag{1.17}$$

$$\frac{\partial\Omega_3}{\partial l} = \Omega_0. \tag{1.18}$$

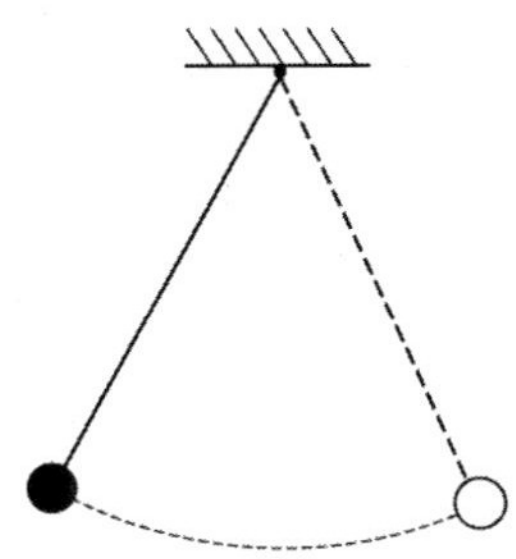

Figure 1.4. An eternal pendulum of which harmonic oscillation is the theme.

How to interpret these relationships? Each mathematical equation must have its corresponding physical process. To specify, when an electron is at the state of ($\Omega_0 + \Omega_1$), its Ω_0 component is transforming into Ω_1 component according to equation (1.15). As this process completes, the integral process of equation (1.16) starts off. Or indeed both differential and integral processes are undergoing simultaneously. In other words, one electron is shifting its state from Ω_0 to Ω_2, losing a time dimension and gaining a space dimension while the other electron is evolving from state Ω_2 to state Ω_0, increasing a time dimension and reducing a space dimension. Hence both electrons in the helium shell are not static, but changing states continuously and periodically.

By simple interpretation, the oscillation of electrons is somewhat similar to a pendulum where a body suspended from a fixed support swings freely back and forth under the influence of gravity (Figure 1.4). One significant difference is that the period of the electronic cycle is very short, say on the order of 10^{-16} s. If each electron were orbiting around the nucleus like planets around the sun kinematically as was suggested by Niels Bohr (1885-1962), then it would emit energy through Larmor radiation due to its high frequency and high acceleration. As a result, the oscillation would be damped quickly and the system could not be conserved for long. To overcome this, both 1s electron must oscillate through changing states so as to maintain total energy conservation and total momentum conservation. In this manner, the system will not lose energy to the outer environment so that the oscillatory cycle proceeds forever. Here the electron is revolving in the sense that it changes physical state continuously and periodically as a point, *A*, orbits around the origin O (Figure 1.3). The circular track of point *A* represents the course of electronic transformation or electromagnetic induction rather than kinematic displacement. This is our new interpretation of simple harmonic oscillation.

Combining equations (1.15) and (1.16) together, and (1.17) and (1.18) together yields:

$$-\frac{\partial\Omega_0}{\partial t}=\frac{\partial\Omega_2}{\partial l}, \tag{1.19}$$

$$-\int\Omega_0 dl=\int\Omega_2 dt, \tag{1.20}$$

which mean that the changing rate of one electron in time is equivalent to the gradient of the other electron in space. The derivative form conforms to Faraday's law of induction between electric field strength **E** and magnetic field strength **B**:

$$\nabla\times\mathbf{E}=-\frac{\partial\mathbf{B}}{\partial t}, \tag{1.21}$$

where the *curl* operator, $\nabla\times$, declines to $\partial/\partial l$ for spherical quantities in one-dimensional space because of electronic harmonic oscillation. Further explanation in this regard is delayed until section 2.2.2. If we treat Ω_0 as a magnetic field strength and Ω_2 as an electric field strength, then equation (1.19) is indeed an expression of Faraday's law. This indicates that electronic oscillation within the atom is an electromagnetic phenomenon. Because Ω_0 and Ω_2 are not necessarily a magnetic field strength and an electric field strength respectively, we may put it in a more general statement that both electrons are in certain states obeying the calculus relationship that defines them at a velocity dimension interval in atomic spacetime. For example, if we treat Ω_0 as a probability density function and Ω_2 as a probability current, then equation (1.19) indicates that a change in the probability density in region l is compensated by a net change in the probability current (aka probability flux) into that region.

This alternative interpretation agrees with the continuity equation for quantum mechanics, too.

Considering $\omega^2 = -1$ and $r^2 = -1$ dimensionally from their imaginary unit property, we may rewrite calculus relationships of (1.17) and (1.18) as:

$$-\frac{\partial \Omega_2}{\partial t} = \Omega_3 , \quad (1.22)$$

$$\int \Omega_3 dl = \Omega_0 . \quad (1.23)$$

The reason for their equivalence is that in the two-dimensional world, reducing a time dimension from a spherical quantity for the second times is equivalent to increasing it while increasing a space dimension from a spherical quantity for the second times is equivalent to reducing it. This is somewhat analogous to the situation of a man who walks along a circle. As he walks forward a distance of half circumference, he arrives at the other end of a diameter; but as he continues to go forward for another half circumference, he returns to the original place, forming a circle. In the same manner, equations (1.15), (1.16), (1.22), and (1.23) form a closed loop.

During the electronic transformation, the process of increasing a space dimension is always accompanied by the process of decreasing a time dimension, or vice versa. In other words, expansion of space is undergoing with release of time wrinkles whereas contraction of space results in condensation of time. When space fully unfolds, it loses all density and wraps back spontaneously according to the stipulated cycle. Space and time components are coupled together in such an intimate way that atomic spacetime is a finite continuum. This spacetime view agrees well with the theory of relativity that spells out time dilation and length contraction.

From the perspective of waves, if we regard each root to the duality equation, Ω_i , as a waveform, then each electronic orbital comprises two adjacent roots, and therefore has two waveforms. Because every pair of adjacent roots are exactly one dimension apart, separated by either a time or space dimension, both waveforms are orthogonal to each other. In other words, an electron manifests as two perpendicular waveforms. An electronic wave propagates from one waveform to another following the differential or integral rule.

How to explain electronic duality? We believe electrons are real particles, but they may not exist as solid particles in the atom all the time. No one has ever captured a single electron in its static particle form as a biologist captures a bacterium under a microscope. If we associate time with a condensed particle and space with an expanded volatile cloud, then electronic motion can be interpreted as oscillation between a point particle and a cloud medium (see section 1.5.3). An electron demonstrates wave and particle behavior because of its varying state between space and time. Space represents the volume, medium, or inertia whereas time represents the particle, density, or propensity. Thus wave and particle duality of an electron is a manifestation of space and time oscillation. A helium shell is a conservative system where energy conservation, momentum conservation, and angular momentum conservation apply for both electrons as a whole, but spacetime dimension conservation applies for each 1s electron individually even though each electron is transforming between two states sinusoidally (cf. Section 9.2).

1.3.3. Duality in an *LC* Circuit

One will not be satisfied with the foregoing abstract description of harmonic oscillation for electrons. Neither quantum mechanics nor the standard model of particles and forces provides more concrete explanation of electronic motion in the inert atom. *LC* oscillations or electromagnetic oscillations are general harmonic oscillations in their fundamental nature. We therefore investigate the property of an *LC* oscillator in this section with the hope that the readers will gain better insight into the behavior of electrons and the significance of spherical quantities.

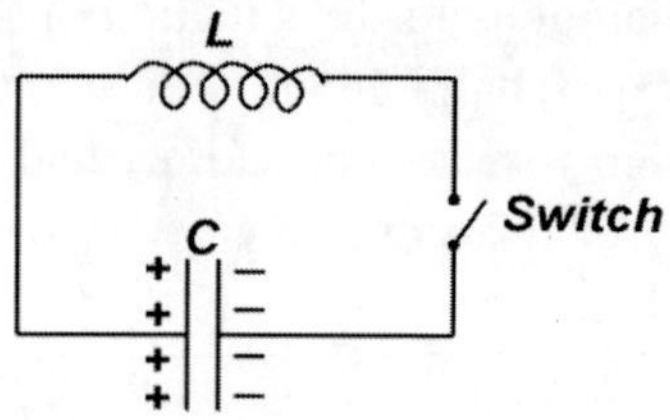

Figure 1.5. An LC circuit.

Consider an idealized circuit shown in Figure 1.5 comprising only a switch, a capacitor C, and an inductor L with N turns of a coil (and then treating N=1 for brevity), ignoring the resistance in the wire. Suppose the capacitor is initially charged so that one plate has positive charge Q and the other plate has negative charge of the same amount. As soon as the switch turns on, the capacitor discharges. The inductor initially opposes the growth of the current in the circuit by creating a change in magnetic flux through itself. The change in the magnetic flux induces an electromotive force in the circuit. After the capacitor discharges completely, the electromotive force drives a current along the circuit further, charging the capacitor in the reverse polarity. The cycle then repeats itself in the opposite direction. In the absence of any electric and magnetic energy loss, the oscillation will continue back and forth indefinitely, which may be characterized by

$$\frac{d^2q}{dt^2} = -\frac{1}{LC}q, \tag{1.24}$$

where q indicates charge in the capacitor varying in sinusoidal form:

$$q = Q\cos\varpi t, \tag{1.25}$$

$$\varpi = \sqrt{\frac{1}{LC}}. \tag{1.26}$$

In such an *LC* circuit, electric current *I*, electromotive force *V*, and magnetic flux Φ have the following basic relationships:

$$I = \frac{dq}{dt}, \tag{1.27}$$

$$V = -L\frac{dI}{dt}, \tag{1.28}$$

$$-N\frac{d\Phi}{dt} = V, \tag{1.29}$$

whence

$$I = -Q\varpi \sin \varpi t, \tag{1.30}$$

$$\Phi = -\frac{LQ\varpi}{N}\sin \varpi t, \tag{1.31}$$

$$V = LQ\varpi^2 \cos \varpi t. \tag{1.32}$$

We may describe the harmonic oscillation by spherical quantities as well as by physical quantities. Instead of electric charge, the following two complex functions satisfy the oscillation equation in general:

$$q - I = Q(\cos \varpi t + \varpi \sin \varpi t), \tag{1.33}$$

$$\Phi + V = -QL\varpi(\sin \varpi t - \varpi \cos \varpi t). \tag{1.34}$$

Here an electron may take the form of a static particle charge, participate in an electric current, transform into a magnetic flux, and build an electromotive force. We also realize that the charge is stored in the capacitor while the magnetic flux is contained inside the inductor, both electronic components being orthogonal in that a capacitor allows alternating current to pass but cuts off direct current whereas an inductor allows direct current to pass but impedes alternating current. Taking the hardware of the circuit into consideration, we may associate $(q - I)$ with $\cos\beta$ for a typical capacitor and associate $(\Phi + V)$ with $\sin\beta$ for a typical inductor, both circular functions being orthogonal.

$$f_1 = Q(\cos \varpi t + \varpi \sin \varpi t)\cos\beta, \tag{1.35}$$

$$f_2 = -QL\varpi(\sin \varpi t - \varpi \cos \varpi t)\sin\beta. \tag{1.36}$$

As an electron undergoes electromagnetic oscillation, it sets off from the capacitor, takes a path along the circuit, enters the coiled wire of the inductor, transforms into a magnetic flux through the inductor, backs into the circuit to build an electromotive force, and returns to the capacitor. Besides temporal sinusoidal waves, spherical quantities f_1 and f_2 have spatial sinusoidal waves in terms of hardware shifting as electrons travel along the wire. Since the temporal wave depends on the spatial transition of the electrons along the circuit, both temporal and spatial signals are synchronized automatically due to the electrons and hardware association along the circuit. Here the electric charge and current constitute a complex function associated with the capacitor while the magnetic flux and electromotive force form another complex function associated with the inductor (cf. Review question 1.10). Careful readers surely found that both f_1 and f_2 functions correspond to both electrons of ($\Omega_0 + \Omega_1$) and ($\Omega_2 + \Omega_3$) within a helium atom pending

$$\frac{1}{L} = -\frac{d\beta}{dl} \tag{1.37}$$

until Chapter 10. We do not imply that both electrons in the helium shell are in the same physical states as those in the idealized *LC* oscillator, but it is certain that they are undergoing analogous harmonic oscillation. Man can design an *LC* oscillator with trivial resistance in a circuit, but an idealized *LC* oscillator without any damping as was described can only be created by nature, in the form of an inert atom, and practically in a DNA molecule as well (see Chapter 7). How wonderful the nature is! How much have we known about its secret?

1.3.4. The Duality Equation vs. Schrödinger's Equation

Our description of electronic orbitals is not only self-consistent but also compatible with well-established quantum mechanics. Schrödinger's equations for the motion of a particle are the starting point for the development of quantum mechanics. The one-dimensional Schrödinger's equation for a free particle is as follows:

$$i\hbar\frac{\partial\xi}{\partial t} = -\frac{\hbar^2}{2m}\frac{\partial^2\xi}{\partial x^2}, \tag{1.38}$$

where ξ is the wave function for the particle. In order to compare this equation with the duality equation, we write down the following basic physical quantity relationships:

$$r = \frac{\lambda}{2\pi}, \omega = 2\pi f, \tag{1.39}$$

$$E = hf, p = \frac{h}{\lambda}, \tag{1.40}$$

$$\hbar = \frac{h}{2\pi}, \tag{1.41}$$

$$E = \frac{p^2}{2m}, \tag{1.42}$$

where h, λ, f, E, and p refer to Planck's constant, wavelength, frequency, energy, and momentum, respectively. In section 4.5, we shall rewrite the energy as a negative value because the electrons are trapped inside the atomic shell, but the final result is the same. For characterizing harmonic oscillation of an entity, parameters ω and r are more descriptive and pertinent than $\hbar$ and m. They are closely related to energy and momentum through the rationalized Planck's constant:

$$\omega = \frac{E}{\hbar}, \quad \frac{1}{r} = \frac{p}{\hbar}. \tag{1.43}$$

After converting parameters $\hbar$ and m into ω and r, equation (1.38) becomes

$$-\omega \frac{i\partial\xi}{\partial t} = \omega^2 r^2 \frac{\partial^2 \xi}{\partial x^2}. \tag{1.44}$$

Treating space dimensions in this equation as spherical quantities yields

$$\omega \frac{i\partial\xi}{\partial t} = v^2 \frac{\partial^2 \xi}{\partial l^2}. \tag{1.45}$$

We cancel out the minus sign because the first derivative and the second derivative of ξ are anti-parallel along dimension *l*. (i.e., $\partial l^2 = -\partial x^2$). Moreover, from wave function Ψ in equation (1.2), we obtain

$$i\Psi = C_1(i\cos\alpha + \sin\alpha) = -\frac{d\Psi}{d\alpha}, \tag{1.46}$$

which indicates that factor i may be regarded as a differential operator when placed before a physical quantity directly. The differential operator transforms its operand to the dimension orthogonal to it. Here we treat Ψ as a physical quantity instead of a spherical quantity so that $i\Psi$ is a conventional multiplication. This is different from the interpretation of factor i as angular velocity ω when placed before a trigonometric function. Based on equations (1.46) and (1.3), we have the following calculations:

$$\begin{aligned}\omega i\xi &= -\omega\frac{\partial \xi}{\partial \alpha} \\ &= \frac{\partial \xi}{\partial \alpha}\cdot\frac{\partial \alpha}{\partial t} \\ &= \frac{\partial \xi}{\partial t}.\end{aligned} \tag{1.47}$$

Substituting $\omega i\xi$ value of this equation into the left-hand side of equation (1.45) yields the duality equation. Thus the one-dimensional Schrödinger's equation is indeed another expression of the duality equation. However, the same wave equation has different significance in the concepts of spherical quantity and of physical quantity. In physical quantities, equation (1.38) is a one-dimensional wave function for a free particle. In spherical quantities, since the second derivative contains the process of the first derivative, the duality equation governs two whole dimensions, i.e., space and time dimensions. The duality equation is a two-dimensional harmonic oscillation equation for both electrons in a helium shell. Thus the duality equation and the one-dimensional Schrödinger's equation are similar in mathematical form even though they have different physical meanings.

1.4. Dynamic Calculus

The foregoing preliminary description of two-dimensional spacetime enables us to make a rigorous definition on dynamic calculus. Dynamic calculus of a spherical quantity defines an entity in harmonic oscillation. Although simple harmonic oscillation equations in spherical quantities and in physical quantities have a similar form, they have different meanings. Spherical quantities have a unique set of rules of operations such as plus, multiplication, and calculus. They also have special dimensional and trigonometric significance that deserves consideration in detail. After discarding any premises along with physical quantities, we need to establish new ground upon which to base spherical quantities firmly.

A central idea of differential calculus is the approximation of a nonlinear function by a linear one. Geometrically, one is approximating a curve or surface or similar object by a tangent line or plane or similar linear object built of straight lines. Through this approximation, questions of the calculus are reduced to ones of the algebra associated with lines and planes – linear algebra. On the contrary, dynamic calculus traverses a diameter by tracing its subtending semicircular arc, regarding the arc as the course of harmonic oscillation to cross a dimension, i.e., using a curve to approach a straight line. Spherical quantities in dynamic calculus complement physical quantities with linear algebra.

1.4.1. Infinitesimal Calculus Definition

Before introducing dynamic calculus, it is helpful for us to review the basic concept of traditional infinitesimal calculus. Calculus includes integral calculus and differential calculus, the former originating from the attempt to determine the area under a curve while the latter to measure the steepness of a curve at a certain point. It is believed that Newton and Leibniz, quite independently of one another, were responsible for developing the ideas of calculus that is hitherto so powerful to solve a variety of problems in science and engineering.

The central idea of differential calculus is the notion of derivative. The derivative of function *f* with respect to *x* is the function f' defined by

$$f'(x) = \lim_{\Delta x \to 0} \frac{f(x+\Delta x) - f(x)}{\Delta x} \tag{1.48}$$

provided the limit exists. It represents the slope of the tangent line to the graph of *f* at any point $(x, f(x))$, or the rate of change of *f* at *x*. The process of obtaining the derivative is called differentiation. Geometrically, differentiation is the approximation of a curve or surface by a tangent line or plane for calculation in linear algebra.

The integral of a continuous function $f(x)$ on the closed interval $[a, b]$ is defined by

$$\int_a^b f(x)dx = \lim_{\Delta x \to 0} \sum_{i=1}^{n} f(x_i)\Delta x \tag{1.49}$$

provided the limit exists. The interval $[a, b]$ is divided into *n* equal subintervals of width Δx, where Δx is $(b - a)/n$, and x_i is the rightmost point in the *i*th interval. According to the fundamental theorem of calculus, the definite calculus can be calculated by

$$\int_a^b f(x)dx = F(b) - F(a), \tag{1.50}$$

where *F* is any antiderivative of *f*. It is equal to the signed area bounded by lines $x = a$, $x = b$, the *x*-axis and the curve defined by function *f*.

If $f(x) = \cos x$, then according to the above definitions of infinitesimal calculus we obtain

$$f'(x) = -\sin x; \int_0^x f(x)dx = \sin x. \tag{1.51}$$

This indicates that the slope of $\cos x$ at any point *x* within the first quadrant is equal to the negative number of the area under the curve from the origin to *x* (Figure 1.6). We may also write

$$f''(x) = -f(x), \tag{1.52}$$

or

$$\frac{d^2 f}{dx^2} = -f, \tag{1.53}$$

which conforms to the equation of simple harmonic oscillation. Performing differentiation twice upon a cosine function returns it to the original cosine with a minus sign, i.e., the circular function wraps back upon repeated calculus operations. The study of cosine and sine conversion in calculus leads to the invention of dynamic calculus in stark contrast to infinitesimal calculus.

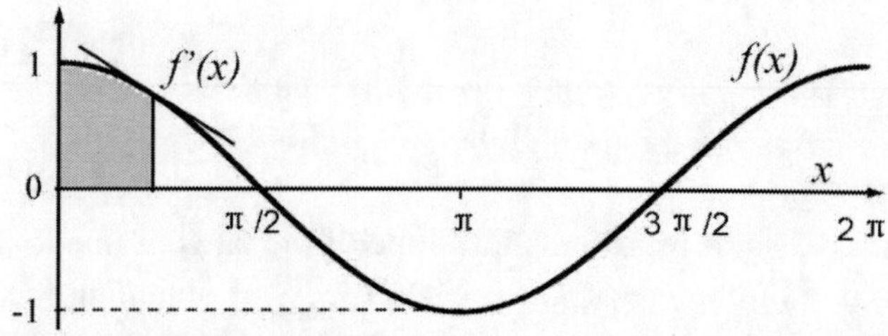

Figure 1.6. The slope of a cosine function is equal to the negative area of its corresponding shaded region in the first quadrant.

1.4.2. Dynamic Calculus Definition

Because in atomic spacetime, electronic motion follows differential and integral operations as demonstrated by equations (1.15) to (1.18) and equations (1.22) and (1.23), here we examine the significance of calculus on spherical quantities. A differential operation upon a trigonometric function with respect to a time dimension, such as that dictated by equation (1.15), does not physically happen in a flash, but it is carried out gradually and smoothly. For example, when $\cos\alpha$ receives a differentiation command, radian angle α is then rotating gradually up to ($\pi/2 + \alpha$), at which point the differential operation completes so that $\cos\alpha$ transforms into $-\sin\alpha$ accordingly. Differential and integral operations upon trigonometric functions can be interpreted as follows:

$$\frac{d}{d\alpha}\cos\alpha = \cos(\alpha + \frac{\pi}{2}), \tag{1.54}$$

$$\frac{d}{d\alpha}\sin\alpha = \sin(\alpha + \frac{\pi}{2}); \tag{1.55}$$

$$\int \cos\beta \, d\beta = \cos(\beta - \frac{\pi}{2}), \tag{1.56}$$

$$\int \sin\beta \, d\beta = \sin(\beta - \frac{\pi}{2}). \tag{1.57}$$

The correctness of these expressions can be easily verified by infinitesimal calculus, but here we regard the plus sign in the parentheses as continuous radian angle rotation up to a displacement of $\pi/2$ and the minus sign as continuous radian angle rotation down to a displacement of $-\pi/2$. In light of the special relationships between calculus and radian angle rotation, here we introduce the concept of dynamic calculus. A typical differential operation upon a trigonometric function $\Psi_0 = C_1 \cos\alpha$ can be expressed as

$$-\frac{d\Psi_0}{dt} = -C_1 \cos\alpha, where\ (\alpha \mapsto \frac{\pi}{2} + \alpha), \tag{1.58}$$

where the arrow $\mapsto$ indicates the process of radian angle rotation from α to ($\pi/2 + \alpha$). A differential operation is realized through the smooth increment of α angle from α to ($\pi/2 + \alpha$), which of course results in the change of $\cos\alpha$ term into $-\sin\alpha$ term, followed by the differential chain rule of $d\alpha/dt$ in the end. Disregarding the differential chain for a moment, we may express a differential operation in terms of radian angle rotation. Since electronic motion is a dynamic process, by treating α as a continuously changing variable, we may omit the subordinate clause in equation (1.58) and express a differential operation by a simple trigonometric function no matter whether the differential process is carried out completely or not. If it is completely done, then α increases a displacement of $\pi/2$ so that $\cos\alpha$ transforms into $-\sin\alpha$ and $-d\Psi_0/dt = -C_1\,\omega \sin\alpha$ by the differential chain rule; otherwise, the differential process is still at a state represented by the magnitude of the radian angle. Because α is a dynamic variable, the difference between both trigonometric terms ($\cos\alpha$ and $-\sin\alpha$) is a matter of difference in the value of α in the cosine function. Thus, both $\cos\alpha$ and $-\sin\alpha$ can be used to express the differentiation on the left-hand side of equation (1.58), even though we traditionally associate the final result of the differentiation with $-\sin\alpha$, and the initial condition with $\cos\alpha$. Thus, we define dynamic calculus as follows.

Definition 1. Given a spherical quantity $f(\alpha) = \cos\alpha$ or $f(\alpha) = \sin\alpha$, a differential operation on $f(\alpha)$ with respect to α means increasing variable α by a displacement of $\pi/2$ in the function.

Definition 2. Given a spherical quantity $g(\beta) = \cos\beta$ or $g(\beta) = \sin\beta$, an integral operation on $g(\beta)$ over β means decreasing variable β by a displacement of $\pi/2$ in the function.

The difference between infinitesimal calculus and dynamic calculus is that the latter is a process covering $\pi/2$ range of radian angle rotation in a trigonometric function whereas the former only captures the terminal state of the trigonometric function at a specific radian value. A rotatory radian angle of a trigonometric function is the underlying implementation of dynamic calculus whereas an infinitesimal limit of a function is the underlying implementation of infinitesimal calculus. Dynamic calculus represents a physical process rather than a mathematical limit. It is a new genus of calculus.

Spherical quantities are wave functions in sine or cosine functions. Dynamic calculus on spherical quantities can be explained in terms of translational movements of a curve. If we plot $f(\alpha) = \cos\alpha$ on the Cartesian plane, then increasing α variable by a displacement of $\pi/2$ in the function is a translational movement of $f(\alpha)$ along the abscissa axis back a distance $\pi/2$ so that $\cos\alpha$ transforms into $-\sin\alpha$ in the graph. Likewise, if we plot $g(\beta) = \cos\beta$ on the Cartesian plane, then reducing β variable by a displacement of $\pi/2$ in the function shifts the whole of $g(\beta)$ forward a distance $\pi/2$ so that $\cos\beta$ becomes $\sin\beta$ in the graph. Thus a calculus operation upon a spherical quantity may be illustrated by a phase shift of $\pm\pi/2$ in the circular function or a translational movement of its curve on the Cartesian plane.

Taking the differential chain rule into consideration beforehand, we may implement equations (1.15) and (1.16) by:

$$-\frac{d}{dt}\cos\alpha = \omega\cos(\alpha + \frac{\pi}{2})\;; \tag{1.59}$$

$$\int \cos\beta \, dl = r\cos(\beta - \frac{\pi}{2})\,. \tag{1.60}$$

As shown in Figure 1.7(a), transformation of the time component of Ω_0 from $C_1\cos\alpha$ to $-C_1\omega\sin\alpha$ can be expressed by dynamic motion of a point C along semicircle ACB. At any specific point C, chord BC denotes $C_1\cos\alpha$ while chord AC denotes $-C_1\omega\sin\alpha$ where factor ω is an imaginary unit indicating a mutually perpendicular relationship between BC and AC. As radian angle α rotates from 0 to $\pi/2$, chord BC disappears while chord AC increases up to the maximum of diameter AB, and the time component of the electron transforms from $C_1\cos\alpha$ into $-C_1\omega\sin\alpha$. This is a trigonometric interpretation of differential operation $-\partial\Omega_0/\partial t$ where the minus sign indicates time component decrement.

Likewise, Figure 1.7(b) illustrates the course of integration of equation (1.60). As radian angle β decreases from $\pi/2$ to 0, point C traces semicircle ACB from A to B. At any specific point C on the course, chord AC denotes $C_2\cos\beta$ while chord BC denotes $C_2 r\sin\beta$, both being perpendicular at any time as noted by factor r as an imaginary unit. Orthogonality means AC$\perp$BC at any moment while both spherical quantities are transforming between each other dynamically. Thus the integration of equation (1.16) is implemented by continuous β angle decrement in Ω_1 . The space component of the electron transforms from $C_2\cos\beta$ into $C_2 r\sin\beta$. Because α and β angles are complementary at any moment, equation (1.19) holds, describing the relationship between Ω_0 and Ω_2 . In this way, we have explained electronic motion by dynamic calculus and the implementation of dynamic calculus by trigonometry in the dimension diagram.

The electronic transformation can be illustrated by another diagram. As shown in Figure 1.8(a), as the electron transforms from point A toward point B along quarter arc ACB, it traverses a time dimension from a dimensionless quantity along the X-axis to the negative angular velocity dimension along the Y-axis. At any specific moment of point C on the course, the time components of both electrons are shown as both vectors $C_1\cos\alpha$ and $-C_1\omega\sin\alpha$ in the graph, respectively. Differential operation $-\partial\Omega_0/\partial t$ transforms $C_1\cos\alpha$ into its orthogonal dimension of $-C_1\omega\sin\alpha$ with factor ω as an imaginary unit. Angular velocity ω from the differential chain rule is consistent with the usage of ω as the imaginary unit that casts an operand into its orthogonal dimension.

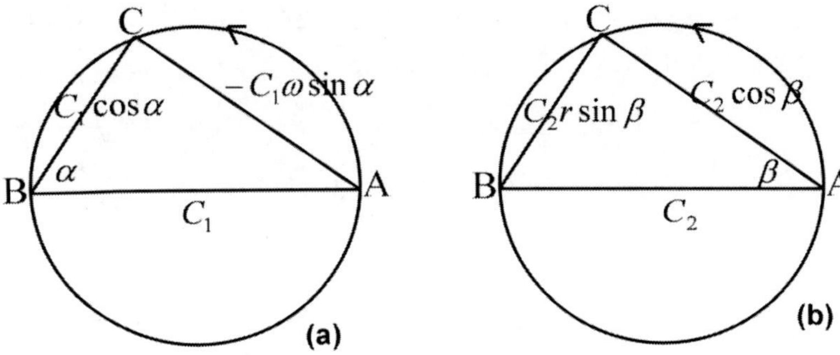

Figure 1.7. Dimension diagram of dynamic calculus of a spherical quantity for the motion of an electron with complementary (a) time and (b) space components.

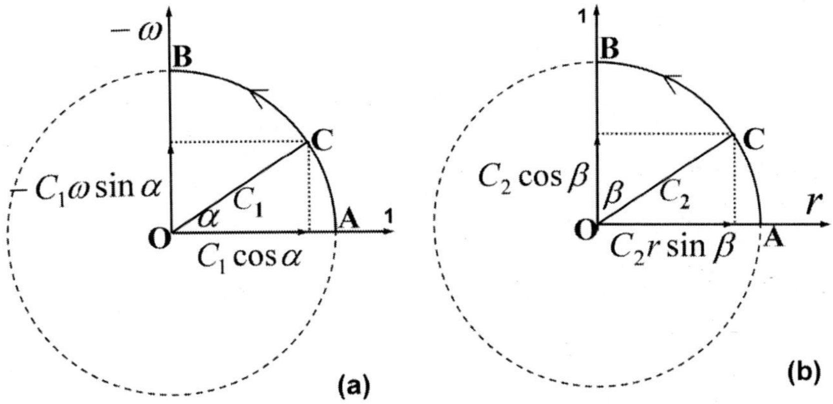

Figure 1.8. Schematic diagram of dynamic calculus of a spherical quantity for the motion of an electron in time and space.

Likewise, as shown in Figure 1.8(b), the quarter circle from A to B represents electronic transformation of $\int \Omega_1 dl$ implemented by the continuous decrement of radian angle β from π/2 to zero in the process. At any specific moment of point C on the course, radian angles α and β are complementary. The integration of $\int \Omega_1 dl$ transforms $C_2 \cos\beta$ into its orthogonal dimension $C_2 r \sin\beta$ with the orbital radius r as an imaginary unit. From points A to B, the electron increases a velocity dimension from spherical quantities Ω_0 to Ω_2 eventually.

1.4.3. Spacetime Continuity

Because a spherical quantity changes by differential and integral operations, there are interesting relationships between a spherical quantity and its first and second derivatives. As shown in Figure 1.9(a) for equation (1.1) in two-dimensional spacetime, dimension $d^2\Psi/dt^2$ must be equivalent to dimension Ψ, i.e., performing differential operations twice upon a spherical quantity returns it to the original quantity provided $\omega^2 = -1$.

Moreover, since operator d^2/dt^2 casting upon Ψ constitutes a complete loop in two-dimensional spacetime, dimensions Ψ and $-d\Psi/dt$ must be symmetric within the cycle. It follows that if Ψ_0 is a valid solution to equation (1.1), then $-d\Psi_0/dt$ must be a valid solution as well for it is an intermediate step for spherical quantity Ψ_0 to evolve toward $d^2\Psi_0/dt^2$. For example, if a solution to equation (1.1) is

$$\Psi_0 = C_1 \cos\alpha , \tag{1.61}$$

then there is another solution in derivative form $-d\Psi_0/dt$:

$$\Psi_1 = -C_1 \omega \sin\alpha . \tag{1.62}$$

Likewise, as shown in Figure 1.9(b) for equation (1.4) in two-dimensional spacetime, the solutions to it can be compactly written in a complex function of equation (1.5).

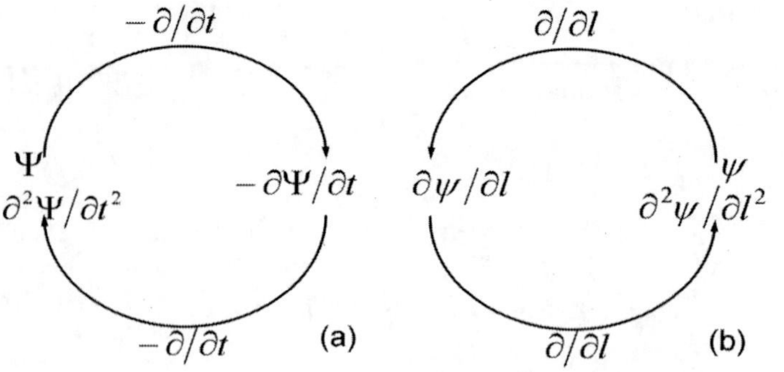

Figure 1.9. The equivalence between a spherical quantity and its second derivative and the symmetry of a spherical quantity and its first derivative for (a) time component and (b) space component of the electrons within a helium atom.

Dynamic calculus of a spherical quantity is a continuous process. As shown in Figure 1.10(a), equations (1.15) (1.16) and (1.22)(1.23) constitute a loop. And if we regard the differential and integral operations as a simultaneous process, then functions Ω_1 and Ω_3 may be deleted from the loop so that Ω_0 and Ω_2 represent both electrons in a helium atom. As shown in Figure 1.10(b), both spherical quantities have an intervening spacetime distance of reducing a time dimension while increasing a space dimension. Performing $\int(-\partial/\partial t)dl$ operation upon Ω_0 produces Ω_2 and performing the same operation upon the latter yields the former, both operations forming a cycle. Each spherical quantity is the symmetric counterpart of the other in the cycle. The implementation of dynamic calculus by smooth angle rotation bridges discrete physical quantities into continuous spherical quantities. Spherical quantities observe the rule of dynamic calculus instead of linear algebra.

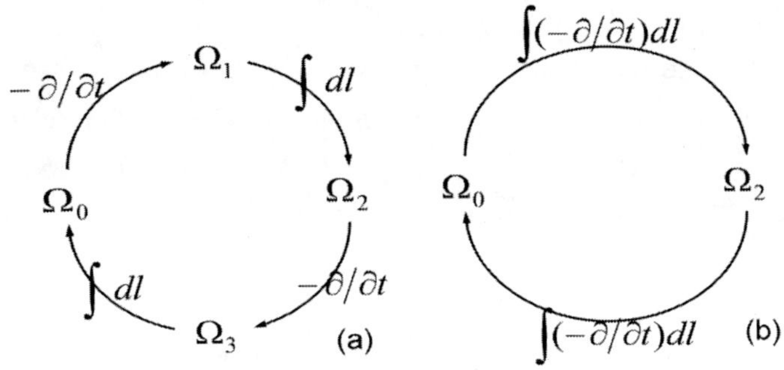

Figure 1.10. Calculus relationships (a) between four characteristic roots to the duality equation and (b) between two representative roots that form a rope structure.

During the course of a dynamic calculus operation on a spherical quantity, there is not any break, or void, or abrupt jump, in the spherical quantity but smooth transition in the physical dimension. Atomic spacetime is continuous in that space and time components are transforming smoothly and continuously from one state to another. Here the spacetime continuity has non-classical meaning. To give an analogy, when we say that a father and a son are spacetime continuous, we do not mean that they sit on the same bench and in close touch, but that the son grows from the seed of the father. Space refers to the biological bodies, not the location or volume of air that they empty off. In the same way, the atomic spacetime continuity does not mean that both electrons occupies two positions closely enough, but that both electrons are transforming from each other smoothly without any interruption even though both electrons do occupy the atomic sphere without a clear physical distinction between them. The spacetime continuity refers to electrons under smooth calculus transformation implemented by trigonometry. This concept is more fundamental and stricter than conventional function continuity at a specific position under Cartesian coordinates.

After introducing dynamic calculus and spacetime continuity concepts, we shall make a modification concerning the general solution to the duality equation. While two representative roots to the duality equation set the dimensional framework for electrons, every point of C along the semicircle (Figure 1.7) or the quarter circle (Figure 1.8) constitutes a valid solution. Because the course of point C is a continuous arc, given a specific pair of complementary α and β values, there is one solution to the duality equation corresponding to it. In other words, there are countless solutions to the duality equation and the arc constitutes the set of all solutions. Among others, two boundary points deserve further scrutiny in mathematics.

Both electrons in a helium atom transform between each other smoothly and continuously by the rule of dynamic calculus. At a specific moment, both electrons constitute two basic dimensions, space and time, separated by a phase of $\pi/2$ in their waveforms. For example, an electron Ω_0 may represent time at ($\beta = \pi/2$; $\alpha = 0$) while the other electron Ω_2 may represent space. At that specific moment, the space component of Ω_0 disappears ($C_2 \cos\beta = 0$) and the entire spherical quantity Ω_0 is equal to the time component of it ($\Omega_0 = C_1 \cos\alpha = C_1$). At the same moment, the time component of Ω_2 disappears ($C_1 \sin\alpha = 0$), and the entire spherical quantity Ω_2 is equal to the space component of it ($\Omega_2 = -vC_1C_2 \sin\alpha \sin\beta = -vC_2$) where factor $-v$ is a dimension identifier. In a spherical quantity, a zero value indicates the disappearance of the current dimension and the beginning of the next one. Dynamic calculus is a physical process. When 0< α <π/2, both electrons contain a mix of space and time components. Since both electrons in the helium shell at a certain phase (e.g., $\beta = \pi/2$; $\alpha = 0$) represent time and space dimensions respectively, they are symmetric and orthogonal as time and space dimensions are.

1.4.4. Properties of Spherical Quantity

The foregoing discussion has demonstrated three ways to express a differential operation on a spherical quantity. The first is by differential form such as $-d\Psi_0/dt$; the second is by a circular function such as $-C_1 \cos\alpha$; and the third is by a shorthand notation such as $\dot{\Psi}_0$, similar to ω or r expression as an imaginary unit or a dimension identifier. These three usages agree with each other in expressing the dynamic transformation of a spherical quantity.

By expressing $-d\Psi_0/dt$ in terms of $-C_1 \cos\alpha$, we deliver at least three messages. First, radian angle α is a dynamic variable rotating from 0 to π/2 smoothly. At any specific moment, the value of the spherical quantity is represented by $-C_1 \cos\alpha$ (Figure 1.11). Second, spherical quantity Ψ_0 changes cosinusoidally corresponding to the uniform rotation of angle α from 0 to π/2, i.e., $-d\Psi_0/dt$ is a special quotient expression with a constant denominator t and a cosinusoidal nominator. Third, operator $-d/dt$ casting on spherical quantity Ψ_0 is equivalent to operator cosine casting on uniform rotatory angle α in expressing the same differentiation. These three meanings agree with each other.

We have encountered many unconventional consequences mathematically with spherical quantities. Firstly, addition has different physical meanings from conventional one. Since electronic motion follows dynamic calculus transforming from one state to another, the plus sign in complex function ($\Omega_0 + \Omega_1$) indicates a spherical quantity at a state shifting from Ω_0 to Ω_1 as was expressed by equation (1.15). The plus sign indicates continuity of the wave functions by calculus operation. In this sense, the addition of spherical quantities ($\Omega_0 + \Omega_1$) has different physical meanings from ($\Omega_1 + \Omega_0$). Hence addition of two spherical quantities does not observe the commutative law. The atomic space is not a linear vector space. Spherical quantities are governed by the rule of dynamic calculus.

Secondly, symmetry has fresh meanings for spherical quantities. There are orbital symmetry and dimension symmetry in atomic spacetime. We consider electronic orbitals Ω_0 and Ω_2 to be symmetric because performing a $\int(-\partial/\partial t)dl$ operation upon Ω_0 results in Ω_2 and performing a $\int(-\partial/\partial t)dl$ operation upon Ω_2 results in Ω_0. This symmetry was illustrated by Figure 1.10(b). We consider space and time to be symmetric because they are two complementary dimensions, each being the counterpart of the other. Dimension symmetry was illustrated by Figure 1.9 and by equations (1.1) and (1.4).

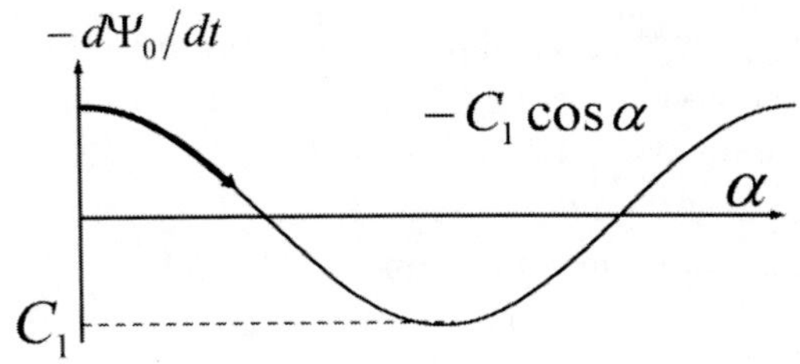

Figure 1.11. The value of a spherical quantity in dynamic differentiation. As α rotates from 0 to π/2 smoothly, the spherical quantity traces the cosinusoidal curve.

Specifically, from equations (1.15) and (1.16), we may write:

$$\int -\frac{\partial\Omega_0}{\partial t}dl = \Omega_2 . \tag{1.63}$$

This indicates that the reduction of a time dimension is accompanied by the increment of a space dimension. Space and time dimensions are complementary and exchangeable in the Ω_0 electron. Moreover, the time component of one electron is equivalent to the space component of the other in differential form if we regard both time and space dimensions as constant denominators. In addition to equation (1.19), we also have:

$$-\frac{\partial\Omega_2}{\partial t} = \frac{\partial\Omega_0}{\partial l} . \tag{1.64}$$

This quotient equivalence may be derived from equation (1.14) with calculations:

$$-\frac{\partial\Omega_2}{\partial t} = -C_1C_2v\omega\cos\alpha\sin\beta ; \tag{1.65}$$

$$\frac{\partial\Omega_0}{\partial l} = -C_1C_2\frac{1}{r}\cos\alpha\sin\beta . \tag{1.66}$$

Because $v^2 = 1$, both differential terms are equal. Equation (1.19) indicates that the time component of the Ω_0 electron is proportional to the space component of the Ω_2 electron; and equation (1.64) indicates that the space component of the Ω_0 electron is proportional to the time component of the Ω_2 electron. Thus, time and space components of both 1s electronic orbitals are symmetric and complementary.

Thirdly, orthogonality has fresh mathematical significance in atomic spacetime. Besides symmetry between Ω_0 and Ω_2, we regard both as orthogonal spherical quantities because they are exactly a time dimension and a space dimension apart as shown in equations (1.15) and (1.16). Orthogonality is defined as an exact differential and/or integral relationship between two spherical quantities. By this definition, we may deduce that four characteristic roots to the duality equation are all mutually orthogonal from equations (1.15) to (1.18).

Orthogonal quantities are coordinative, which permits dot operation between them. As shown in equation (1.8), dot operation is applied to space and time components of an electron to indicate coordination. Orthogonal quantities are interdependent and perpendicular. As was shown in Figure 1.7, chords AC and BC are perpendicular. They form two legs of a right triangle in the dimension diagram. In trigonometric functions, two orthogonal spherical quantities have a $\pi/2$ phase difference, but they are transforming from one spherical quantity to another according to dynamic calculus. This is radically different from orthogonality in Euclidean space where orthogonal vectors are supposed to be linearly independent and inexchangeable.

1.5. Linear Interpretation of Spherical Quantity

We have described the motion of 1s electrons by spherical quantities in harmonic oscillation equation, trigonometry, and dynamic calculus. The readers naturally wish to know how to relate spherical quantities in atomic spacetime to variables in Euclidean space that we are familiar with. The answer is not so straightforward as the common translation between spherical polar coordinates and Cartesian coordinates because it is about relating two systems that are rooted in fundamentally different concepts. Nonetheless, the translation between both systems is critical to the success of the atomic spacetime theory because only when it is properly integrated with conventional knowledge can it be recognized and gain its place in science. Spherical quantities must correspond to physical quantities in a certain way; and the electrons in atomic space must have their geometric shapes in Euclidean space. This section tries to establish such correspondences.

1.5.1. Spherical Quantity as a Rotatory Vector

A spherical quantity is a wave function in harmonic oscillation. Harmonic oscillation can be treated as uniform circular motion or the projection of uniform circular motion onto a certain profile. Figure 1.12 shows the vibration of a ball under the action of springs. The displacement of the ball from the equilibrium position is governed by both Newton's second law and Hooke's law

$$m\frac{d^2x}{dt^2} = -kx, \tag{1.67}$$

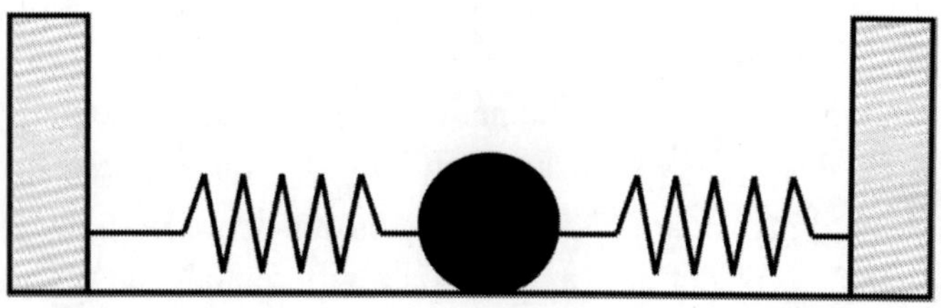

Figure 1.12. Harmonic oscillation of a ball under the influence of springs after shifting the ball a displacement from the central equilibrium position and letting it go without friction.

where k is the spring constant and m is the mass of the ball. This equation is equivalent to equation (1.1) in noting

$$\omega^2 = \frac{k}{m}. \tag{1.68}$$

Harmonic oscillation represents a large class of natural phenomena whose far-reaching implications have not been recognized yet. Regarding harmonic oscillation as uniform circular motion, we may adopt a rotatory vector **u** around a central point to describe the state of the oscillating entity. As shown in Figure 1.13, a circle is situated at a Cartesian coordinate system whose abscissa axis is of dimension C and ordinate axis of its orthogonal dimension $-iC$. Any point P located at the circle can be expressed by complex function ($C\cos\alpha - iC\sin\alpha$). Vector OP, denoted by **u**, rotates from dimensions $C\cos\alpha$ to $-iC\sin\alpha$ as α increases from 0 to π/2, i.e., the rotation transforms the vector along the abscissa axis to one along the ordinate axis. Here we are concerning about the direction change rather than the magnitude change of the vector. The length of the vector is fixed while the direction of the vector varies continuously. The direction shift represents dimension transformation in the wave function. Thus the rotatory vector represents a spherical quantity. We define the rotatory process by dynamic differentiation upon the spherical quantity with respect to a time dimension:

$$\frac{dC\cos\alpha}{dt} = -iC\sin\alpha \tag{1.69}$$

with the assignment of $i = \omega$ as a spherical quantity dimension for the system. The inherent physical property, $\sqrt{k/m}$, determines the magnitude of the spherical quantity dimension. This differential process rotates the vector from the abscissa axis to the ordinate axis on the Cartesian plane. It is implemented by the continuous counterclockwise rotation of radian angle α from 0 to π/2.

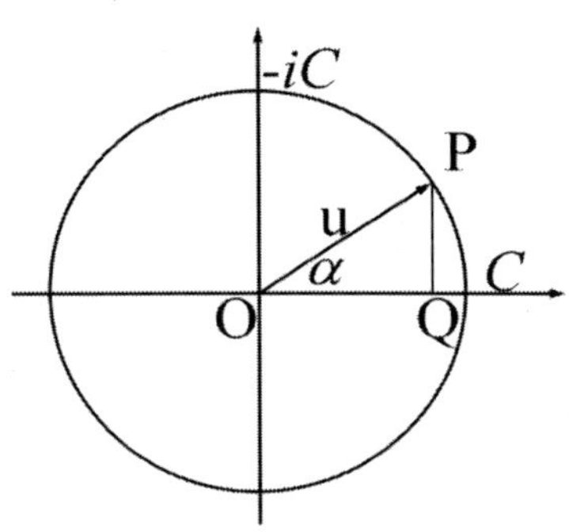

Figure 1.13. Schematic illustration of a rotatory vector to denote a spherical quantity in harmonic oscillation.

Defining differential calculus of a spherical quantity as a rotatory vector, we may express harmonic oscillation with a rotatory vector. As the vector, **u**, rotates counterclockwise, at

$\alpha = 0$, vector **u** is at its initial state corresponding to spherical quantity $\Psi = C\cos\alpha$; at $\alpha = \pi/2$, the rotatory vector reaches the ordinate axis position corresponding to the first derivative of the spherical quantity $-d\Psi/dt = -Ci\sin\alpha$; at $\alpha = \pi$, the rotatory vector returns to the abscissa axis corresponding to the second derivative $d^2\Psi/dt^2 = C\cos\alpha$, which is equal to the original function in terms of time dimensions (Figure 1.9a). In the range of $0 < \alpha < \pi/2$, the spherical quantity is transforming from Ψ toward $-d\Psi/dt$ with the tip of vector **u** sweeping along the arc in the first quadrant; between $\pi/2 < \alpha < \pi$, the spherical quantity is transforming from $-d\Psi/dt$ toward $d^2\Psi/dt^2$ with the tip of vector **u** sweeping along the arc in the second quadrant as was shown in Figure 1.14. From the trigonometric perspective, the differentiation is a gradual course or a smooth process rather than a single step jump. Thus, although $-d\Psi/dt$ is conventionally associated with the final calculation result of the differentiation, it should be understood here as the process of a position vector rotating around the origin with its tip along the arc, bridging the gap between two dimensions smoothly.

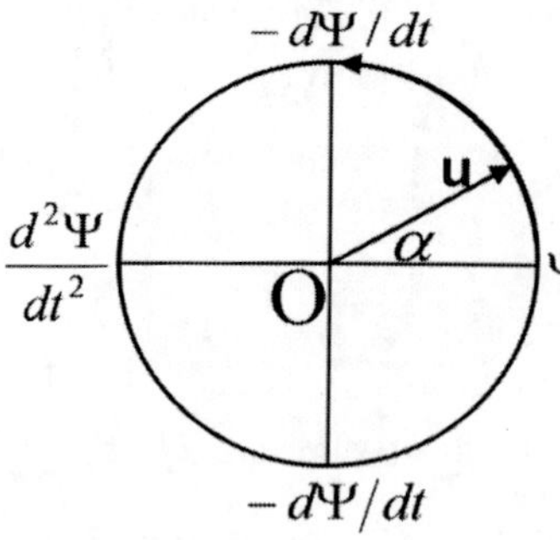

Figure 1.14. A rotatory vector u at various positions corresponding to a spherical quantity and its derivatives.

Since rotatory vector **u** indicates the state of a spherical quantity during harmonic oscillation, the physical significance of the spherical quantity can be interpreted by the vector. If a physical quantity x_0 denotes the initial maximum displacement of the ball from its equilibrium position (Figure 1.12), then the magnitude of x_0 is the norm of the vector ($|\mathbf{u}| = x_0$ in Figure 1.13) and displacement is the initial dimension of the spherical quantity ($\Psi = C = meter$ in SI unit). As the ball moves to a certain position with displacement *x*, vector **u** rotates to a position that forms an angle $\alpha = \arccos(x/x_0)$ with the abscissa axis, and the spherical quantity becomes $\Psi = (C\cos\alpha - iC\sin\alpha)$. As the ball reaches the equilibrium position (*x*=0), vector **u** reaches the ordinate axis, and the spherical quantity becomes $-d\Psi/dt = -C\omega\sin\alpha$ with $\alpha = \pi/2$. In terms of SI unit, the spherical quantity is $\Psi = -\omega C = meter/second$. At this moment, the spherical quantity denotes velocity instead of displacement. When α rotates from 0 to π/2, displacement is decreasing while velocity is increasing (see Figure 1.7a). A spherical quantity in dynamic calculus, represented by a rotatory vector **u**, describes both trends. In SI unit, the spherical quantity is

$\Psi = meter \cdot second^{-\sin\alpha}$ when $0 < \alpha < \pi/2$ with the minus sign indicating contrary alignment of ω relative to the time dimension orientation. Here we use second^{-1} to denote ω for simplicity, but actually ω is a time unit characteristic of the harmonic oscillation system. Dimension transformation of a spherical quantity from the abscissa axis to the ordinate axis accounts for the transformation between physical quantities of displacement and velocity, or the conversion between potential energy and kinetic energy of the ball under the influence of springs.

So is a spherical quantity measurable? Yes. The value of a spherical quantity is always expressed as a sine or cosine function, which is the ratio of two physical quantities. For example, in the above case, the time component of the spherical quantity is $-C\omega\sin\alpha$ with $\alpha = \arccos(x/x_0)$ where x_0 is the maximum displacement off the equilibrium position and x is the current displacement. In this case, the spherical quantity varies in its time component only, with a constant space component. A spherical quantity can be calculated from physical quantities.

Table 1.1. A property comparison between physical quantity and spherical quantity

Quantity	Physical quantity	Spherical quantity
Variable	Magnitude changes, dimension fixed	Dimension transforms, magnitude fixed
Calculus	Infinitesimal calculus	Dynamic calculus
Complex notation	The imaginary unit i	Dimension factor ω
Favorable spacetime	Euclidean space Newtonian time	Atomic spacetime
Applicable laws	The laws of physics	The law of nature
Quantification	Measurable	As a ratio of two physical quantities

Factor ω as a physical quantity of angular velocity is also a dimension factor in the spherical quantity. As expressed in equation (1.68), it describes the interaction between the ball and the spring, the struggle between the guided and the guider, and the unity of Newton's second law and Hooke's law that govern the system. As a dimension factor, $\omega^2 = -1$ indicates to and fro of an entity in harmonic oscillation, the restoration of the initial dimension after two consecutive differential operations. The differences between physical quantity and spherical quantity are summarized in Table 1.1.

1.5.2. Dimension Concepts of Space and Time

This section discusses the connection between a dimension of spherical quantity and a dimension in linear algebra. The number of dimensions in linear algebra is the number of independent vectors in space as the X, Y, and Z axes typically represent three dimensions. In our daily life, when we talk about a three-dimensional ball, it generally means that there are three mutually perpendicular orientations within it, namely the X, Y, and Z directions in Cartesian coordinates. However, if this ball is isotropic, i.e., if it contracts or expands uniformly along all directions, then we may use a radius to describes the size of the ball. Hence a radius variable may replace three variables along the X, Y, and Z axes. If this ball is a round balloon, we only have to specify its radius to define its size. Of course, we also wish to know the density or air pressure inside the balloon. Thus we need another parameter in addition to a radius. Assuming that the air inside the balloon is an ideal gas that distributes homogeneously over the entire balloon at any moment, we may introduce a density variable to describe the pressure of the air. When the balloon expands or contracts continuously in a sinusoidal way, there is a one-to-one point correspondence between the density and time duration. So we may use the density variable to represent time during the continuous process. Thus we have exactly two variables, a radius and a density, to describe a balloon satisfactorily. The radius denotes space while the density denotes time. Can we say that the radius and the density constitute two dimensions of the balloon?

What is a dimension? Traditionally, it is assumed that three axes in a Cartesian coordinate system represent three space dimensions, and hence a real number axis along the X direction indicates a dimension. If we can establish that a radius or density variable has a one-to-one point correspondence with a real number axis, then the variable would qualify as a dimension. For an electron within a helium atom, a radius and a density meet that strict condition. An electron cloud is not an ideal gas. As shown in Figure 1.15, suppose within a helium atom, an electron cloud is expanding and contracting harmonically in a periodic cycle where the length of its radius OB changes sinusoidally with point B tracing the circle Q. Segment OA is the maximum value of the radius.

If we set up a real number axis tangential to circle Q at point O, which is the origin of the axis, and draw a line AB and extend it to the axis at point B', then we may establish a one-to-one point correspondence between the circle OBA and the real number axis. For each point B along the circle, there is a point B' at the axis corresponding to it, and vice versa. When B is at the position of O, it corresponds to the origin of the axis; and when B is at the position of A, it corresponds to ∞ or $-\infty$ at the real number axis, both positive and negative infinities being interconnected.

For each value of the radius OB, there is a segment OB' on the axis corresponding to it, and vice versa. Given a radian angle α , there are a unique value $OB' = OA \tan \alpha$ and a unique value $OB = OA \sin \alpha$ that correspond each other. Thus the radius is a space dimension because it corresponds to the real number axis, even though the scale of the axis tends to be linear and the radius OB is sinusoidal along circle Q.

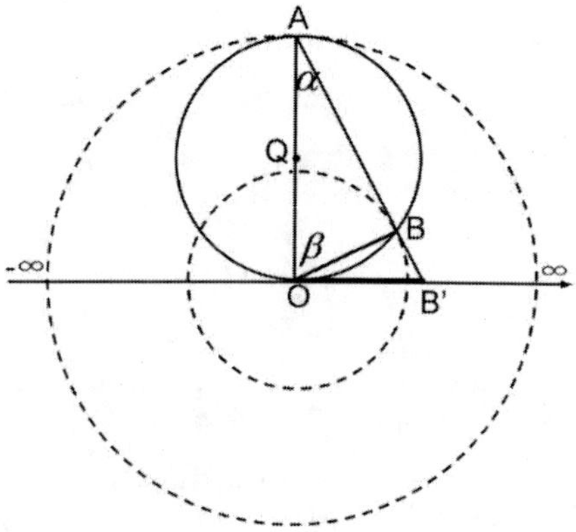

Figure 1.15. Schematic diagram of a one-to-one point correspondence between a circle Q except standpoint A and a real number axis OB'.

However, we define a dimension of the radius as the diameter OA of circle Q that can be approached to and fro by its semicircular arcs. The semicircular arc OBA corresponds to the positive part of the real number axis, and the other semicircle corresponds to the negative part of the axis, both traversing the diameter OA back and forth. In a dimensional diagram (Figure 1.7), the diameter AB is a space dimension. An electron traverses the dimension through both semicircular arcs. A dimension is the span of an electron in harmonic oscillation. If the density of the electron cloud also obeys harmonic oscillation, then the density would qualify as a time dimension as well. The dimensions of space and time are antiparallel. If OA direction is gaining half a space dimension, then OA direction is also losing half a time dimension. A pair of space and time spherical quantities for an electron are orthogonal and complementary because in Figure 1.15, $AB \perp OB$ and $\alpha + \beta = \pi/2$. The lapsing of a time dimension is naturally compensated by the gaining of a space dimension in the meanwhile, and vice versa.

It so happens that both electrons in a helium shell are symmetric and complementary in space and time, i.e., they are in space and time reversal. The space component of one electron is equivalent to the time component of the other electron; and the time component of one electron is equivalent to the space component of the other electron. We may regard both electrons in a helium shell as living in a space dimension and a time dimension (equations 1.2 and 1.5), respectively. However, we normally exchange a space semicircular arc for a time semicircular arc for the electron in space, and consequently exchange a time semicircular arc for a space semicircular arc for the other electron in time. This equivalence allows us to swap corresponding space and time components of equations (1.2) and (1.5) in the formulation of spherical quantities Ω_0 and Ω_2 in equation (1.14). In this way, each electron still maintains a one-dimensional spacetime entity in simple harmonic oscillation. Thus both electrons can be characterized by the same radius and density variables but with space and time reversal. This will become apparent as we define the geometric transformation of their spherical quantities in Euclidean space in the next section. We shall describe a system of electronic orbitals within inert atoms step by step that eventually changes the traditional spacetime worldview.

1.5.3. Spherical Quantities in Euclidean Space

As shown in Figure 1.16, if we interpret harmonic oscillation of 1s electrons as spherical expansion and contraction in Euclidean space, then as electron Ω_0 is undergoing sinusoidal expansion, electron Ω_2 experiences cosinusoidal contraction in the meanwhile. At any specific moment, the spherical radius of Ω_0 depends on OB as point B traces semicircle OBA, and the spherical radius of Ω_2 depends on OC as point C traces semicircle ACO, both being synchronized by complementary radian angles β in Figure 1.16(a) and α in Figure 1.16(b). When point B reaches position A, electron Ω_0 attains a maximum volume and begins to wrap back following the case of Ω_2 in Figure 1.16(b). On the other hand, when Ω_2 diminishes to minimum space at point O, it starts to expand following the case of Ω_0 in Figure 1.16(a).

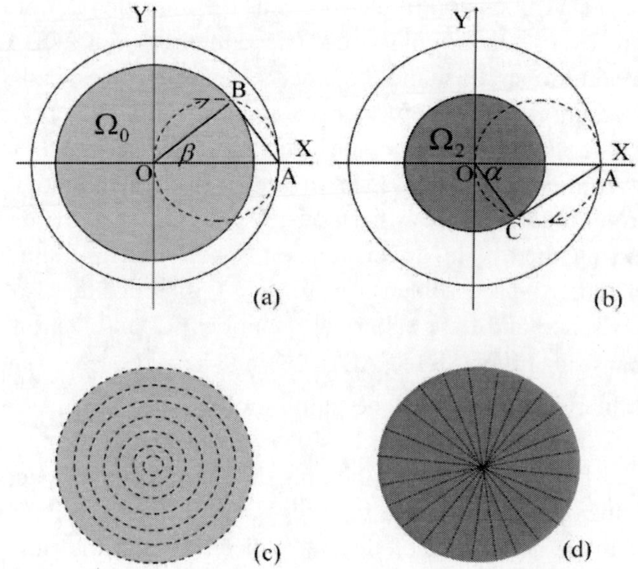

Figure 1.16. Interpretation of both 1s electrons in Euclidean space where shaded solids of (c) and (d) are electron clouds with electromagnetic properties. According to the Maxwell relationship (equation 1.19), two terms in the whole differential formula $d\Omega = \Omega_0 dl - \Omega_2 dt$, upon dynamic integral, correspond to the synchronized magnetic and electric fields respectively. The radius in the helium sphere is treated as one-dimensional in space as well as in time.

Both electrons inflate and deflate their spatial spheres harmonically with their centers at the origin and their radii stretching out or drawing back via the dashed circle, like two cranks driving the dashed wheel. Since the time and space components of Ω_0 are equivalent to the space and time components of Ω_2 respectively, Ω_0 and Ω_2 are spacetime symmetric. The space of Ω_0 and the time of Ω_2 (symbolically represented by density) are oscillating at the same phase from the center to the whole atomic sphere. However, due to their dimensional difference, both electrons are orthogonal and not in the same physical state during the oscillation. For example, they might exist as a pair of mutually inductive magnetic and electric fields (Figure 1.16c and 1.16d). The magnetic field Ω_0 is represented by space chord OB while the electric field Ω_2 by time chord AC, obeying $\int \Omega_0 dl = -\int \Omega_2 dt$. Space and time dimensions have an interval of a velocity dimension that gives the magnetic and electric fields an orthogonal relationship. Thus, in the helium shell, both electrons are actually expanding outwards from the nucleus to the whole sphere, one being the manifestation of space, the other of time (cf. Figure 11.1). They are mutually inductive and one is the root of the other. As shown in Figure 1.16(c) (d), the electronic waves are created, progressing outwards and disappearing one wave after another sinusoidally. This space and time symmetric phenomenon is counterintuitive, which we shall revisit in Figures 3.6 and 3.9 later on.

Given the foregoing scenario, we may calculate the density distribution of a time dimension of electron Ω_0 within the spatial sphere in the case of Figure 1.16(a). At any specific moment, the spatial volume of the electron is:

$$V = \frac{4}{3}\pi R^3 , \tag{1.70}$$

where radius R corresponds to chord OB along the dashed circle. If we designate the maximum radius OA as R_0, then

$$R = R_0 \cos\beta , \tag{1.71}$$

where radian angle β is rotating from π/2 to 0 uniformly. Supposing the electron cloud distributes homogeneously within the occupied sphere, then the electronic density must vary with the increment of the spherical volume as follows:

$$dP = -D\frac{4\pi R^2 dR}{V} , \tag{1.72}$$

where dP means the differential of electronic density, and D is a constant. Integrating both sides of the equation yields

$$P = \int -D\frac{4\pi R^2}{V} dR . \tag{1.73}$$

Substituting equation (1.70) into (1.73) gives the density distribution of a time dimension at a specific R value as

$$P = -3D\ln(R) , \tag{1.74}$$

which, upon normalizing D value and ordering R_0=1, transforms into

$$P = -\ln(\frac{R}{R_0}) = -\ln(\cos\beta) , \tag{1.75}$$

which indirectly involves two transcendental numbers: π in trigonometry and e in natural logarithm. From an alternative perspective of two-dimensional spacetime, when the space component of the electron is $R_0 \cos\beta$, its time component is $R_0 \sin\beta$ so that

$$dP = -\frac{R_0 \sin\beta}{R_0 \cos\beta} d\beta , \tag{1.76}$$

which also leads to equation (1.75) upon integration. The atomic sphere is two-dimensional in spacetime because it is homogeneous and isotropic at any moment, sufficiently expressed by a radius of space and a density of time. And it just so happens that the time component of one electron is the space component of the other electron, and the space component of one electron is the time component of the other electron. Figure 1.17 shows the logarithmic function with radial ratio R/R_0 in the range of (0,1). The graph indicates that the electronic density is zero at the outer boundary of $R=R_0$ so that the electron never actually reaches radius R_0.

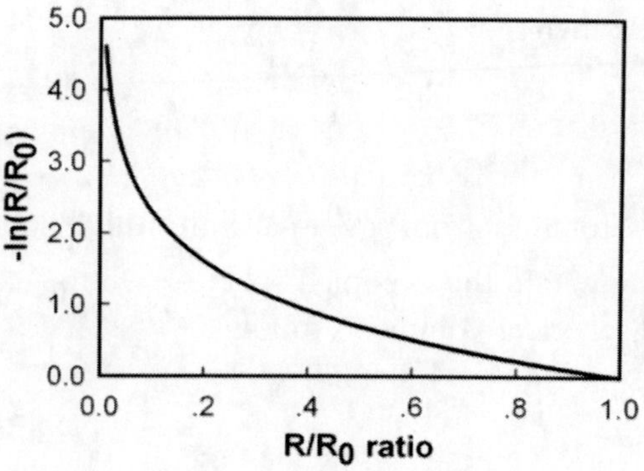

Figure 1.17. Natural logarithmic function $-\ln(R/R_0)$ as the density distribution of a time dimension of a 1s electron at R distance from the nucleus where R_0 denotes the maximum radius.

On the other hand, as R approaches 0 at the origin, the electronic density becomes infinitely large, but R cannot be assigned to 0 because it is a fraction denominator in equations (1.72) and (1.73). We assume that the electron is neutralized or annihilated at the nuclear center. Thus the sphere of the electron is an open set, or a domain, in Euclidean space. A domain is an open region without including any boundary points. Each electron cloud is homogeneous within its volume and spherically isotropic at any specific moment even though its density changes with time continuously. This is a strict requirement of a sphere of one-dimensional space with one-dimensional time.

Conceptually, the time component of spherical quantity Ω_0 is $C_1 \cos\alpha$, which is indicative of time density without taking the space component into consideration. Function (1.75) is the density distribution of a time dimension in space of a 1s electron, taking the space component into consideration. On the other hand, the product of time and space components of an electron is equal to the probability density or the chance of finding the electron within a unit volume upon a certain reliable physical conversion from the continuous electron cloud to discrete particle counts.

$$P = C_1 \cos\alpha \times C_2 \cos\beta = \frac{C_1 C_2}{2} \sin 2\beta \quad (1.77)$$

This probability density of the 1s electron is plotted in Figure 1.18 as two areas, the positive value corresponding to β from π/2 to 0 whereas the negative area corresponding to β from 0 to -π/2. The probability density is proportional to the energy density of the electron (cf. equation 9.7). The probability density indicates the most probable location of finding an electron that defines an electronic orbital in quantum mechanics. However, according to equation (1.77), an electron has a clear spherical boundary and a helium atom has a definite atomic radius, contrary to the point of view of modern quantum mechanics. When β is at the boundary point of $\pm\pi/2$ or 0 value, the probability density is zero. Thus the electron cloud is distributed in a domain of $\beta \in (\pi/2, 0)$ or $(0, -\pi/2)$.

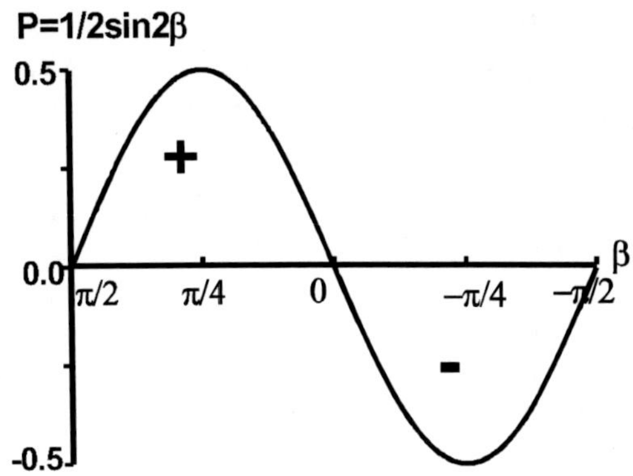

Figure 1.18. The probability density of a 1s electron. The most probable chance of finding the electron is at $\beta = \pi/4$ or $-\pi/4$. The negative area indicates the wrapping back of the electron cloud after reaching the maximum radius. The points at $\beta = \pi/2$ and $-\pi/2$ are connected forming a close cycle.

With the vivid geometrical picture (Figure 1.16), this chapter reaches a climax in the characterization of both electrons within a helium atom by definite orbitals. The motion of an electron is totally deterministic in space and time, contrary to the probability doctrine of quantum mechanics, which asserts that the indetermination is a property inherent in nature and not merely a profession of our temporary ignorance. The double-slit interferences were often given as an example to mystify the wave-particle duality. However, with spherical quantities in dynamic calculus, there is not any difficulty in explaining the duality of electrons. An electron exhibits wave and particle properties because of its oscillation between space and time. This duality model explains the results of the familiar Young's double-slit experiments reasonably well.

When a parallel beam of electrons falls on two slits whose sizes are comparable to the wavelength, an interference pattern is observed on the screen behind. Since the interference pattern would occur even when electrons passed through the slits one at a time, it is clear that the interference pattern would not arise from the interaction of one electron with another, but arise from the electron that passed through both slits at the same time, interfering with itself. This is so because an electron is oscillating between an electronic cloud (space) and a particle (time) continuously. In Figure (1.16), radian angle α may vary from 0 to $\pi/2$ and β value from $\pi/2$ to 0, which determine the density and size of the electron. Because this duality oscillation cycle is shorter than the duration for electrons to travel from the source to the double slits, the radian phases take random values between 0 to $\pi/2$ at the slits. There is wave behavior associated with the spacetime oscillation, and a wave could certainly travel through both slits at once. With spherical quantities in dynamic calculus, one can certainly follow the path of a single electron precisely through space and time during a double-slit experiment. Since each electron with a different radian phase will pass the double slits in a different way, a large amount of electrons with their random radian angles dictating their wave behavior will statistically produce an interference pattern on the screen. The successful explanation of experimental phenomena is the basic evidence of duality theory.

Conclusion

Based on the mere assumption that electrons observe general harmonic oscillation, we have given an orbital interpretation of space and time orthogonality in the two-dimensional world. Time and space are inherently associated with physical entities that instantiate them. They are symmetric and relative spherical quantities referring to two modes or two dimensions in spacetime as were represented by two electrons in a helium shell. Time and space of both electrons in a helium shell are symmetric; and time and space of each electron are orthogonal and complementary. In atomic spacetime, space is no more a conventional three-dimensional volume, and time is no more a unidirectional flow. The meanings of space and time are different from their usual meanings. Space still has X, Y, and Z orientations in Euclidean space, so does time, but it is more proper to adopt sine and cosine functions in various dimensions (1, ω, ν, r) to characterize their intricacy than to use the linear X, Y, and Z coordinates independent of real objects. Spherical quantities are dimensional quantities. Spherical quantity time is true physical existence that transforms in a certain dimension sinusoidally as a chord subtended by the arc of uniform circular Newtonian time in the dimension diagram; and spherical quantity space is true physical existence that transforms in a certain dimension sinusoidally as a chord subtended by the arc of uniform circular motion in the dimension diagram. Both time and space are sinusoidal quantities pegged to that specific dimension. Atomic spacetime is continuous electron clouds that transform by differential and integral operations. This new outlook of spacetime is termed the spherical view, beyond our usual mental concept. The spherical view is the scientific approach of adopting spherical quantities in dynamic calculus to characterize natural phenomena.

We have characterized electronic motion by harmonic oscillation, defined harmonic oscillation by spherical quantities in the duality equation, explained the equation by dynamic calculus, implemented dynamic calculus by trigonometry, and interpreted the distribution of an electron cloud in terms of probability density in Euclidean space. Atomic spacetime is about spherical quantities in dynamic calculus. It is not linear vector space with uniform time, neither Euclidean space nor Hilbert space. A spherical quantity corresponds to a wave function that describes the motion of an electron. The product of its time and space components indicates the probability density of the electron that traditionally defines the region of the electronic orbital. However, within an atomic shell, an electronic orbital can be simply understood as an electron in motion, or a spherical quantity in dynamic calculus.

As infinitesimal calculus and complex number notation drive mathematics toward idealism farther and farther, our new definition of dynamic calculus and dimensional interpretation of complex numbers bring it back to reality. However, granting the new interpretation of space and time does not mean to overthrow existing successful theories. Just as Einstein's relativity does not invalidate Newtonian mechanics, but sees objects in a wider scope and effectively extends it, our example here supplements physical quantities with spherical quantities for the description of electronic orbitals in atomic spacetime where classical mechanics ceases to be effective and the power of quantum mechanics is so limited.

Because of the great success of Newtonian mechanics, the biosphere of the earth is customarily assumed to be a medium of three-dimensional space with one-dimensional time from the perspective of human beings, but it might not be the case when observed from the perspective of an animal, say a sessile barnacle on a bridge pier (which only cares about the ebb and flow of the tide), or from the perspective of the planetary god Gaia (if exists at all) who senses every pain on the earth. We never know how a virus sees its living environment, or how a plant seed feels about itself in the air and in the soil. In order to achieve real objectivity in the quest for truth, one must try to perceive the world from an alternative subject, at a different angle, and by a new standard. Such original perceptions and tastes are this book all about.

By viewing space and time in a helium atom differently from conventional three-dimensional space with one-dimensional time, we are actually examining the world from a new perspective and by a fresh mindset. Just as things may take various shapes from different angles, it is not surprising that the results we get might be quite different from those of quantum mechanics. For instance, a Chinese sees that an apple is red in color whereas an Englishman sees that the apple is round in shape. In this case, both are right and compatible. It is easy to understand that different perspectives and approaches can be taken to attack a problem. In analytic geometry, an equation in polar coordinates is normally different from that in Cartesian coordinates. The question is not which one is correct but which one is more informative and elegant to tackle the problem under consideration. Indeed, our theory could be established without challenging any successful theories, but adding fresh insight to them instead.

Although this introductory chapter looks deceptively naïve, it is merely the beginning of a series of discoveries with the two-dimensional helium shell as the simplest case of atomic spacetime. A unique perspective by a single mind is required to develop a complex theory coherently. A crowd never does without a unified scheme of thought. The fundamental law of nature will not be easily explained in a couple of pages because it actually relates to every corner of science. We shall unfold this great beauty step by step leading to its full feather, but we still cannot exhaust every feature of it, nor can we integrate it with established disciplines in all aspects. Our solutions and logic may seem radical or even perverse on occasion, which may attract many criticisms from conservative physicists. The readers should grasp the heart and soul of the spherical view for integrity instead of probing the details impatiently even though every specific issue is believed to be important. We therefore urge you to put up with the rough until the end of the book.

Chapter 2

The Theory of Quaternity

Electronic motion is not like mechanical movement in Euclidean space with Newtonian time where linear algebra is applied to physical quantities. Electronic motion obeys electromagnetism in atomic spacetime where dynamic calculus of spherical quantities is the rule. To extend the two-dimensional spacetime model in a helium shell to four dimensions in a neon shell, here we put forward a novel spacetime theory, defining four space dimensions of a sphere in geometry, trigonometry, and calculus consistently in the context of harmonic oscillations. Four-dimensional spacetime is explained and instantiated by the 2s2p electron octet, which is built mathematically upon the spherical quantities of both 1s electrons in the helium shell. Ten electrons within a neon atom constitute a multi-dimensional spacetime continuum where wave radiation from 1s electrons toward 2s2p electrons is realized by dimension encapsulation mechanism, and wave reflection inwards from the neon shell to the helium shell follows the chain rules of dynamic calculus.

2.1. Introduction to Four Space Dimensions

In Chapter 1, we introduced spherical quantities for describing electronic orbitals within a helium atom. Filled by both 1s electrons, the helium shell is a perfect two-dimensional spacetime continuum. However, the most important spacetime is of four dimensions, as will be found in a neon shell with the 2s2p electron octet. We think that the best way to delineate four-dimensional space is to define four geometric elements corresponding to four dimensions. This approach is markedly different from modern geometry, which tends to give pure abstraction (such as n-dimensional vector space) instead of intelligible pictures. When mathematical formulation dissociates from reality, it is losing ground and value and sailing to nowhere. Therefore, we adhere to laying out quaternity spacetime on a solid basis through geometric and physical instantiations [4, 5]. Frankly speaking, accurately identifying four orthogonal geometric elements in a sphere was a milestone in the early development of the atomic spacetime theory. There are definitely four orthogonal shapes or elements in Euclidean space corresponding to four dimensions in a sphere, but they are not so obvious at first glance or at first thought. This is because we are not taught in traditional geometry at school in that way. So please keep an open mind in searching the basic elements of geometry in the following sections.

2.1.1. Four Geometric Elements in a Sphere

How to divide a spatial sphere into four dimensions logically is the subject of this section. Euclidean geometry is the de facto foundation of all knowledge of space, so we shall begin our discussion from it. Nowadays, every middle school student makes acquaintance with the basic elements of geometry: points, lines, planes, and solids. Two points determine a one-dimensional straight line and two parallel lines determine a flat plane. In addition to straight lines and flat planes, there are circular arcs and spherical surfaces. The relationship between a circle and straight lines deserves further scrutiny, so does that between a spherical surface and flat planes.

Because the center of a circle is the unique and pivotal point of it, we shall study a circle from that standpoint. As shown in Figure 2.1, a circle corresponds to two lines and two points. For every point *a* in the circle, we can always find a point *A* in the lines corresponding to it except at points P_1 and P_2. Conversely, for every point *B* in line L_1 or L_2, we can always find a point *b* in the circle corresponding to it. There are projective correspondences between the circle and two lines plus two points. Points P_1 and P_2 and lines L_1 and L_2 are two basic elements of the circle. If we regard the two points as a geometric dimension and the two lines as another dimension, then the circle contains two dimensions. Both dimensions are, of course, not the same. Just as the number 77 has two digits, the first one referring to seventy and the second to seven only, a circle has two dimensions of different geometric structures: two points and two straight lines. Thus a circle has two dimensions in the same way that we would say that the number 77 has two digits.

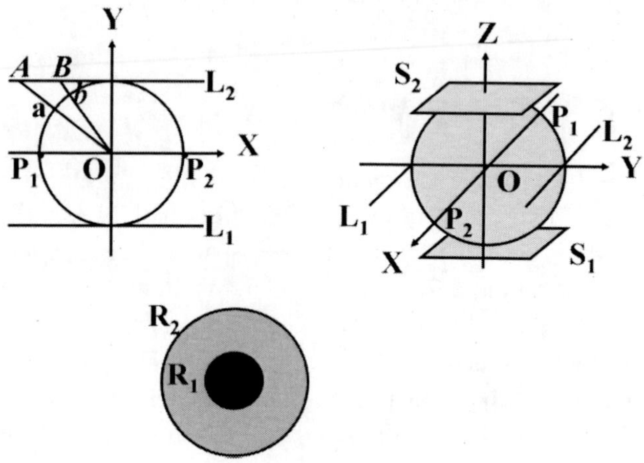

Figure 2.1. Geometric analyses of dimensions in a spherical layer. A circle is projectively decomposed into two lines and two points; a spherical surface into two planes S_1 and S_2, two lines L_1 and L_2, and two points P_1 and P_2; and a spherical layer is an outer solid sphere R_2 minus an inner solid sphere R_1.

In the same way, a spherical surface is composed of two flat planes, two straight lines, and two points. There is a one-to-one point projective correspondence between a spherical surface and three mutually exclusive elements: P_1 and P_2, L_1 and L_2, and S_1 and S_2. In other words, these three geometric elements are a distorted expression of the spherical surface. Thus a spherical surface possesses three dimensions of various geometric structures, the third one being planes S_1 and S_2.

Because two points, P_1 and P_2, might determine a straight line, we regard these two points as a one-dimensional shape; because two parallel lines, L_1 and L_2, might define a flat plane, we treat both parallel lines as a two-dimensional shape; and likewise, we regard parallel planes S_1 and S_2 combined together as a three-dimensional geometric element. Thus a spherical surface has three dimensions in the sense that it is composed of three structural elements: one-dimensional double points, two-dimensional double lines, and three-dimensional double planes. This is analogous to say that the number 777 has three digits: ones, tens, and hundreds.

A solid sphere has one additional dimension in the radial direction compared with a spherical surface. So we regard it as a four-dimensional shape. A radius represents a dimension, the addition of which to a spherical surface yields a solid sphere. A spherical layer is an outer sphere minus a concentric inner sphere or the medium between two concentric spherical surfaces. As we study a spherical layer, the outer sphere is treated as a four-dimensional space whereas the inner sphere as a zero-dimensional point. In traditional geometry, a point is assumed to be infinitesimal in size, but here a point has the size of the inner sphere. Since the inner sphere might be treated as a point, the spherical layer might be treated as a solid sphere as well, ignoring the existence of the inner sphere. Throughout this book, a sphere always means a solid spherical object rather than a spherical surface.

Table 2.1. Various geometric elements within a spherical layer analogous to various digits in a decimal number

Geometries		Number expression	
An inner sphere	Zero-dimensional	0	Starting point
Two points	One-dimensional	7	First digit
Two lines	Two-dimensional	70	Second digit
A Circle	Has two dimensions	77	A two digit number
Two planes	Three-dimensional	700	Third digit
A spherical surface	Has three dimensions	777	A three digit number
A radius (of outer sphere)	Four-dimensional	7000	Fourth digit
An outer sphere	Has four dimensions	7777	A four digit number

The notion of four space dimensions within a spherical layer is called quaternity space. The four space elements (spheres, planes, lines, and points) that we have defined possess various spatial meanings. This is somewhat analogous to number expression. For example, in the number 7777, the first 7 means seven thousand, the second means seven hundred, and so forth. Four digits actually denote different magnitudes. In fact, it is such a kind of structural grading that makes it feasible to express a large number. Quaternity space follows the same logic. Four space elements form four rungs in a ladder (i.e., four, three, two, one, and zero) in terms of dimensions, contrary to Cartesian coordinates where dimensional properties of the X, Y, and Z orientations are assumed to be the same. Owing to their grading, four different dimensions are able to express properly a spherical layer. The logic of combining various dimensions of geometric elements to account for a spherical layer is similar to that of adopting various digits to express a large number (Table 2.1). We focus on the geometry of a spherical layer because it represents a large class of atomic shells, such as the 2s2p electron octet within a neon atom.

2.1.2. Orthogonality and Scalability of Four Space Dimensions

We have characterized a spherical layer by four dimensions: two spheres, two planes, two lines, and two points. As was shown in Figure 2.1, points P_1 and P_2 are along the X-axis; the perpendiculars to lines L_1 and L_2 are along the Y-axis; and the normal lines to planes S_1 and S_2 are along the Z-axis. These three geometric elements are mutually exclusive in their positions and mutually perpendicular in their orientations. Together they form a spherical surface. As the fourth dimension, the radius of the sphere is perpendicular to the spherical surface. Thus, four space dimensions are mutually orthogonal in Euclidean geometry.

In atomic spacetime, orthogonality is defined as being an exact dimension apart in spacetime. As was mentioned previously, points P_1 and P_2 determine a one-dimensional line; lines L_1 and L_2 determine a two-dimensional plane; planes S_1 and S_2 determine a three-dimensional surface; and spheres R_1 and R_2 determine a four-dimensional solid. Thus these four geometric elements are orthogonal because of the whole dimension interval between them (Figure 2.2).

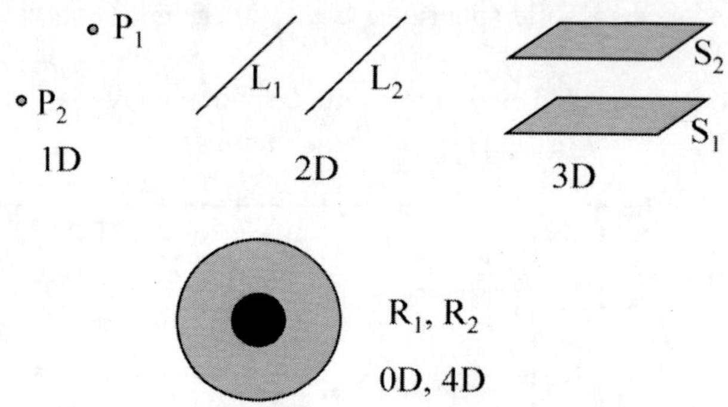

Figure 2.2. Four orthogonal elements of geometry for characterizing a spherical layer with reference to Figure 2.1.

There are interesting differential and/or integral relationships between various geometric structures within a spherical layer. As shown in Figure 2.3, if we start with a four-dimensional outer sphere E_0 and reduce it a dimension, then we get a spherical surface E_1. The dimension reduced is the highest dimension of the geometric structure, a space dimension oriented in the radial direction. Such an action is like taking away 7000 from the number 7777 because a radius is four-dimensional. We may write this process in differential form as:

$$\frac{dE_0}{dl_0} = E_1, \qquad (2.1)$$

where l_0 denotes a space dimension in the radial direction. If we further deduct a dimension from the spherical surface, we get a circle E_2. The immediate space dimension taken away is represented by both parallel planes S_1 and S_2, which has a three-dimensional space property. The action is analogous to subtracting 700 from the number 777. Likewise, reducing a subsequent dimension from the remaining circle leads to two symmetric points E_3 about the center; and reducing the last dimension from E_3 exposes the central point E_4, a zero-dimensional inner sphere. These geometric transformations may be expressed in differential form as

$$\frac{dE_1}{dl_1} = E_2; \ \frac{dE_2}{dl_2} = E_3; \ \frac{dE_3}{dl_3} = E_4 \tag{2.2}$$

where l_1 is a three-dimensional geometric element represented by two parallel planes, l_2 is a two-dimensional geometric element represented by two parallel lines, and l_3 is a one-dimensional geometric element represented by two points dislocated from the spherical center. Differentiation upon a geometric structure is always operated with respect to the highest outstanding dimension, analogous to taking away the leftmost digit from a number.

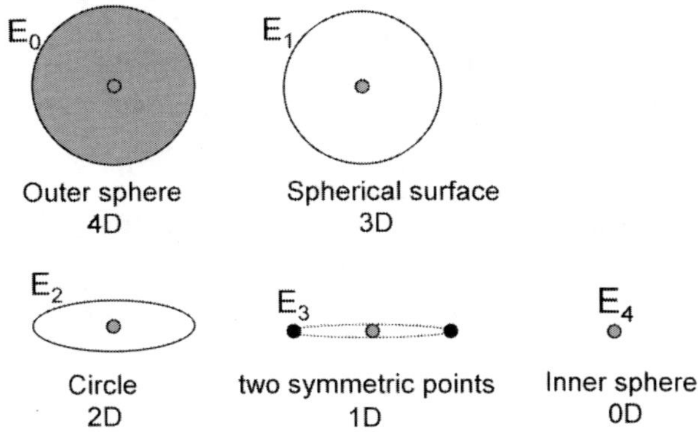

Figure 2.3. Calculus relationships between geometric structures within a spherical layer where the inner sphere hides at the geometric center until exposure in the last one.

It is interesting to find that a sphere reduced four dimensions consecutively produces the same shape of a different size. The four space dimensions reduced are a radius, two planes, two lines, and two points in sequence, and the final result is a smaller sphere represented by an inner radius. These geometric elements correspond to $2s2p_z2p_y2p_x2s$ orbital shapes as will be described in sections 2.2 and 2.3. They are contiguous in atomic space. A sphere is divided into four space dimensions, which is the basis of orbital quantization.

As we study a well-organized structure with multiple spherical layers, the dimension of a specific sphere depends on whether it is the outer sphere or the inner sphere within the spherical layer of concern. If it is the outer sphere, then it must be regarded as one containing four dimensions in space; and if it is the inner sphere, then it is a zero-dimensional point encapsulating all complexity of itself. That performing differential operations upon a sphere four times consecutively produces another smaller sphere gives scalability for the four dimensions, which can be peeled off in sequence repeatedly from a large and well-structured sphere, such as an argon atom. As an exception, a one-dimensional helium sphere, after being reduced one dimension, results in a zero-dimensional point immediately. This case has been explained in Chapter 1.

2.2. Quaternity Spacetime

We have analyzed the dimensional structure of a sphere in space. We shall further explain the dimensional structure of a sphere as time is added into the scene. The complexity of four-dimensional spacetime is remarkably greater than that of two-dimensional spacetime as a neon atom is compared with a helium atom. The fundamental law of atomic spacetime is spherical quantities in dynamic calculus. The appropriate initial concept on which to base the theory of quaternity is that of four-dimensional structure in a sphere. The most difficult part of quaternity is perhaps dimension encapsulation mechanism that connects the helium and neon shells. If dynamic calculus and four-dimensional structure are two cornerstones of the theory of quaternity, then dimension encapsulation mechanism is the keystone of the mathematical architecture. So the readers are encouraged to digest the following material for grasping the intricate structure of a neon atom.

2.2.1. Symmetry and Complementarity of Time and Space

Geometric analyses indicate that a spherical layer possesses four orthogonal space dimensions. Because we view time and space as two relative and symmetric spherical quantities, a spherical layer must contain four time dimensions to complement its space dimensions. Traditionally we take it for granted that time is one-dimensional and space is three-dimensional. After giving up this presumption and treating it as a special case in atomic spacetime, we obtain four kinds of space and time dimension combinations in a four-dimensional sphere (Table 2.2).

Table 2.2. Quaternity spacetime in a four-dimensional sphere based on four orthogonal dimensions

Quaternity axis	Space dimension	Time dimension	Projective geometries	Spherical layer correspondences
1	0 or 4	4 or 0	Two radii	Inner and outer spheres
u	1	3	Two points	Two symmetric poles
u^2	2	2	Two lines	Two semicircular arcs
u^3	3	1	Two planes	Two hemispherical surfaces

Space and time dimensions are complementary. When the total number of dimensions is set for a given entity, reducing a dimension in space inevitably increases a dimension in time accordingly, and vice versa. If we use a real number 1 to indicate a spherical quantity of four-dimensional time with zero-dimensional space, then by reducing a time dimension while increasing a space dimension, we obtain a velocity dimension u to indicate a spherical quantity of three-dimensional time with one-dimensional space. Dimension u refers to velocity in four-dimensional spacetime whereas v refers to velocity in two-dimensional spacetime, both belonging to different spherical layers.

In the same manner, we use u^2 to represent a spherical quantity of two-dimensional time with two-dimensional space and u^3 to represent a spherical quantity of one-dimensional time with three-dimensional space. These four distinct combinations of time and space dimensions form four orthogonal axes in quaternity coordinates (Figure 2.4). The home axis indicates the current spacetime dimension of a spherical quantity. The purpose of the quaternity coordinate system is to illustrate the intricate dimensional relationships between various spherical quantities in dynamic calculus instead of facilitating metric measurements between positions in the Cartesian coordinate system.

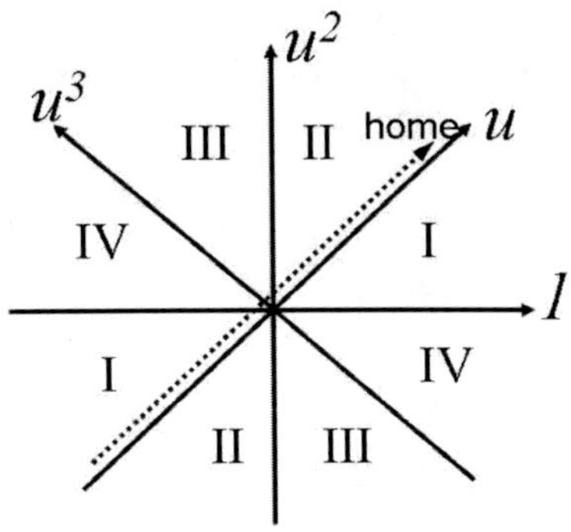

Figure 2.4. Four quaternity axes dividing a spacetime sphere into four spherical quadrants.

As shown in Figure 2.4, four quaternity axes divide a four-dimensional spacetime continuum into four spherical quadrants. The spacetime dimension between the *1* and u axes is the first spherical quadrant; the spacetime dimension between the u and u^2 axes is the second spherical quadrant; the spacetime dimension between the u^2 and u^3 axes is the third spherical quadrant; and the spacetime dimension between the u^3 and *1* axes is the fourth spherical quadrant. The negative directions of quaternity axes (1, u, u^2, and u^3) may be alternatively marked as u^4, u^5, u^6, and u^7 dimensions, respectively. Spherical quantities in these four spherical quadrants are termed distinctively: a polor is a spherical quantity of the first spherical quadrant, a metor of the second, a vitor of the third, and a scalor of the fourth. For the literal meanings of these four spherical quantities, please refer to section 11.3. Quaternity means four spherical quadrants in one sphere.

Time and space are not necessarily in symmetry with each other in each spherical quadrant, but when we consider four spherical quadrants together, time and space are indeed symmetric. For example, the polor is in symmetry with the vitor with regard to the space and time switch. Four spherical quadrants reflect the symmetry and complementarity of space and time in the whole sphere. We are not accustomed to adopting multiple time dimensions. One way to grasp it is to ponder over the physical quantity of acceleration with a unit of m/s^2 that contains two dimensions of time, getting the feel of propensity. In the spherical view, multiple time dimensions define harmonic oscillations in multiple orientations consecutively.

The proposition of quaternity spacetime carries a further step beyond Minkowski's four-dimensional events where time is one-dimensional. Quaternity is not framed by that traditional worldview. To illustrate, suppose that we are observing the far away heavenly bodies from the earth, which is at the center of the imaginary celestial sphere, receiving the light from the stars. The distance between the earth and a star is best characterized by the time interval that light travels between them. The unit of light-year is normally used in this sense. So at the center of the celestial sphere, we are watching the stars located on a spherical surface, the radial direction being the coming light. The light showering on us indicates the flow of time. If the universe is a medium of one-dimensional time with three-dimensional space, then a spherical surface must be three-dimensional in space because it is exactly one dimension less along the radial direction than the whole celestial sphere, the dimension along the radial direction being the time flow. However, since the time flow can be expressed by a distance, the celestial sphere is indeed four-dimensional in space. Thus a celestial sphere may be characterized by one-dimensional time with three-dimensional space or by four-dimensional space only. If a dimension in time and a dimension in space can be traded according to a certain perceptual interpretation, then the complementarity of time and space dimensions listed in Table 2.2 is conceivable.

2.2.2. Rotatory Operations in the Quaternity Equation

When the total number of spacetime dimensions is fixed for an electron, the electron decreasing a dimension in time inevitably increases a dimension in space accordingly, and vice versa. Because time and space dimensions are antiparallel, an electron undergoes concurrent differential and integral operations, i.e., differentiation upon a spherical quantity with respect to a time dimension must be accompanied by integration over a space dimension. This can be expressed in quaternity coordinates by a counterclockwise rotation of the home axis that reduces a time dimension while increasing a space dimension for a spherical quantity. If Φ is a spherical quantity, then its counterclockwise rotation Φ^0 is defined as

$$\Phi^0 \equiv \int\left(-\frac{\partial \Phi}{\partial t}\right)dl, \tag{2.3}$$

or

$$\frac{\partial \Phi^0}{\partial l} = -\frac{\partial \Phi}{\partial t}, \tag{2.4}$$

where l represents the highest space dimension within Φ^0, t represents its corresponding time dimension within Φ, and the minus sign indicates that space and time dimensions are anti-parallel, aligning in opposite directions. The principle of rotatory operations applies to spherical quantities in various spherical quadrants. For example, if the home axis of Φ is at the location of three-dimensional time with one-dimensional space, then a rotatory operation brings it to the location of two-dimensional time with two-dimensional space as Φ^0. The rotation of a polor results in a metor.

It is important to realize that the principle of rotatory operations is not invented out of pure imagination, but deeply rooted in Faraday's law, one of Maxwell's equations, as was mentioned in Chapter 1. We may relate equation (1.19) in dynamic calculus to equation (1.21) in infinitesimal calculus closely in a logical way. Equation (1.19) governs spherical quantities Ω_2 and Ω_0 and defines an electron along a hypothetical semicircle uniformly through electromagnetic induction (see Figure 1.7). While both time and space components are sinusoidal, uniform circular motion along the semicircle guarantees that radian angles of time and space are progressing uniformly. Equation (1.21) governs physical quantities of electric field strength **E** and magnetic field strength **B** in Euclidean space with Newtonian time. Newtonian time is uniform time corresponding to uniform circular motion along a semicircle, but Euclidean space as measured by the X, Y, and Z axes is rectilinear and must be converted into circular by using the *curl* operator in order to achieve uniform progress along the semicircle. Quantities on both sides of the equation must be equivalent in time and space whirling. A circular function with a uniformly varying radian angle, or dynamic calculus, expresses best the process of a *curl* operation. For example, a *curl* operation signifies the sinusoidal change of chord OC along the dashed circle (Figure 1.16b). As chords OC and OB vary sinusoidally along the dashed circle, equations (1.19) and (1.21) can be expressed as $\cos\alpha = \sin\beta$. Thus, Faraday's law spells out the same thing as the principle of rotatory operations does in describing an electron in harmonic oscillation or electromagnetic induction. Please refer to section 10.3 for more details.

We may explain the principle of rotatory operations in a simply way. For example, a boy grows into a man over years. A seed develops into a tall tree over a certain time period. In both natural phenomena, time elapsing is in return for size gaining of the organisms. A spherical quantity losing a dimension in time is always accompanied by gaining a new dimension in space.

It can be seen from Figure 2.4 that rotating the home axis eight times consecutively brings it to the same axis position; and rotating the home axis four times consecutively brings it to the same axis but in the opposite direction. Because a neon shell can be regarded as a four-dimensional spherical layer, a proper spherical quantity describing the electronic oscillation within it must satisfy the following equations:

$$\Phi^{08} = \Phi\,, \tag{2.5}$$

$$\Phi^{04} = b\Phi\,, \tag{2.6}$$

where Φ^{08} and Φ^{04} denote rotating counterclockwise eight times and four times from spherical quantity Φ respectively, and b is a dimension parameter. Since the distance between the Φ^{04} and Φ dimensions is four axes apart, b actually bears a spacetime factor of u^4 so that equation (2.6) can also be expressed in partial derivative form as

$$\frac{\partial^4\Phi}{\partial t^4} = u^4\,\frac{\partial^4\Phi}{\partial l^4}\,, \tag{2.7}$$

which we shall call the quaternity equation.

The physical significance of quaternity equation can be understood in a simple way. We have established the equivalence between the duality equation and the one-dimensional Schrödinger's equation in Chapter 1. Quantum mechanics naturally extends one-dimensional Schrödinger's equation into three dimensions by introducing the Laplacian operator, which sees three dimensions of space in the X, Y, and Z orientations so that linear algebra works. In contrast to this, quaternity views the space represented by the X, Y, and Z axes as possessing different levels of dimensions. As shown in Figure 2.5, supposing that we are trying to locate a particular seat in a great auditorium, we would first locate the floor number as is represented by the Z direction, and then within that proper floor search the tier number as is represented by the Y-axis, and finally look for the X position in that tier. Here, three-dimensional Z direction refers to the arrangement of all the floors, each of which contains multiple tiers; two-dimensional Y direction refers to the sequence of the tiers, each of which contains multiple seats; and one-dimensional X direction refers to the order of seats. This simple logic is not beyond our everyday living experience, but mathematicians have not distinguished the dimensional meanings between these three axes. The wave functions in three orientations must represent different meanings and constitute three orthogonal dimensions. The natural shifting of the dimensions from floor order, to tier order, and to seat order in locating a particular seat in the auditorium reflects the rule of dynamic differentiation, which is always operated with respect to the most immediate or highest dimension. In formulating multi-dimensional wave equations, quaternity climbs to higher ranks of derivatives than the second and embodies four space dimensions in $\partial/\partial l$, $\partial^2/\partial l^2$, $\partial^3/\partial l^3$, and $\partial^4/\partial l^4$ forms instead of adopting the Laplacian operator, $\frac{\partial^2}{\partial x^2}+\frac{\partial^2}{\partial y^2}+\frac{\partial^2}{\partial z^2}$. Thus a spherical quantity obeys dynamics calculus instead of linear algebra.

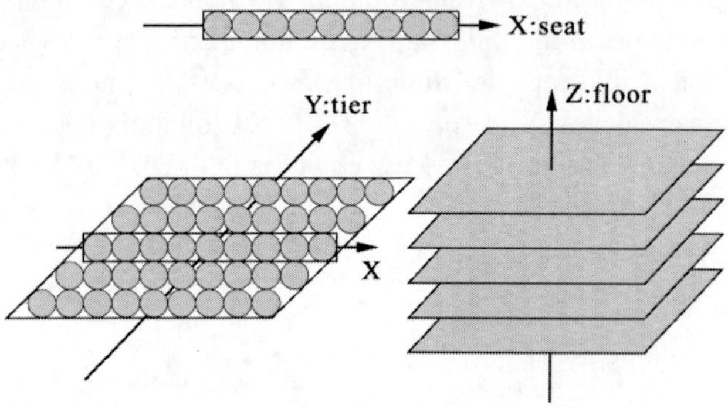

Figure 2.5. Even in a theater ticket, the meanings of the X, Y, and Z axes denote different levels of space dimension.

Because the fourth derivative $\partial^4\Phi/\partial l^4$ contains the operations of the third, second, and first derivatives, the quaternity equation governs electronic oscillations in four varying dimensions. The quaternity equation holds true no matter where the home axis of a spherical quantity is. For example, if the home axis is at the four-dimensional time with zero-dimensional space position, then the equation governs the scalor. Likewise, if the home axis is at the three-dimensional time with one-dimensional space location, then the equation is for that specific polor as well. Spherical quantities in four spherical quadrants have four different kinds of space and time combinations. They satisfy the quaternity equation individually and simultaneously in their respective dimensions. In other words, the quaternity equation is an invariant formula for four spherical quantities in four spherical quadrants. It is not a one-dimensional linear differential equation, but a four-dimensional harmonic oscillation equation, the solution to which will give spherical quantities for eight electrons in a neon shell. Within the neon shell, four electrons colonize four spherical quadrants in the positive direction of quaternity axes whereas another quartet colonizes spherical quadrants in the negative direction of the same axes.

2.2.3. Electronic Orbitals in a Neon Shell

Since electrons in a neon shell occupy a spherical layer outside a helium shell, spherical quantities for describing 2s2p are closely related to those for 1s electrons. We return to the first chapter where time and space components of 1s electrons are represented by

$$\Psi = C_1(\cos\alpha - \omega\sin\alpha)\,; \tag{2.8}$$

$$\psi = C_2(\cos\beta + r\sin\beta)\,. \tag{2.9}$$

As shown in Figure 2.6, semicircle ACB represents the course of dynamic calculus upon spherical quantities Ψ and ψ. Both $C_1\cos\alpha$ and $-C_1\omega\sin\alpha$ terms together determine a time state point C on semicircle ACB; and both $C_2\cos\beta$ and $C_2 r\sin\beta$ terms together determine a space state point C on the same semicircle. As point C travels from A toward B along the semicircle, one 1s electron transforms into another by a rotatory operation:

$$\Omega_0{}^0 = \Omega_2\,, \tag{2.10}$$

which indicates orthogonality between both 1s electrons. When treated as a spherical quantity, Ψ is measured by chord BC= $C_1\cos\alpha$ and chord AC= $-C_1\omega\sin\alpha$; but when treated as a radian angle, Ψ is measured by arc AC (or $\angle DAC$), which increases from 0 to π/2 during the differentiation. As a radian angle, the whole semicircle ACB corresponds to circumferential angle π/2 in the dimension diagram. Arc Ψ equals radian angle α provided that the differential course is a strict semicircle, but they measure radian angles in different spherical layers. Radian angle α belongs to the helium shell while radian angle Ψ belongs to the neon shell. As point C travels from A toward B, arc BC is retreating while arc AC is progressing. In the scope of the neon shell, functions Ψ and ψ represent both arcs AC and BC respectively, which are complementary in terms of radian angles:

$$\Psi + \psi = \pi/2\,. \tag{2.11}$$

Electronic orbitals are in ever-changing dynamic states. At any moment, a state point C is codetermined by four segments: $C_1\cos\alpha$, $-C_1\omega\sin\alpha$, $C_2\cos\beta$, and $C_2 r\sin\beta$ in right triangle ACB. This redundancy check on the electronic orbital reinforces synchronization of time and space components and complementarity of radian angles Ψ and ψ during the rotatory operation.

With Ψ and ψ representing time and space radian angles respectively, we shall construct eight spherical quantities to represent $2s^2 2p^6$ electrons in the neon shell. Given an initial function of

$$\Phi_{00} = A_1 A_2 \cos\Psi \cos\psi\,, \tag{2.12}$$

where A_1 and A_2 are time and space dimension constants, we may perform a rotatory operation on it as follows. The rationale of constructing equation (2.12) will be explained in section 2.2.4. Suffice it to say here that the equation is a bridge connecting the helium and neon shells.

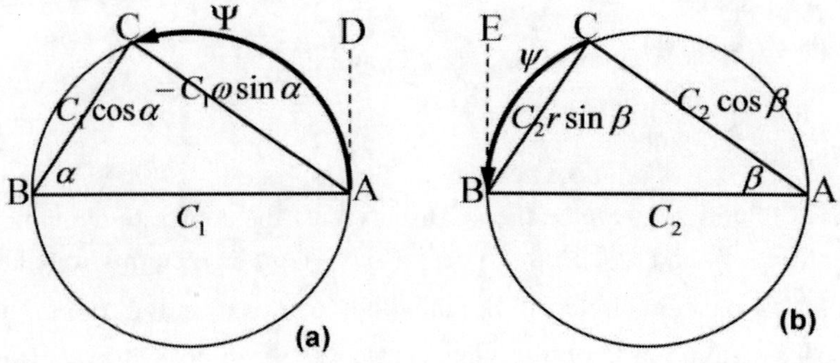

Figure 2.6. Spherical quantities Ψ and ψ as complementary radian angles in bold arcs in (a) time and (b) space dimension diagram where $\Psi = \angle DAC$ and $\psi = \angle EBC$. Here an angle formed by a chord and a tangent that intersect on a circle equals the measure of the intercepted arc.

$$
\begin{aligned}
\Phi_{00}^{0} &= -\frac{dA_1\cos\Psi}{dt}\int A_2\cos\psi\ dl \\
&= (A_1\frac{d\Psi}{dt}\sin\Psi)\cdot(-A_2\frac{1}{\psi'}\int\cos\psi\ d\psi) \\
&= -A_1A_2(-\frac{1}{\psi'}\dot{\Psi})\sin\Psi\sin\psi \\
&= -A_1A_2u\sin\Psi\sin\psi,
\end{aligned}
\tag{2.13}
$$

where $\dot{\Psi}$ and $-1/\psi'$ are the angular velocity and orbital radius, both being constants in the neon shell.

$$\dot{\Psi} = -\frac{d\Psi}{dt}, \tag{2.14}$$

$$\psi' = -\frac{d\psi}{dl}, \tag{2.15}$$

The product of $\dot{\Psi}$ and $-1/\psi'$ is a velocity dimension u in the neon shell, which manifests as a rotatory operation upon a spherical quantity in the helium shell.

$$
\begin{aligned}
u &= -\frac{1}{\psi'} \cdot \dot{\Psi} = \frac{dl}{d\psi} \cdot (-\frac{d\Psi}{dt}) = \Omega_0^0 \\
&= \int C_2 \cos\beta dl \cdot (-\frac{dC_1 \cos\alpha}{dt}) \\
&= (r\int C_2 \cos\beta d\beta) \cdot (-C_1 \omega \sin\alpha) \\
&= -r\omega C_1 C_2 \sin\alpha \sin\beta \\
&= -v C_1 C_2 \sin\alpha \sin\beta \\
&= \Omega_2,
\end{aligned}
\tag{2.16}
$$

where $dl/d\psi$ expands as an integral operation in the helium shell. After the execution of $\frac{1}{\psi'}\int \cos\psi \, d\psi = \frac{1}{\psi'}\sin\psi$ in equation (2.13), $dl/d\psi$ follows the integral chain rule. The integral chain rule parallels the differential chain rule in dynamic calculus. We made use of equations (1.3), (1.6), and (1.7) in the above calculation. Through consecutive rotatory operations, we obtain four spherical quantities for $2s2p_x2p_y2p_z$ in the positive direction of quaternity axes:

$$
\begin{pmatrix} \Phi_0 \\ \Phi_1 \\ \Phi_2 \\ \Phi_3 \end{pmatrix} = \begin{pmatrix} \Phi_{00} - \dfrac{\partial \Phi_{00}}{\partial t} \\ \Phi_{00}^{0} - \dfrac{\partial \Phi_{00}^{0}}{\partial t} \\ \Phi_{00}^{02} - \dfrac{\partial \Phi_{00}^{02}}{\partial t} \\ \Phi_{00}^{03} - \dfrac{\partial \Phi_{00}^{03}}{\partial t} \end{pmatrix},
\tag{2.17}
$$

where Φ_{00}^{02} and Φ_{00}^{03} indicate performing counterclockwise rotations upon function Φ_{00} twice and three times respectively. Substituting the initial value of Φ_{00} into the matrix produces

$$
\begin{pmatrix} \Phi_0 \\ \Phi_1 \\ \Phi_2 \\ \Phi_3 \end{pmatrix} = A_1 A_2 \begin{pmatrix} \cos\Psi\cos\psi - \dot{\Psi}\sin\Psi\cos\psi \\ -u(\sin\Psi\sin\psi + \dot{\Psi}\cos\Psi\sin\psi) \\ u^2(\cos\Psi\cos\psi - \dot{\Psi}\sin\Psi\cos\psi) \\ -u^3(\sin\Psi\sin\psi + \dot{\Psi}\cos\Psi\sin\psi) \end{pmatrix}.
\tag{2.18}
$$

These four spherical quantities belong to four different spherical quadrants. They are the polor, metor, vitor, and scalor in quaternity spacetime. After carrying out four consecutive rotatory operations upon Φ_{00}, we arrive at function

$$
\Phi_{40} = A_1 A_2 u^4 \cos\Psi\cos\psi.
\tag{2.19}
$$

At this moment, time and space radian angles switch their roles, i.e., Ψ represents a space radian angle while ψ denotes a time radian angle for further rotatory operations. For example,

$$\begin{aligned}\Phi_{40}^{0} &= -u^{4}\frac{dA_{2}\cos\psi}{dt}\int A_{1}\cos\Psi dl \\ &= A_{1}A_{2}u^{4}\frac{d\psi}{dt}\sin\psi\cdot\left(-\frac{1}{\Psi'}\int\cos\Psi d\Psi\right) \\ &= -A_{1}A_{2}u^{4}\left(-\frac{1}{\Psi'}\dot{\psi}\right)\sin\psi\sin\Psi \\ &= -A_{1}A_{2}u^{5}\sin\psi\sin\Psi,\end{aligned} \tag{2.20}$$

where

$$\dot{\psi} = -\frac{d\psi}{dt}, \tag{2.21}$$

$$\Psi' = -\frac{d\Psi}{dl}, \tag{2.22}$$

$$\begin{aligned}u &= -\frac{1}{\Psi'}\cdot\dot{\psi} = \frac{dl}{d\Psi}\left(-\frac{d\psi}{dt}\right) \\ &= \left(-\int C_{1}\omega\sin\alpha dt\frac{dl}{dt}\right)\cdot\left(-\frac{dC_{2}r\sin\beta}{dl}\frac{dl}{dt}\right) = \Omega_{2}^{0} \\ &= (C_{1}\cos\alpha)\cdot(C_{2}\cos\beta)\left(\frac{dl}{dt}\right)^{2} \\ &= v^{2}C_{1}C_{2}\cos\alpha\cos\beta \\ &= \Omega_{0}\end{aligned} \tag{2.23}$$

In the same manner, we may derive spherical quantities for another $2s2p_x2p_y2p_z$ quartet from

$$\begin{pmatrix}\Phi_{4}\\ \Phi_{5}\\ \Phi_{6}\\ \Phi_{7}\end{pmatrix} = \begin{pmatrix}\Phi_{40} - \dfrac{\partial\Phi_{40}}{\partial t}\\ \Phi_{40}^{0} - \dfrac{\partial\Phi_{40}^{0}}{\partial t}\\ \Phi_{40}^{02} - \dfrac{\partial\Phi_{40}^{02}}{\partial t}\\ \Phi_{40}^{03} - \dfrac{\partial\Phi_{40}^{03}}{\partial t}\end{pmatrix}, \tag{2.24}$$

which expands into trigonometric expression as

$$\begin{pmatrix}\Phi_4\\ \Phi_5\\ \Phi_6\\ \Phi_7\end{pmatrix} = A_1A_2u^4\begin{pmatrix}\cos\psi\cos\Psi - \dot{\psi}\sin\psi\cos\Psi\\ -u(\sin\psi\sin\Psi + \dot{\psi}\cos\psi\sin\Psi)\\ u^2(\cos\psi\cos\Psi - \dot{\psi}\sin\psi\cos\Psi)\\ -u^3(\sin\psi\sin\Psi + \dot{\psi}\cos\psi\sin\Psi)\end{pmatrix}. \tag{2.25}$$

These four electrons, $2s2p_x2p_y2p_z$, are in the negative direction of quaternity axes. The quaternity axis *1* refers to a four-dimensional time with zero-dimensional space quantity whereas factor u^4 denotes a four-dimensional space with zero-dimensional time sphere, both being at the same quaternity axis but in opposite directions. It has been stated that the home axis rotated four times consecutively returns to the same axis but in the opposite direction; and the home axis rotated eight times consecutively returns to its original position in quaternity coordinates so that

$$u^8 = 1. \tag{2.26}$$

After the electron Φ_4 reaches four-dimensional space at the quaternity axis of u^4, radian angles Ψ and ψ switch their time and space roles so that a counterclockwise rotation begins anew from zero-dimensional space with four-dimensional time and produces a polor Φ_5 that falls in the spherical quadrant I. Further counterclockwise rotations in the negative direction of quaternity axes give spherical quantities Φ_6, Φ_7, and finally Φ_0 in sequence. The space dimension increases while the time dimension decreases during the rotation processes. At spherical quantity Φ_0, the electron contains four-dimensional space with zero-dimensional time, but it immediately becomes zero-dimensional space with four-dimensional time upon switching radian angles Ψ and ψ to their default roles again. And another cycle begins from this point. Spherical quantities in equations (2.18) and (2.25) form four pairs of electrons with reversal roles of space and time radian angles, each pair being located at the same quaternity axes but in opposite directions. They are the polors, metors, vitors, and scalors, respectively, in quaternity coordinates.

Every spherical quantity contains two trigonometric terms at a time dimension interval. This indicates that each electron has two mutually orthogonal waveforms or two interrelated properties that form a complex function. Since a rotatory operation is in essence a Maxwell's equation in quaternity spacetime, an electron may transform by electromagnetic induction, for example, a changing magnetic field induces an electric field.

Among eight electrons, there are invariant rotation relationships between adjacent ones, e.g., between Φ_0 and Φ_1, Φ_1 and Φ_2, Φ_7 and Φ_0, and so on. For instance, if Φ_{00} and Φ_{01} represent the first and second terms of spherical quantity Φ_0 respectively, then there is a partial differential relationship between them:

$$-\frac{\partial\Phi_{00}}{\partial t} = \Phi_{01}, \tag{2.27}$$

and Φ_{01} has an integral relation with Φ_{10}, the first term of spherical quantity Φ_1.

$$\int \Phi_{01} dl = \Phi_{10} . \tag{2.28}$$

An electron may shift its state from one waveform to another and further migrate to the second electronic state according to the rotation principle. Equations (2.27) and (2.28) represent the pathway or mechanism of electronic transformation from state Φ_0 to state Φ_1. Four spherical quantities in equation (2.18) follow the similar counterclockwise rotation pathway whereas the other four in equation (2.25) switch time and space roles of radian angles Ψ and ψ in differential and integral operations, and lastly the rotation of Φ_7 results in Φ_0. At this moment the time and space roles of Ψ and ψ switch back to the default settings.

Within the neon shell, eight electrons, $2s^2 2p^6$, circulate according to the principle of rotatory operations. An electron of Φ_0 is traveling in quaternity spacetime and evolving into Φ_1 while Φ_1 is changing into Φ_2, and Φ_2 into Φ_3, and so on. Each electron is gradually shifting its state in the following link direction:

$$\Phi_0 \mapsto \Phi_1 \mapsto \Phi_2 \mapsto \Phi_3 \mapsto \Phi_4 \mapsto \Phi_5 \mapsto \Phi_6 \mapsto \Phi_7 \mapsto \Phi_0 , \tag{2.29}$$

where arrow $\mapsto$ indicates a rotation from one spherical quantity to another. Wave function Φ_0 is initially a 2s orbital of zero-dimensional space with four-dimensional time. It goes through three 2p-orbitals and then returns to a 2s orbital Φ_4 step by step. The number of space dimensions is increasing while that of time dimensions is decreasing, i.e., electronic space occupation is expanding gradually while time wrinkles are releasing during the rotations. The spherical quantity at u^4 is a medium of four-dimensional space with zero-dimensional time. The 2s-orbital Φ_4 with a u^4 factor means that radian angles Ψ and ψ must switch time and space roles so that counterclockwise rotations may proceed for spherical quantities in the negative direction of quaternity axes. After going through eight electronic states, the electron returns to its original waveform, and another cycle begins. Within the neon shell, a factor of u^8 is equivalent to a dimensionless number *1*. The electronic circulation involves eight harmonic oscillations prettily expressed by the quaternity equation. Eight electrons fulfill a complete cycle during the rotations, which explains satisfactorily why a full octet configuration of a neon shell is a stable system.

2.2.4. Dimension Encapsulation Mechanism

As we derived spherical quantities for electrons in a neon shell, we simply treated spherical quantities Ψ and ψ as two radian angles. What is the basis for this usage? Of course, to treat a quantity as an angle, it is necessary to use a sine or cosine operator to cast on it. What is the physical significance of circular functions that implement the conversions of Ψ and ψ from spherical quantities to radian angles? This section will examine these critical questions thoroughly.

It has been established that dynamic calculus is implemented by trigonometry. Upon the time component of a 1s electron, we may perform a differential operation

$$-\frac{d\Psi}{dt} = -C_1 \cos\alpha\,, \tag{2.30}$$

where the left-hand side may be interpreted as a dynamic fraction whose denominator is the magnitude of a full time dimension t and numerator is the current time quantity Ψ. In Figure 2.6, this fraction is proportional to ratio BC/AB where C moves from A toward B along the semicircle. The differential notation indicates cosinusoidal reduction of spherical quantity Ψ from a full dimension to vanishing, which corresponds to α angle rotation from 0 to π/2. Thus, when expressing a spacetime component changing from a full dimension to vanishing cosinusoidally, we may logically cast a cosine operator upon a proper radian angle. This is the fraction interpretation of a circular function and hence of a differential operation. Recognizing that electrons are in harmonic oscillations so that their spherical quantities are transforming in a sinusoidal pattern, we set forth two theorems with regard to dynamic calculus.

Theorem 1. For a spherical quantity f, the derivative form $-\partial f/\partial t$ indicates cosinusoidal change of its time component from a full dimension t to vanishing such as

$$-\frac{\partial f}{\partial t} = -B_1 \cos\alpha\,, \tag{2.31}$$

where B_1 is a dimension constant orthogonal to $\cos\alpha$, t is a fixed time dimension, and the cosinusoidal decrease of the time component in the range of $f \in [t,0]$ corresponds to the rotation of the radian angle in the range of $\alpha \in [0, \pi/2]$.

Theorem 2. For a spherical quantity g, the integral form $\int g dl$ indicates cosinusoidal change of its space component from naught to a full dimension l such as

$$\int g dl = B_2 \cos\beta\,, \tag{2.32}$$

where B_2 is a dimension constant orthogonal to $\cos\beta$, l is a fixed space dimension, and the cosinusoidal increase of the space component in the range of $g \in [0, l]$ corresponds to the rotation of the radian angle in the range of $\beta \in [\pi/2, 0]$.

By interpreting calculus in terms of a circular function, and the circular function in terms of a varying fraction or numerator, we have provided fresh physical significance for calculus. Trigonometrically, the cosinusoidal value of differential process $-\partial f/\partial t$ can be explained by a chord in a dimension diagram. As shown in Figure 2.7(a), in a circle with diameter PR of dimension t, as Q moves from P toward R along the semicircle, radian angle α rotates from

0 to $\pi/2$ and chord QR reduces from a full dimension t to vanishing following the cosine function. The minus sign indicates the reducing tendency of the spherical quantity.

On the other hand, integral process $\int gdl$ is illustrated by Figure 2.7(b). The integral is neither an indefinite integral nor a definite integral, but a dynamic integral implemented by trigonometry. As Q moves from P toward R along the semicircle, radian angle β rotates from $\pi/2$ to 0 so that chord PQ increases from naught to a full diameter dimension l following the cosine function. As state point Q traces along arc PQR, the time and space components of the electron are measured by chords QR and PQ, respectively.

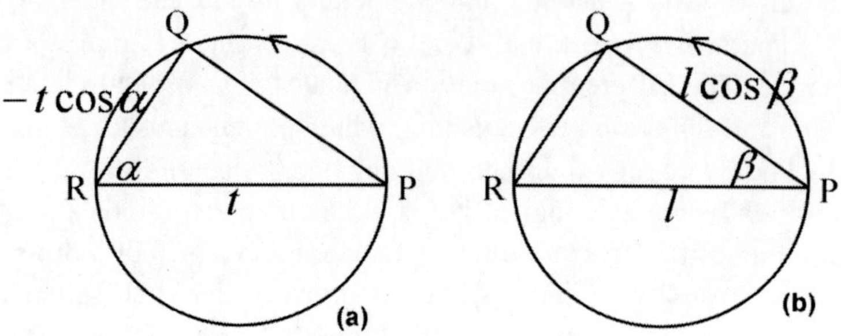

Figure 2.7. Trigonometric interpretation of dynamic calculus by cosinusoidal change of chords QR and PQ with point Q orbiting along the semicircle PQR.

To express cosinusoidal change of the time component within 1s electron Ω_0, we had equation (2.30); and to express cosinusoidal change of the time component within 2s electron Φ_0, we may write in the same manner

$$-\frac{\partial \Phi_0}{\partial t} = -A_1 \cos\Psi\,, \tag{2.33}$$

where the space component is omitted for brevity. The connection between equations 2.30 and 2.33 may be explained by rotatory vectors. As shown in Figure 2.8, vector C_1 rotates around origin O with its tip moving from A to B along arc APB, transforming spherical quantity $C_1\cos\alpha$ into $-C_1\omega\sin\alpha$ as radian angle α increases from 0 to $\pi/2$. Arc APB records the course of electronic transformation from Ω_0 toward Ω_2 in a time component represented by $-C_1\cos\alpha$ in equation (2.30). The tangential vector A_1 is another time dimension orthogonal to vector C_1. Shifting the perspective scope from the helium shell to the neon shell, we examine the motion of vector A_1 from the standpoint of P. Even though P is a moving point at Figure 2.8(a), it is treated as a stationary point at Figure 2.8(b), where vector A_1 rotates around P with its tip traveling from E toward F along arc EGF. The perspective of point P shifts from the periphery (Figure 2.8a) to the center (Figure 2.8b). When treated as a spherical quantity in the helium shell, Ψ is represented by vector C_1, whose tip is coordinated by $C_1\cos\alpha$ and $-C_1\omega\sin\alpha$; but when treated as a radian angle, Ψ is $\angle EPG$ or arc AP, which is increasing from 0 to $\pi/2$. So the cast of a cosine operator on Ψ

in the neon scope is justified mathematically. A cosine operator may be treated as a differential operator or an integral operator as was demonstrated in equations (1.54) and (1.56). If a radian angle in a circular function increases uniformly by a displacement of π/2, then its cosine operator is a differential notation. From Figure 2.6, Ψ equals to chord $-C_1\omega\sin\alpha$ in the helium shell and equals to $\angle DAC$ or arc AC subtending the chord in the neon shell, thus the differentiation of $-C_1\omega\sin\alpha$ with respect to a time dimension is equal to $A_1\cos\Psi$ upon the scope shift. And likewise the integration of $C_2 r\sin\beta$ over a space dimension is equal to $A_2\cos\psi$. The arc subtending chord $-C_1\omega\sin\alpha$ is the radian angle in spherical quantity $A_1\cos\Psi$.

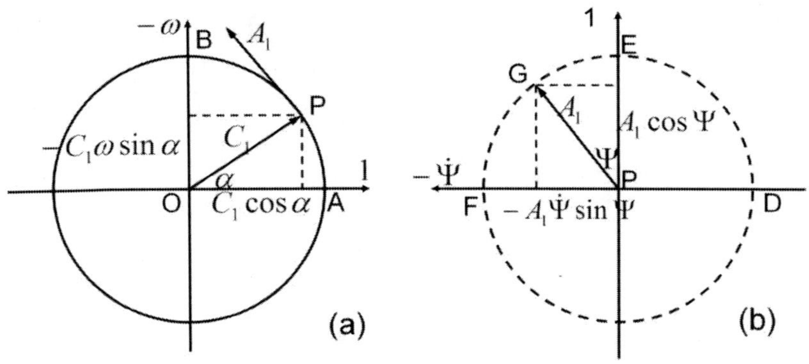

Figure 2.8. Rotatory vector interpretation of time component Ψ as (a) a spherical quantity $C_1(\cos\alpha-\omega\sin\alpha)$ and as (b) a radian angle encapsulating 1s electrons upon a standpoint shift.

After rotating counterclockwise around the vertex B for π/2, the initial and terminal sides of $\angle ABC$ parallel those of $\angle DAC$ (Figure 2.6a); and after rotating counterclockwise around the vertex B for π/2, the angle $\angle EBC$ parallels the circumferential angle $\angle BAC$ (Figure 2.6b). The Cartesian coordinate system rotates π/2 counterclockwise from Figure 2.8(a) to 2.8(b) so that spherical quantity $A_1A_2\cos\Psi\cos\psi$ coincides with spherical quantity $C_1C_2\cos\alpha\cos\beta$ in spacetime at the moment $\Psi = \alpha = 0$ (cf. Review question 2.9). Dimension encapsulation by equation (2.12) is actually a rotatory operation upon spherical quantity $-C_1C_2v\sin\alpha\sin\beta$.

Spherical quantity Ψ is the solution to simple harmonic oscillation equation of 1s electrons in time; and spherical quantity ψ is the solution to simple harmonic oscillation equation of 1s electrons in space. Simple harmonic oscillation is a projection of uniform circular motion. So Ψ and ψ define a circle in time and a circle in space respectively with their two legs corresponding to the diametrical dimension, the hypotenuse in the right triangle ABC (Figure 1.7). They represent a time dimension and a space dimension of the helium shell respectively. After casting cosine operators upon Ψ and ψ, we treat them as radian angles in dynamic rotation (Figure 2.6). Trigonometric functions $\cos\Psi$ and $\cos\psi$ encapsulate both time and space dimensions of the helium shell, respectively.

As A_1 rotates from PE to PF, the time component of spherical quantity Φ_0 transforms from $A_1\cos\Psi$ to $-A_1\dot{\Psi}\sin\Psi$ as Ψ rotates from 0 to $\pi/2$. The rotation of A_1 from PE to PF corresponds to the movement of point P tracing arc APB, surrounding the boundary of 1s electrons. So both 1s electrons are enclosed by arc APB. As was shown in Figure 2.6(a), diameter C_1 is subtended by arc ACB; and as was shown in Figure 2.8(a), vector C_1 is wrapped by arc APB dimensionally. Allowing radian angles α and Ψ to rotate a displacement of $\pi/2$, $\cos\Psi$ is dimensionally encapsulating the helium shell in the neon scope. The rotation step from Φ_0 to Φ_1 in the neon shell is geared with the rotation step from Ω_0 to Ω_2 with coinciding spherical quantities Φ_0 and Ω_0 in spacetime (cf. Figure 4.3). And the rotation step from Φ_7 to Φ_0 is geared with the rotation step from Ω_2 to Ω_0. The inner shell drives the outer shell. Such a dimension encapsulation mechanism is the reverse operation of the chain rules of dynamic calculus.

Geometrically, function Ψ in the helium shell and $\cos\Psi$ in the neon shell together determine a time dimension of 1s electrons. As was shown in Figure 2.6, a given length of chord $-C_1\omega\sin\alpha$ with subtending arc Ψ (implemented by $\cos\Psi$) in a circle reflects the curvature of the circle and hence is an indicator of circular diameter C_1. To specify, when the arc subtending a fixed-length chord is in the radian range of (0, $\pi/2$), the more severely it curves, the smaller is its circular diameter, the smaller is $\cos\Psi$ value, and vice versa. The magnitudes of a chord and its subtending arc codetermine the diameter of the circle. In this case, chord $-C_1\omega\sin\alpha$ and arc Ψ codetermine the time dimension of the helium shell (cf. Review question 2.9).

Since from the perspective of the neon shell, $\cos\Psi$ and $\cos\psi$ are two orthogonal diametrical indicators of a coiled time and space structure, the coordination of them encapsulates the spacetime block of both 1s electrons. As we built spherical quantity Φ_{00} based on $A_1A_2\cos\Psi\cos\psi$, we were actually dealing with a larger spatial sphere, i.e., using a helium shell as the block to construct a neon shell by dynamic calculus. In other words, we treated $A_1A_2\cos\Psi\cos\psi$ as zero-dimensional space with four-dimensional time in the scope of the neon shell and derived spherical quantities for 2s2p electrons according to rotatory operation principle beginning from that spherical point. Such a trigonometric transformation reveals the mechanism of dimension curling up.

Since Theodor Kaluza (1885-1954) proposed his fifth dimension curling up hypothesis in 1919, dimension encapsulation mechanism has not been demonstrated mathematically despite the popularity of trigonometric functions. Our wave function construction has confirmed the phenomena of dimension curling up and instantiated the mechanism for hiding a sphere or two dimensions of the helium shell from the neon shell.

Dimension encapsulation mechanism has far-reaching implications in philosophy (cf. Review question 2.11). However, dimension encapsulation phenomena need not be esoteric. The instances of hidden dimensions are ubiquitous in our daily life. For example, as we are counting the members of the United Nations, the United States is one of them. We disregard the fact that there are 50 states within the United States because we treat the nation as a unit. The details of 50 states are hidden from direct visualization from the standpoint of the United

Nations. Nonetheless, as we turn our focus toward the United States, we are interested in the 50 states that it comprises, but we still ignore the fact that the state of Delaware has three counties, namely Sussex, Kent, and New Castle, which are too detailed and irrelevant to the subject of study. Thus dimension-hiding phenomena are closely associated with the scope of concern. For example, when we are talking about John's family, we usually refer to John and his wife, his daughters and sons. The details about the family of a John's daughter including her husband and her kids are not of direct concern, which can be considered to be curled up dimensions in the scope of John's family. It is by the same logic that solid state physicists don't care about quarks.

2.2.5. The Differential Chain Rule

The role conversion of Ψ and ψ from spherical quantities of 1s electrons to radian angles for 2s electrons is implemented by dimension encapsulation mechanism. Conversely, the role conversion of Ψ and ψ from radian angles in the neon shell to spherical quantities in the helium shell reflects the chain rules of dynamic calculus. For example, considering the time component only, we have

$$\begin{aligned} -\frac{\partial\Phi_0}{\partial t} &= -A_1 \cos\Psi,\ \Psi \mapsto \Psi + \frac{\pi}{2} \\ &= -A_1\dot{\Psi}\sin\Psi \\ &= -A_1 \sin\Psi \cdot (-C_1 \cos\alpha),\ \alpha \mapsto \alpha + \frac{\pi}{2} \\ &= -A_1 \sin\Psi \cdot (-C_1\omega \sin\alpha). \end{aligned} \tag{2.34}$$

where arrow $\mapsto$ indicates smooth rotation of the radian angle up to a displacement of $\pi/2$. Both differential operations are implemented by cosine functions in two simultaneous steps. The first step of the differentiation $-\partial\Phi_0/\partial t$ is implemented by $-A_1\cos\Psi$ for radian angle Ψ to increase $\pi/2$, which results in $-A_1\dot{\Psi}\sin\Psi$. This process is illustrated in Figure 2.8(b) through the rotation of vector A_1 around point P. Function $-A_1\cos\Psi$ of the neon shell encapsulates the time dimension of the helium shell. The second step of differentiation $-d\Psi/dt$ is hidden in the neon shell by shorthand expression $\dot{\Psi}$. While the dimension of $-A_1\cos\Psi$ transforms into $-A_1\dot{\Psi}\sin\Psi$, $\dot{\Psi}$ is implemented by $-C_1\cos\alpha$ for radian angle α to increase $\pi/2$, which results in $-C_1\omega\sin\alpha$ finally. This second process is illustrated in Figure 2.8(a) through the rotation of vector C_1 from 0 to $\pi/2$. Both differential steps are coupled together by the differential chain rule and proceed simultaneously. In the first step, Ψ is a radian angle; but in the second step, it is treated as a spherical quantity. The role changes due to a perspective shift.

It is noteworthy that $\dot{\Psi}$ is different from ω because $\dot{\Psi}$ contains a hidden dimension of $-C_1\cos\alpha$ to spread out by the differential chain rule whereas ω is only a dimension constant. Angular velocity $\dot{\Psi}$ controls the pace of $-A_1 \cos\Psi$ oscillation while radian angle Ψ is the arc subtending chord $-C_1\omega\sin\alpha$ (Figure 2.6). As shown in equation (2.13), performing a rotatory operation upon Φ_{00} leads to spherical function $\Phi_{10} = -A_1 A_2 u \sin\Psi \sin\psi$ in the neon shell where the dimension factor u is implemented by Ω_0^0 in the helium shell (see equation 2.16). Rotatory operations in both shells are coupled together just as two cogwheels are geared to transmit circular motion (Figure 2.9).

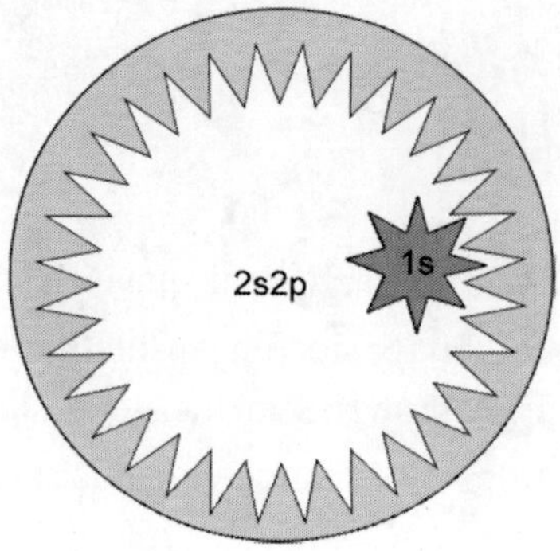

Figure 2.9. Conceptual illustration of the differential chain rule between 2s2p and 1s electrons as cogwheels in mesh.

The helium shell is inside the neon shell and their rotatory directions are all counterclockwise. The cogwheels in mesh reinforce the wave signals in the helium shell. When the 2s2p octet completes a full cycle, both 1s electrons have looped four cycles because radian angles Ψ and α are synchronized, so are radian angles ψ and β. The cycling of 1s electrons is the engine, each cycle delivering a rotatory operation upon 2s2p electrons. Four dimensions of time in the neon shell indeed derive from four 1s cycles. The circular motion is transmitted from the helium shell to the neon shell by dimension encapsulation mechanism. The oscillatory motion of 1s electrons drives the undulation of 2s2p electrons, which reflects inwards to reinforce the rhythm of 1s electrons by the chain rules of dynamic calculus.

Because electrons in both helium and neon shells are geared with each other (Figure 2.9), when the 2s2p electron octet completes a cycle, the 1s electrons have completed four cycles. Thus the angular velocity of 1s electrons is four times as quick as the angular velocity of the 2s2p electrons:

$$\omega = 4\dot{\Psi}, \tag{2.35}$$

so that the energies E_1 and E_2 for electrons in helium and neon shells within a neon atom have the same relationship.

$$E_2 = -\hbar\dot{\Psi} = -\frac{\hbar\omega}{4} = \frac{E_1}{2^2}. \tag{2.36}$$

This result agrees with the classical Bohr model and De Broglie's hypothesis on the wave-particle duality for energy quantized electrons within atoms in general (cf. Section 9.2), disregarding the fine differences between the energy levels of 2s, $2p_x$, $2p_y$, and $2p_z$ orbitals. But we have extended the particle-wave duality of electrons to quaternity because 2s2p electrons may take the forms of points, arcs, surfaces, and spherical clouds. As continuous spherical quantities are converted into discrete physical quantities, dimension stairs in the atomic structure take whole dimensions, resulting in energy quantization of electrons.

Bridging both differentiation steps, $\dot{\Psi}$ actually acts as a gateway between helium and neon shells. The copulation of both shells observes the differential chain rule mathematically. Such an elegant relationship between both 1s and 2s spherical layers is a demonstration of the entangled atomic structure. Strictly speaking, although $\dot{\Psi}$ is expressed by equation (2.14), there is a subtle difference between $\dot{\Psi}$ and $-d\Psi/dt$ under the context of dynamic differentiation. The former represents the initial state of the differential operation with respect to a time dimension while the latter term is traditionally associated with the final state of the differentiation. The former hides a time dimension of the helium shell in the neon scope while the latter is an unfolding time dimension of helium under focus. The observer's perspective shift accompanies the differential chain rule and the observer's standpoint determines the physical meaning of a function. In the neon shell, $\dot{\Psi}$ is only a dimension identifier like an imaginary unit as was shown in Figure 2.8(b) because differential process $-d\Psi/dt$ does not unfold yet. The outspread of the curled up dimension manifests only in the helium shell, where $-d\Psi/dt$ propagates through equations (1.15) to (1.18), describing both 1s electrons.

We call $\dot{\Psi}$ a gateway function for it connects both 1s and 2s electrons into a spacetime continuum within a neon atom. It really serves as a portal between the outer waves (2s2p electrons in this case) and the inner waves (1s electrons). The gateway function is also a unit imaginary number like i. This becomes evident when we compare spherical quantity Φ_0 with that of a 1s electron having the form of

$$\Omega_0 + \Omega_1 = C_1 C_2 (\cos\alpha\cos\beta - \omega\sin\alpha\cos\beta), \qquad (2.37)$$

where ω is another expression of i as indicated by equation (1.10). In the same way, we call ω a gateway function too because it communicates between 1s electrons and the nucleus. Thus, the derivative term $\dot{\Psi}$ is indeed a dimension identifier, an imaginary unit, and a gateway function. Spherical quantities of 1s and 2s2p electrons are spacetime continuous. They are in waveforms that feature sine and cosine functions, complex numbers, and gateway communications that synchronize waves between both spherical layers. In these ways, we have plotted a unique version of multi-dimensional hyperspace where ten spacetime dimensions represented by 1s2s2p orbitals are continuous via consecutive rotatory operations. The coupling of 1s and 2s2p electron waves is by dimension encapsulation mechanism in the outward direction and by the chain rules of dynamic calculus inward.

By the way, we have discussed the differential chain rule of time components in dynamic differentiation. For space components in dynamic integration, there is an integral chain rule parallel to the differential chain rule. The integral chain rule is a unique feature of dynamic calculus even though it does not exist in infinitesimal calculus. It is different from the substitution rule or the formula of integration by parts. Calculus chain rules include the chain rules for both differentiation and integration.

2.3. Four-Dimensional Harmonic Oscillations

We have characterized electronic orbitals in a neon shell by the quaternity equation and provided solutions to the equation in calculus and trigonometry. We shall further examine electronic transformation from one state to another in geometry and establish the connection between spherical quantities and four geometric elements in a spherical layer as was discussed in section 2.1. Because the electron octet as listed in equations (2.18) and (2.25) comprises four types of orbitals, $2s2p_x2p_y2p_z$, the transformation of spherical quantities includes four distinct geometric paths. This section illustrates the geometry of four electrons in rotatory operations of $\Phi_0 \mapsto \Phi_1 \mapsto \Phi_2 \mapsto \Phi_3 \mapsto \Phi_4$ in detail. On the negative direction of quaternity axes, another four rotation steps $\Phi_4 \mapsto \Phi_5 \mapsto \Phi_6 \mapsto \Phi_7 \mapsto \Phi_0$ follow similar geometric paths but with radian angle space and time roles reversal (Figure 2.10). The space and time roles reversal in radian angles in the counterclockwise rotation of Φ_4 indicates that the electron rotations from a new state of zero-dimensional space with four-dimensional time.

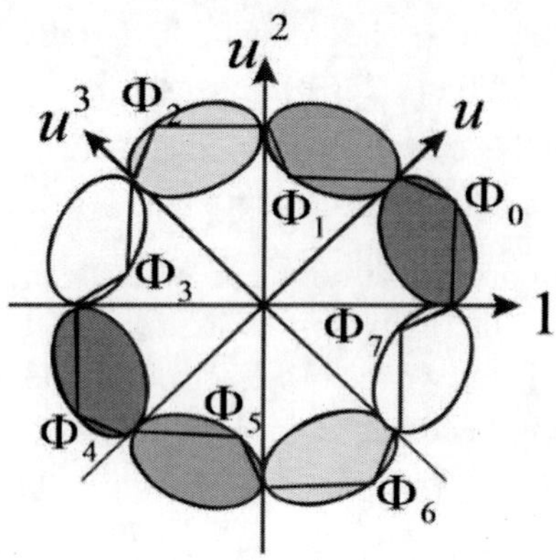

Figure 2.10. Dimension diagrams of 2s2p electronic orbitals with four pairs of similar orbital types transforming along four closed circles in quaternity coordinates, each pair orbiting along both semicircles of the same shaded circle.

Every two adjacent electrons have a rotatory operation relationship. For brevity, here we only consider the relationship between the first terms of spherical quantities Φ_0 and Φ_1 because differential and integral processes are undergoing concurrently so that their second terms may be omitted. Given that

$$-\frac{\partial \Phi_{00}}{\partial t} = \frac{\partial \Phi_{10}}{\partial l}, \tag{2.38}$$

where Φ_{00} and Φ_{10} refer to the first terms of Φ_0 and Φ_1 in equation (2.18) respectively, the rate of time component reduction of one electron from a full time dimension to vanishing is equivalent to the rate of space component change of its adjacent electron from a full space dimension to vanishing. The equation may be expressed in trigonometry as

$$\cos\Psi = \sin\psi, \tag{2.39}$$

where variables Ψ and ψ constitute two acute angles of a right triangle so that a rotatory operation is actually an alternative expression of the Pythagorean theorem.

Since both Ψ and ψ represent complementary time and space radian angles respectively, together they determine the course of an electron along a semicircle in a dimension diagram (Figure 2.11). As Ψ increases from 0 to π/2, time component $A_1 \cos\Psi$ transforms into $-A_1\dot{\Psi}\sin\Psi$; and as ψ reduces from π/2 to 0, space component $A_2 \cos\psi$ transforms into $-A_2 \frac{1}{\psi'}\sin\psi$. Spherical quantities of 2s2p electrons are derived from the coordination of time and space components, such as $A_1 \cos\Psi$ and $A_2 \cos\psi$. At any moment, the Pythagorean theorem synchronizes the rotation of radian angle variables Ψ and ψ for the electrons in harmonic oscillations.

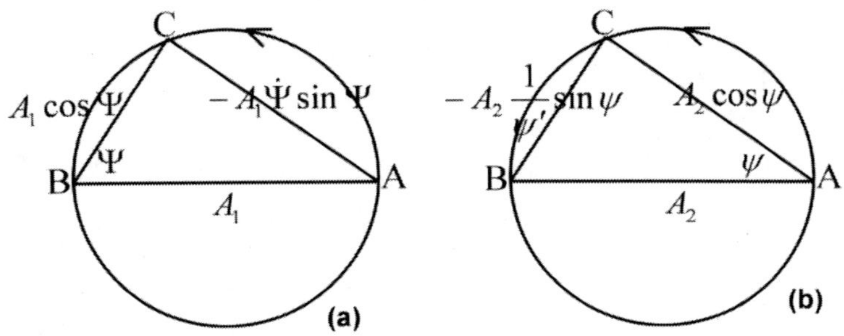

Figure 2.11. Dimension diagram of synchronized differential and integral operations of $\Phi_0 \mapsto \Phi_1$ in (a) time and (b) space.

Figure 2.11 shows the dimension diagram of electronic transformation from Φ_0 to Φ_1. Dimension diagrams for other electrons follow a similar pattern with semicircles as the courses of dynamic calculus (Figure 2.10). These courses have diverse correspondences in geometric paths. We shall discuss four typical cases characteristic of 2p$_x$, 2p$_y$, 2p$_z$, and 2s electrons in a neon shell or a carbon atom. The elucidation of the electron configuration of a carbon atom with these four types of outer electrons has far-reaching implications in organic chemistry as will be seen in Chapter 6.

2.3.1. Harmonic Oscillation in One Dimension

The spherical quantity for a 1s electron is $C_1C_2 \cos\alpha \cos\beta$; and the spherical quantity for the initial 2s electron is $A_1A_2 \cos\Psi \cos\psi$. Both are electronic particles with the same geometric radius at the phase of ($\beta = \pi/2, \psi = \pi/2$), but the 1s sphere is a one-dimensional sphere with a variable isotropic radius only while the neon sphere is a four-dimensional one with four variable anisotropic orientations. At this moment, both electrons coincide. The 1s electron reaches a minimum size and begins to expand while the 2s electron arrives at the nuclear center and begins to skew into a polor. This is a chance for the neon shell to exchange matter and energy with the helium shell because the 1s and 2s electronic orbitals are spacetime continuous within the atom.

The initial spherical quantity $A_1A_2\cos\Psi\cos\psi$ built by the encapsulation of the helium shell represents a sphere of four-dimensional time and zero-dimensional space in the neon shell. Sphere O is zero-dimensional in space because its center is located at the nucleus; it is also four-dimensional in time because of its hypothetical maximum temporal size. Spherical quantity $A_1A_2\cos\Psi\cos\psi$ encapsulates both 1s electrons.

In Chapter 1, we have indicated that space defines the scale of a 1s electron and time the density for convenience. However, time has a vivid tangible shape indicative of its magnitude here. When space reaches a minimum size, time attains a maximum sphere. In the helium shell, we regard the full time 1s sphere as one-dimensional along all radial direction; but in the neon shell, the same 2s sphere has four time dimensions that must be traversed by an electron through four consecutive rotations.

As shown in Figure 2.12(a), the electron cloud of sphere O starts to condense and its center is gradually displaced from the nucleus. The 2s electron, which initially permeates the entire sphere as an electron cloud, gradually contracts its activity sphere as its spherical center moves from point O toward point B, which represents a solid particle. This polarization corresponds to the rotatory operation of spherical quantity $A_1A_2\cos\Psi\cos\psi$. As the electron gradually appears from a misty cloud to a solid particle, time reduces from four dimensions to three while space increases from zero dimension to one, i.e., from center O to position B with a one-orientational displacement from the nucleus. We characterize the electron by an electron cloud because of its four-dimensional time property within the atom even though it is always a particle when actually detected instrumentally.

At any specific moment, the temporal radius of sphere O reduces from OB to AB as its center O moves to center A, which is the vertical projection of point E onto OB. Hypothetical point E is tracing uniformly along a semicircle OEB. In other words, spherical center A is traveling along the straight line OB but follows the harmonic oscillation principle as if it were tracing smoothly along a hypothetical semicircle OEB.

In right triangle OEB, side BE measures the time component of the electron, corresponding to the product of hypotenuse BO and $\cos\Psi$ whereas side OE is the space displacement from the origin O, corresponding to the product of hypotenuse OB and $\cos\psi$. Hypotenuse OB represents a time dimension and a space dimension that are antiparallel. With the rotation of Ψ angle from 0 to $\pi/2$, ψ rotates from $\pi/2$ to 0 so that time component $BO\cos\Psi$ reduces while space component $OB\cos\psi$ increases. From point O to point B, spherical quantity $A_1A_2\cos\Psi\cos\psi$ changes from dimension A_1A_2 to dimension $-A_1A_2u$. Spherical quantity $-uA_1A_2$ is located downstream of A_1A_2 at a full velocity dimension interval. Semicircle OEB is the hypothetical course for the electron to traverse diameter OB uniformly, the first space dimension. The course represents electronic transformation from the 2s state to the full $2p_x$ state. At point B, the electron has one-dimensional space OB and three-dimensional time as a solid particle B.

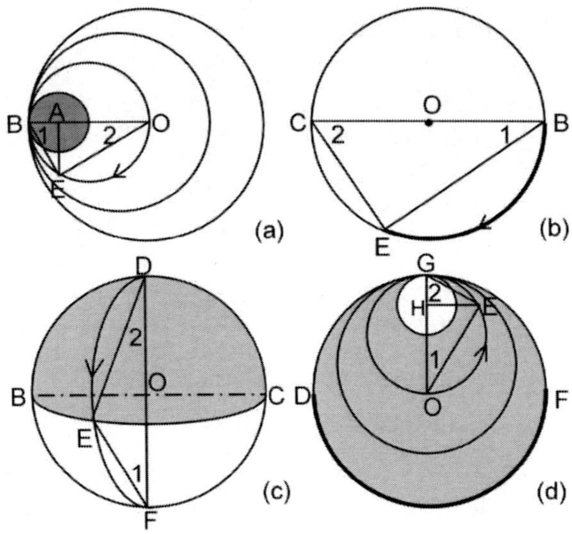

Figure 2.12. The geometric paths of electrons in various transformations: (a) 2s$\mapsto$ $2p_x$, (b) $2p_x \mapsto 2p_y$, (c) $2p_y \mapsto 2p_z$, and (d) $2p_z \mapsto$ 2s where $\angle 1$ denotes radian angle Ψ and $\angle 2$ denotes radian angle ψ.

2.3.2. Harmonic Oscillation in Two Dimensions

As shown in Figure 2.12(b), semicircle BEC represents the course as well as the geometric path of an electron transforming from the $2p_x$ state to the $2p_y$ state. A full $2p_x$ electron is a point particle at B whereas a full $2p_y$ electron is a semicircle of BEC in geometry. As the electron stretches along the semicircle from point B to point C, radian angle Ψ increases from $\pi/2$ to π (i.e., decreases from $\pi/2$ to 0 in right triangle BEC). The geometrical metamorphosis reflects the rotatory operation of spherical quantity $-uA_1A_2\sin\Psi\sin\psi$. The initial point B denotes a spherical quantity at dimension $-uA_1A_2$. After point E reaches the end point C, the spherical quantity becomes $u^2A_1A_2$. Semicircle BEC is the actual path for the electron to attain the second space dimension BC.

At any specific moment along the course, chord CE indicates the time component of the electron while chord BE is the space component of it. In right triangle BEC, side CE equals $\text{CB}\sin\Psi$ and side BE equals $\text{BC}\sin\psi$. The time component is decreasing while the space component is increasing with the decrease of radian angle Ψ from $\pi/2$ to 0. Even though semicircle BEC is the course of the electron, time and space components of it are measured by their corresponding chord lengths, CE and BE, for any point E along the course. As the electron reaches the end point C, it has traversed the second space dimension denoted by diameter BC and has drawn the shape of a full $2p_y$ orbital as a semicircle on the flat plane. A $2p_y$ electron is two-dimensional in space because its center is away from the central nucleus in two orientations, an orientation along semicircle BEC in addition to the $2p_x$ electron along a radial direction OB. Throughout this book, a semicircle always refers to a semicircular arc.

2.3.3. Harmonic Oscillation in Three Dimensions

Figure 2.12(c) explains the motion of an electron transforming from a $2p_y$ orbital to a $2p_z$ orbital. It is the rotation of semicircle BEC around axis BC in geometry. The position of whole semicircle BEC is determined by radian angle Ψ in right triangle DEF. With the increment of radian angle Ψ from π to $3\pi/2$ (i.e., from 0 to $\pi/2$ in right triangle DEF), semicircle BEC smears over a hemispherical surface, which gives the shape of a full $2p_z$ orbital. In other words, the electron spreads over from a semicircle into a hemispherical surface in geometry during the life of $u^2 A_1 A_2 \cos\Psi \cos\psi$. The initial position of semicircle BEC at BDC denotes a spherical quantity at dimension $u^2 A_1 A_2$. After point E reaches point F, the spherical quantity becomes $-u^3 A_1 A_2$ in spacetime.

At any specific moment along the course of DEF, the interval between semicircles BDC and BEC, as measured by side DE in right triangle DEF, represents the space displacement of the electron along the third dimension. Side DE defines the progress of a curved surface shown in Figure 2.12(c) as a shaded area while side FE represents the time component of the electron. DE=DF $\cos\psi$ and FE=FD $\cos\Psi$. As Ψ rotates from 0 to $\pi/2$ in right triangle DEF, space increases while time dwindles. After the electron traverses through diameter DF in time and space, its spherical quantity tacks on another velocity factor. The geometrical shape of a full $2p_z$ is a hollow hemisphere such as BDCFE bra, which is three-dimensional in space. A $2p_z$ electron has potential energy higher than a $2p_y$ electron because it spreads over another space dimension in DF orientation in addition to the initial $2p_y$ electron.

2.3.4. Harmonic Oscillation in Four Dimensions

Figure 2.12(d) indicates electronic transformation from orbital type $2p_z$ to type 2s. The initial $2p_z$ electron is a hemispherical surface DF, which gradually grows into a crescent. The waxing of the crescent involves the inward expansion of the hemispherical surface into a solid sphere as the volume of empty sphere H diminishes from the maximum sphere O into point G along the radial direction. Diameter OG is the fourth space dimension for the electron to traverse harmonically via hypothetical semicircle OEG (cf. Section 3.3.4 for an alternative description).

At any specific moment, side OE in right triangle OEG represents the space component of the electron with OE=OG $\sin\psi$; and side GE represents the time component of the electron with GE=GO $\sin\Psi$. As radian angle Ψ increases from $3\pi/2$ to 2π (i.e., decreases from $\pi/2$ to 0 in right triangle OEG), the time component reduces while the space component increases. The process is the rotatory operation of spherical quantity $-u^3 A_1 A_2 \sin\Psi \sin\psi$. As the shaded crescent develops into a full sphere, the spherical quantity transforms from $-u^3 A_1 A_2$ into $u^4 A_1 A_2$. Diameters OG and GO are antiparallel space and time dimensions, and their ratio OG/GO equals $-u$. Once again, the full 2s sphere of $\Phi_{40} = A_1 A_2 u^4 \cos\Psi \cos\psi$ coincides with a 1s electron $\Omega_2 = -C_1 C_2 v \sin\alpha \sin\beta$ at the center O with $\psi = \pi/2$ and

$\beta = \pi/2$. The 1s electron has attained maximum space and is ready to contract spherically while the 2s electron has reached the center and is ready to skew into a polor. The 2s electronic orbital is spacetime continuous with the 1s orbital at the center. This handshake exchanges matter and energy between s-orbitals in the helium and neon shells. The 2s electron occupies the whole sphere O. It is a four-dimensional space with zero-dimensional time sphere. Sphere O is four-dimensional in space because it is the maximum spatial scale ever attained by the electron; and it is also zero-dimensional in time because of its center at the nucleus.

At this state, time and space radian angles (Ψ and ψ) switch their roles so that further rotatory operations upon the spherical quantities begin anew from the state of zero-dimensional space with four-dimensional time. The space and time roles switch is materially realized through the exchange of both electrons Ω_0 and Ω_2 at the coincidence with 2s sphere. This becomes evident as we compare equation (2.16) with equation (2.23). We perform counterclockwise rotation upon $\Phi_{40} = A_1 A_2 u^4 \cos\Psi \cos\psi$ from the starting point of $\Psi = 0$ as the radian angle for the space component of the spherical quantity (Equation 2.20). The radian angles switch is due to entering the negative direction of quaternity coordinates.

Further oscillatory processes of $\Phi_4 \mapsto \Phi_5 \mapsto \Phi_6 \mapsto \Phi_7 \mapsto \Phi_0$ are in the negative direction of quaternity axes. We have to switch time and space roles of radian angles Ψ and ψ for spherical quantities in the counterclockwise rotation series. They constitute another quartet of $2s2p_x2p_y2p_z$ orbitals. Each pair of similar orbital type electrons transform along the course of two semicircles in the opposite directions of quaternity axes. Each pair make up a closed circle as was drawn by the same shade (Figure 2.10). They are symmetric in spacetime about the origin of quaternity coordinates. For example, the time component of the Φ_0 electron is equivalent to the space component of the Φ_4 electron; and the space component of the Φ_0 electron is equivalent to the time component of the Φ_4 electron. They are two 2s orbitals. Spherical quantity Φ_3 with one-dimensional time and three-dimensional space is in symmetry with spherical quantity Φ_7 of one-dimensional time and three-dimensional space, both being $2p_z$ electronic orbitals located at the opposite directions of the quaternity axis. After eight consecutive rotatory operations, an electron returns to its original state and initiates another cycle.

The foregoing description has demonstrated the principle of right triangles, i.e., the Pythagorean theorem as expressed by equation (2.39). Four distinct diameters OB, BC, DF, and OG in Figure 2.12(a), (b), (c), and (d) represent four space dimensions respectively for an electron to traverse during $\Phi_0 \mapsto \Phi_1 \mapsto \Phi_2 \mapsto \Phi_3 \mapsto \Phi_4$ transformation processes via their subtending semicircles. We label these space dimensions as l_3, l_2, l_1, and l_0 corresponding to 1D, 2D, 3D, and 4D orientations, respectively. We may generalize l_3, l_2, l_1, or l_0 dimension as l, which refers to the current space dimension in transformation or the dimension of home axis. And time dimension t follows similar usage. This general usage has been adopted in the text above and so below.

As an electron undergoes $2s \mapsto 2p_x \mapsto 2p_y \mapsto 2p_z \mapsto 2s$ rotations, the number of time dimensions decreases from 4 to 0 while the number of space dimensions increases from 0 to 4 sinusoidally (Figure 2.13). Four time dimensions lost are t_0, t_1, t_2, and t_3 in sequence while four space dimensions gained in sequence during the rotations are l_3, l_2, l_1, and l_0. These dimension sequences are arranged for an electron rotating counterclockwise in the same order in the negative direction of quaternity axes. Eight 2s2p electrons move one wave after another, forming a continuous medium of four spacetime dimensions.

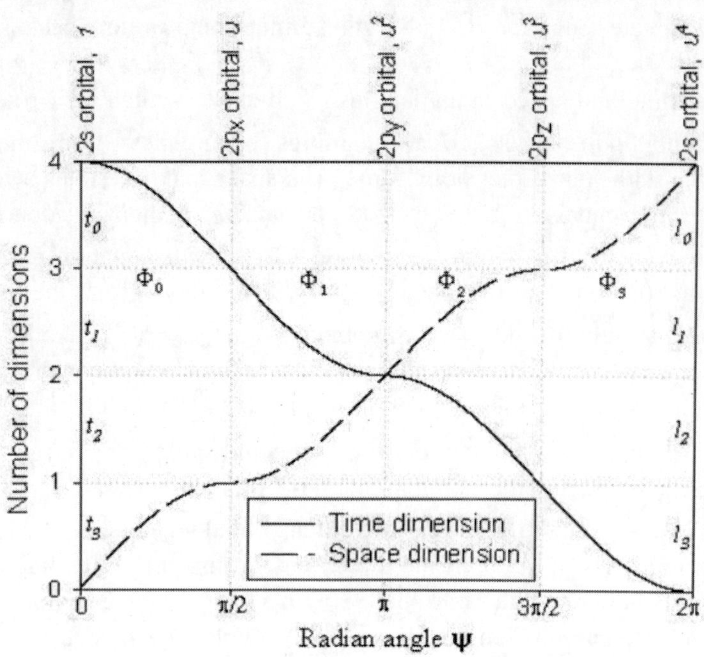

Figure 2.13. Harmonic oscillations of electrons in various space and time dimensions in correspondence with orbital types and quaternity axes.

2.3.5. Four Kinds of Orbital Geometries

From the above discussion on the geometries of electrons in various harmonic oscillations, it is clear that a $2p_x$-orbital is moving along a radial direction; a $2p_y$-oribital is stretching the particle into a two-dimensional semicircular arc; a $2p_z$-electron is spreading over a three-dimensional hemispherical surface; and a 2s-orbital is permeating the entire four-dimensional sphere. As shown in Figure 2.14, both $2p_x$ orbitals are represented by two symmetric points corresponding to points P_1 and P_2 in Figure 2.2; both $2p_y$ orbitals are represented by two complementary semicircles corresponding to lines L_1 and L_2 in Figure 2.2; both $2p_z$ orbitals are represented by two hemispherical surfaces corresponding to flat planes S_1 and S_2 in Figure 2.2; and both 2s electrons are represented by two concentric spheres corresponding to spheres R_1 and R_2 in Figure 2.2. The electron octet takes the shapes of four geometric elements in a spherical layer.

Under quaternity coordinates, a pair of $2p_x$ electrons are located at opposite directions of the u-axis. They are spherical quantities of one-dimensional space with three-dimensional time, constituting two symmetric points. A pair of $2p_y$ orbitals are located at opposite directions of the u^2-axis. They are spherical quantities of two-dimensional space with two-dimensional time, constituting two complementary semicircles. A pair of $2p_z$ orbitals are located at opposite directions of the u^3-axis. They are spherical quantities of three-dimensional space with one-dimensional time, constituting two complementary hemispherical surfaces. These three pairs of 2p electrons have an equal distance from the nuclear center and hence have an equal potential energy. A pair of 2s orbitals are located at opposite directions of the *1*-axis. They are two concentric spheres with space and time dimensions reversal (Table 2.3). Both 2s electrons are closer to the nucleus than 2p electrons and hence have a lower energy than 2p electrons.

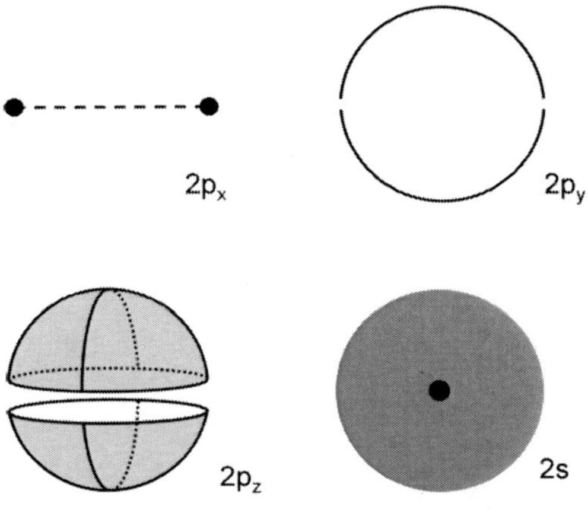

Figure 2.14. The geometrical shapes of four orbital types.

Table 2.3. Relationships between spherical quantities, space and time dimensions, and orbital types within a neon shell

Quaternity axis	Quantity types	Spherical quantities	Space dimension	Time dimension	Electronic orbital
1	Scalors	Φ_0, Φ_4	0, 4	4, 0	2s
u	Polors	Φ_1, Φ_5	1	3	$2p_x$
u^2	Metors	Φ_2, Φ_6	2	2	$2p_y$
u^3	Vitors	Φ_3, Φ_7	3	1	$2p_z$

There is a node between electron clouds of each orbital pair. A pair of $2p_x$ orbitals are separated by the inner 2s sphere; a pair of $2p_y$ orbitals are separated by both $2p_x$ points; a pair of $2p_z$ orbitals are separated by the $2p_x$ and $2p_y$ circle; and both 2s electrons are channeled by 1s electrons. We paint one sphere to represent a 2s electron and a concentric black point to indicate another 2s sphere at the other end of the tunnel. A polor includes the u-axis, but not the *1*-axis whereas a 2s-orbital (of wave function Φ_0) includes the *1*-axis, but not the u-axis, and the dimension between both axes is the first spherical quadrant. Other spherical quantities and orbital types follow the same denotation method.

Under spherical coordinates (ρ, ϕ, θ), ρ is the distance from the origin, ϕ is the polar angle from the Z-axis, and θ is the azimuthal angle in the X-Y plane. A 2s orbital is defined as $\rho = 0$ or $0 < \rho < R_0$ where R_0 is the maximum spatial sphere of the neon shell; a $2p_z$ orbital is defined as $\rho = R_0$ and ($0 \leq \phi < \pi/2$ or $\pi/2 < \phi \leq \pi$); a $2p_y$ orbital is defined as $\rho = R_0$ and $\phi = \pi/2$ and ($0 < \theta < \pi$ or $\pi < \theta < 2\pi$); and a $2p_x$ orbital is defined as $\rho = R_0$ and $\phi = \pi/2$ and (

θ=0 or θ=π). Four types of orbitals are contiguous in a neon shell. It becomes apparent that the anisotropic 2p electronic orbitals account for the chiral property of a carbon atom bonded to four different substituents. Strong evidence of these 2s2p shapes can be found in Chapter 6 as we discuss the structure of organic molecules.

2.4. Compatibility with Classical and Quantum Mechanics

Dynamic calculus of spherical quantities is the theme of the theory of quaternity that runs through the entire book. It is a revolutionary idea, but we find that it integrates with well-established physics perfectly. First, the application of quaternity equation and rotatory operations to electronic orbitals actually indicates that electrons are oscillating among various states. The quaternity equation is a well-known one-dimensional oscillation equation in classical mechanics. However, in a neon shell, all electrons are undergoing harmonic oscillations around the nucleus simultaneously, each of them complying with the quaternity equation in its own dimension. The quaternity equation governs four dimensions indeed. Thus, classical physics can be applied to electronic orbitals with proper new interpretation without changing its mathematical formula. In this sense, quaternity integrates with classical mechanics and effectively extends it to the atomic world.

Second, electronic oscillations follow the principle of rotatory operations. From spherical quantities, it is easy to verify that every two adjacent electrons are orthogonal such as Φ_0 and Φ_1, Φ_1 and Φ_2, Φ_2 and Φ_3, Φ_3 and Φ_4, Φ_7 and Φ_0, etc. Since Faraday's law is equivalent to the principle of rotatory operations, electrons are interacting and transforming according to the electromagnetic law. Moreover, consider the relation of

$$\frac{\partial P(x,t)}{\partial t} = -\frac{\partial J(x,t)}{\partial x}, \tag{2.40}$$

where $P(x,t)$ represents a probability density function of an electron in region x, and $J(x,t)$ denotes the flux or probability current into that region. Quantum mechanics states that a change in the density in x is compensated by a net change in flux into that region. If we replace x with l in quaternity spacetime, then equation (2.40) is actually a rotatory operation. Thus quaternity comfortably adapts to quantum mechanics in the interpretation of the relationship between the probability density and flux.

Third, the theory of quaternity is compatible and complementary with quantum mechanics. Harmonic oscillations of electrons within four spherical quadrants are in correspondence with four quantum numbers in quantum mechanics. The scalors correspond to the principal quantum number that accounts for the energy level; the vitors correspond to the azimuthal quantum number that determines the magnitude of orbital angular momentum; the metors in the equatorial plane are associated with the magnetic quantum number; and the polors, as two spacetime symmetric points, describe intrinsic angular momentum perfectly. This comparison demonstrates the existence of four typical physical properties in a spherical layer. With regard to the inert property of a neon atom, the quaternity theory indicates that 2s2p electrons are a set of valid spherical quantities fulfilling a complete cycle in their rotatory operations and hence explains satisfactorily why a full octet configuration is conservative.

The definitions of four varying spherical quantities and four quantum numbers are not exactly the same, but there are also similarities between them. In a neon shell, four spherical quadrants belong to the same spherical layer of principal quantum number 2. Quaternity defines a spherical layer between *1* and u^4 to account for this energy band. All 2p-electrons have an azimuthal quantum number 1 while 2s orbitals have an azimuthal quantum number 0. With regard to three pairs of 2p electrons, quaternity uses dimension factors of u, u^2, and u^3 to account for their different magnetic angular momentums, among which u^2 is in space and time symmetry by itself corresponding to magnetic quantum number 0 while u and u^3 are a pair of space and time reversal quantities corresponding to magnetic quantum numbers of +1 and −1. The spin numbers, ±1/2, indicate the opposite electronic spins, which are distinguished in quaternity by the pairs of scalors, polors, metors, and vitors that colonize the same quaternity axes but in opposite directions. Thus, four types of quantum numbers in a neon shell are well expressed in quaternity.

Finally, quaternity is advantageous over quantum mechanics in formulating a wave equation for electrons because it adopts a new spacetime framework instead of Cartesian coordinates. Schrödinger's time-dependent equation, a cornerstone in quantum mechanics, takes the following form given wave function Φ (x, y, z, t) and potential energy $V(x, y, z)$:

$$i\hbar \frac{\partial \Phi}{\partial t} = -\frac{\hbar^2}{2m} \nabla^2 \Phi + V\Phi \,. \tag{2.41}$$

By using the Laplacian operator ∇^2, it implies that electronic orbitals distribute equally in the X, Y, and Z directions, i.e., space is isotropic in three axis directions. Notwithstanding this impression, quantum mechanics handles Schrödinger's equation by transforming the isotropic Cartesian coordinates into anisotropic spherical coordinates pertinently and then tries to derive quantum information under the constraints that the equation must have solutions and that the solutions must be normalizable. Only via this transformation is quantum mechanics successful in getting much useful information on electronic orbitals corresponding to discrete quantum numbers, but the information is far from complete. By this approach, even the solution for the electron in a hydrogen atom is a formidable mathematical task. This indirectly illustrates the merits of quaternity spacetime over Euclidean space. As a matter of fact, the formulation of Schrödinger's equation and the approach to solve it have predetermined its outcome. Spherical quantities in quaternity spacetime obey dynamic calculus implemented by trigonometry. Because of the close relationship between space dimensions in a spherical layer and spherical coordinates, the solutions to the quaternity equation are very neat in terms of spherical coordinates as were described in section 2.3.5. The simplicity and beauty of quaternity reveal that it describes nature correctly. For the rigorous mathematical proof of duality as a solution to one-dimensional Schrödinger's equation and quaternity as a solution to three-dimensional Schrödinger's equation, please refer to section 4.5.

Conclusion

We have described quaternity spacetime using geometry, trigonometry, and calculus in a coherent manner with the instantiation of electronic orbitals within a neon shell. To be specific, we have identified four distinct but continuous space dimensions in a spherical layer, discovered the mechanism for dimension curling up for the first time, examined rotatory operations as the rule of electronic transformation, and defined the geometrical shapes of 2s2p orbitals in the context of harmonic oscillations. Quaternity space is spherical space; quaternity axes are dimensional axes; quaternity coordinates measure spherical quantities in dynamic calculus. A semicircle connecting two axes in quaternity coordinates (Figure 2.10) is the course of an electron in electromagnetic induction rather than the trajectory of an object in kinematic movement.

Electrons within a neon shell satisfy the quaternity equation. They are oscillating simultaneously and individually in harmonic ways in four dimensions: a $2p_x$ electron is point displacement, moving along a one-dimensional radius; a $2p_y$ electron is line stretching, progressing along a two-dimensional semicircular arc; a $2p_z$ electron is surface spreading, smearing over a three-dimensional hemispherical surface; and a 2s electron is solid expansion, permeating the entire sphere. The neon shell is four-dimensional in space. This original perception is remarkably different from the traditional worldview of spacetime. It directly challenges the traditional isotropic 2p orbital model and the theory of orbital hybridization based on this model. We shall discuss this important issue in Chapter 6.

The quaternity description of 2s2p electrons by smooth trigonometric functions in four dimensions is a logical extension of classical physics, different from quantum mechanics that abruptly deviates from our common sense. Four space dimensions in a neon shell, expressed as l_3, l_2, l_1 and l_0, are four diameters in dimension diagrams which electrons traverse via their corresponding semicircles uniformly during harmonic oscillations. It is interesting to note that electrons trace along the semicircles in diverse geometric manifestations, but their space and time components at any moment are measured by their subtended chords consistently. The semicircular track is uniform Newtonian time while its subtended chord is sinusoidal time in a spherical quantity. In a dimension diagram, two orthogonal time or space components as two legs and the diametrical dimension as a hypotenuse always maintain a right triangle so that the Pythagorean theorem indeed governs electronic motion in all cases. It is striking that Faraday's law, the Pythagorean theorem, and the principle of harmonic oscillations boil down to dynamic calculus of spherical quantities that we have formulated for electronic motion. The unification of classical physics and particle physics is effectively underway.

Chapter 3

Dynamic Calculus in Vector Calculus

This chapter is an exploration of adopting vector calculus to express dynamic calculus of spherical quantities. As an electron transforms from one state to another, its geometrical shape evolves continuously. Vector calculus is introduced to account for the geometric evolution in Cartesian coordinates during electronic transformation. Spherical quantities are represented by vectors with variable X, Y, and Z components in Euclidean space. Spherical quantities in dynamic calculus are expressed by vector calculus with respect to point displacement, arc length, surface area, and spherical volume. This is an advanced topic intended to deepen the subject discussion. A thorough understanding of conventional vector calculus and the general Stokes' theorem is a prerequisite for studying this chapter. Here we present the structure of electron arrangement in a neon shell from an alternative but unified approach. It is hoped that this chapter will help the readers get better insight into the property of spherical quantities and the structure of four-dimensional space and master the mindset of quaternity essentially.

3.1. Introduction

On the Cartesian plane, exponential function $e^{-i\alpha}$ or complex function $\cos\alpha - i\sin\alpha$ defines a point tracing from coordinates A(1, 0) to B(0,1) along a quarter circle as radian angle α rotates from 0 to $\pi/2$, but in quaternity coordinates $\cos\alpha - i\sin\alpha$ represents differential transformation of a spherical quantity from $\cos\alpha$ to $-i\sin\alpha$ as α rotates from 0 to $\pi/2$ (Figure 3.1). The differentiation involves time dimension transformation from 1 to $-i$ (= $d\alpha/dt$) instead of position movement. Although the meanings of the complex function in Cartesian coordinates and in quaternity coordinates are different, both coordinate systems coincide, which allows us to translate dynamic calculus of a spherical quantity into vector calculus along an arc [6, 7].

It has been established in section 1.5.1 that dynamic differentiation of a spherical quantity may be represented by a unit rotatory vector, **u**, on a two-dimensional plane. In Figure 3.1, a rotatory vector **u** is the directed segment OC rotating around origin O. We may also use a directed curve AC starting from point A and extending along the quarter arc to represent the path of circular motion in Cartesian coordinates as well as the course of dynamic calculus in quaternity coordinates. The rotation of vector **u** around the origin with its vector tip along arc AC corresponds to dynamic differentiation of a spherical quantity from 1 to $(\cos\alpha - i\sin\alpha)$.

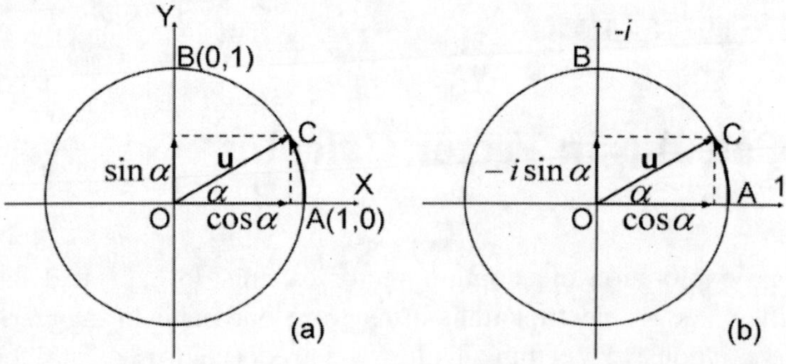

Figure 3.1. Comparison of a complex function in (a) Cartesian coordinates and (b) quaternity coordinates.

A spherical quantity in dynamic calculus is shown in a dimension diagram (Figure 3.2) where a state point C traces along semicircle ACB uniformly. Diameter AB represents a time dimension as well as a space dimension during the rotatory operation. The dynamic differential process can be expressed by smooth rotation of α for a displacement of $\pi/2$ followed by the differential chain rule:

$$-\frac{dC_1\cos\alpha}{dt} = -C_1\cos(\alpha+\frac{\pi}{2})\cdot\frac{d\alpha}{dt}. \tag{3.1}$$

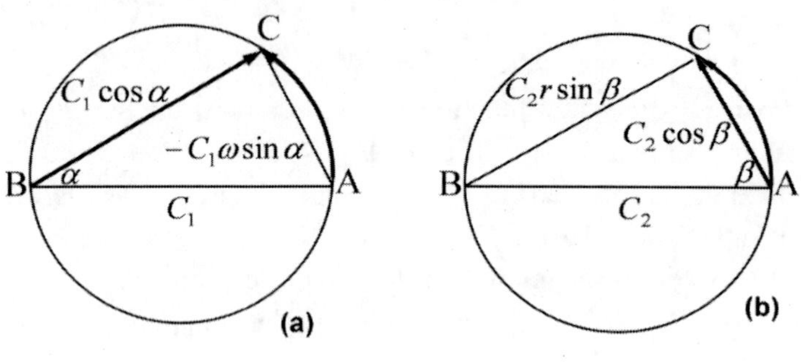

Figure 3.2. Dimension diagram of a spherical quantity in dynamic calculus as vector calculus with respect to arc length.

As radian angle α increases from 0 to $\pi/2$, point C moves from A toward B so that chord $C_1\cos\alpha$ dwindles from a diameter dimension C_1 to vanishing. Arc AC is the course of spherical quantity $C_1\cos\alpha$ in dynamic differentiation; and the diameter is the dimension with respect to which dynamic differentiation is performed. The electron is tracing along the semicircle, but the spherical quantity of it is measured by chord BC at any moment. In other words, the spherical quantity in dynamic calculus defines the course of an electron along the semicircle. This process may be expressed by a rotatory vector BC, which rotates around point B with its vector tip along arc AC.

Equation (3.1) may be regarded as the differentiation of vector BC with respect to time. Because arc AC represents the uniform time flow, the vector differentiation with respect to time is along the path of arc AC. Here we regard the differential path as an essential parameter parallel to an integral path in vector calculus. For example, if vector BC is a physical quantity of displacement, then the vector differentiation along the path of arc AC yields a velocity vector.

During a rotatory operation, differentiation is undergoing concurrently with integration. As chord BC decreases, chord AC increases accordingly. The gaining of a space dimension can be represented by dynamic integral of

$$\int C_2 \cos\beta dl = C_2 \cos(\beta - \frac{\pi}{2}) \cdot r\ , \tag{3.2}$$

where

$$\frac{1}{r} = \frac{d\beta}{dl}\ . \tag{3.3}$$

As β decreases from π/2 to 0, chord AC extends over a space dimension of diameter AB when $C_2 \cos\beta$ changes into $C_2 r \sin\beta$ according to equation (3.2). Here radian angles α and β are always complementary. Changes in $\cos\alpha$ and $\cos\beta$ values as the radian angles rotate smoothly represent a synchronized differential and integral process. The time dimension of a spherical quantity trades for the space dimension during the rotatory operation. Spherical quantity $C_2 \cos\beta$ is represented by chord AC and may be treated as a rotatory vector AC around point A with its vector tip along arc AC. The dynamic integral may be interpreted as vector integral over a space dimension along the integral path of arc AC. For example, if vector AC is a force field, then the vector integral along the path of arc AC represents the work done by this force.

Spherical quantities in dynamic calculus may be expressed by vector calculus. In the dimension diagram, we have translated spherical quantities in dynamic calculus into vector calculus. However, because spherical quantities in dynamic calculus have multifarious geometric manifestations in Cartesian coordinates, their expressions in vector calculus entail consideration case by case for electrons within helium and neon shells. In ancient Chinese culture, Dao and De indicate roads or paths. Dao refers to the general law, and De refers to the individual quality or manifestation of Dao, both corresponding to dynamic calculus and vector calculus, respectively. The proposition of DaoDe as social morality reflects the ancient Chinese good understanding of natural law.

3.2. Vector Calculus in One-Dimensional Space

Electrons within a helium atom transform between space and time dimensions. Their motion has been described in dynamic calculus, trigonometry, geometry, and probability in Chapter 1. Here we characterize the motion of the electrons in vector calculus. We only have to concern with the space component of both 1s electrons because the time component can be understood by its symmetry and complementarity to the space component. As shown in Figure 3.3, spherical center O denotes the position of the nucleus, and the shaded spheres OP and ON represent the space components of both 1s electrons, Ω_0 and Ω_2, within a helium atom, respectively. As points P and N trace along the dashed semicircles, the radius of one sphere OP increases while that of the other ON decreases. Line segments PQ and QN are tangent to both spheres respectively so that profile NOPQ is a rectangle at any moment. Let R_0 represent the maximum spatial sphere of electron Ω_2 so that R_0=OQ. As point N traces from Q toward O along the dashed semicircle, the space component of Ω_2 reduces cosinusoidally (Figure 3.3a). This space shrinking process can be expressed by spherical quantity

$$\Omega_2 = R_0 \sin\beta = R_0 \cos\alpha, \tag{3.4}$$

$$\frac{\partial \Omega_2}{\partial l} = \cos\alpha , \tag{3.5}$$

where l denotes space dimension R_0 and α increases from 0 to $\pi/2$ smoothly. Under Cartesian coordinates, the dynamic differentiation can be expressed by vector calculus as

$$\frac{\partial \Omega_2}{\partial l} = \frac{\partial \Omega_2}{\partial x}\mathbf{i} + \frac{\partial \Omega_2}{\partial y}\mathbf{j} + \frac{\partial \Omega_2}{\partial z}\mathbf{k} , \tag{3.6}$$

or in a concise gradient expression as

$$\frac{\partial \Omega_2}{\partial l} = \nabla \Omega_2 , \tag{3.7}$$

where Ω_2 on the left-hand side is a spherical quantity and on the right-hand side is a conventional scalar represented by a sphere. This equation establishes the relationship between dynamic calculus of a spherical quantity and the gradient of a scalar sphere. When calculus is performed upon l, a spherical quantity in dynamic calculus is assumed; and when calculus is performed upon x, y, and z, conventional calculus is assumed as above, so below. To prove equation (3.6), we write a scalar

$$\Omega_2 = \sqrt{x^2 + y^2 + z^2}\cos\alpha = R_0 \cos\alpha \tag{3.8}$$

to represent a one-dimensional solid sphere so that

$$\frac{\partial \Omega_2}{\partial x} = \frac{\partial \sqrt{x^2 + y^2 + z^2}}{\partial x}\cos\alpha = \frac{x}{R_0}\cos\alpha, \tag{3.9}$$

$$\nabla \Omega_2 = \frac{x\mathbf{i} + y\mathbf{j} + z\mathbf{k}}{\sqrt{x^2 + y^2 + z^2}}\cos\alpha = \cos\alpha = \frac{\partial \Omega_2}{\partial l} . \tag{3.10}$$

The proof has used R_0 to represent a one-dimensional solid sphere and as a spherical surface in the Cartesian coordinate system without distinction because it is the surface boundary that defines the motion of the spatial sphere. Equation (3.6) actually establishes the connection between one-dimensional spherical dimension l and Cartesian coordinates x, y, and z. Dimension l is the diameter OQ subtended by semicircular arc QNO or the radius of sphere O. Spherical quantity Ω_2 is the shrinking radius or one-dimensional sphere, which is treated as a scalar sphere in Cartesian coordinates.

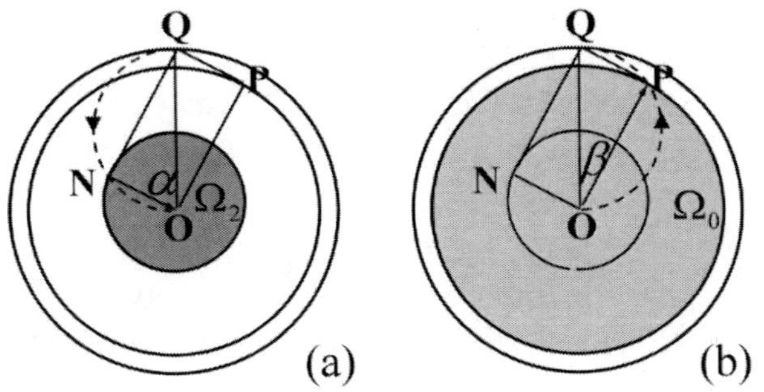

Figure 3.3. Spatial spheres of both 1s electrons within a helium atom represented by shaded circles.

While one 1s electron condenses cosinusoidally, the other experiences spatial expansion with spherical quantity Ω_0 increasing sinusoidally (Figure 3.3b). Under the perspective of one-dimensional sphere,

$$\Omega_0 = \cos\beta , \tag{3.11}$$

where β decreases from $\pi/2$ to 0 smoothly. This process can be expressed as dynamic integral:

$$\int \Omega_0 dl = R_0 \cos\beta , \tag{3.12}$$

which is translated under Cartesian coordinates into

$$\int \Omega_0 dl = \iiint_V \left(\frac{\partial \Omega_{0x}}{\partial x} + \frac{\partial \Omega_{0y}}{\partial y} + \frac{\partial \Omega_{0z}}{\partial z}\right) dxdydz , \tag{3.13}$$

or written in a divergence operator as

$$\int \Omega_0 dl = \iiint_V (\nabla \cdot \Omega_0) dxdydz , \tag{3.14}$$

where Ω_0 on the right-hand side is regarded as a vector of

$$\Omega_0 = \Omega_{0x}\mathbf{i} + \Omega_{0y}\mathbf{j} + \Omega_{0z}\mathbf{k} , \tag{3.15}$$

and volume V is the sphere of

$$\sqrt{x^2 + y^2 + z^2} \leq R_0 \cos\beta. \tag{3.16}$$

To prove equation (3.14), we apply Gauss's theorem

$$\iiint_V (\nabla \cdot \Omega_0) dxdydz = \oiint_S \Omega_0 dA, \tag{3.17}$$

where S is the spherical surface enclosing the shaded sphere V, and the area integral on the right-hand side indicates the growing of the spherical radius by appending spherical surfaces layer upon layer as β decreases from $\pi/2$ to 0. The extending radius OP represents a vector whose tip flips through the spherical surface stack along the dashed semicircle (Figure 3.4). This is a vivid explanation of the orthogonal relationship between a radius and a spherical surface notwithstanding the continuous variable property of the radius. The integral of the radius over the area of the spherical surface results in a solid sphere:

$$\oiint_S \Omega_0 dA = \oiint_S \cos\beta dA = R_0 \cos\beta = \int \Omega_0 dl. \tag{3.18}$$

Vector Ω_0 is the radius, representative of the sphere, growing from zero to R_0 sinusoidally. Equation (3.13) establishes the relationship between a spherical quantity in dynamic integration and volume integration of the divergence of a radial vector. The spherical quantity and the radial vector are synonymous in defining the spatial sphere in this case.

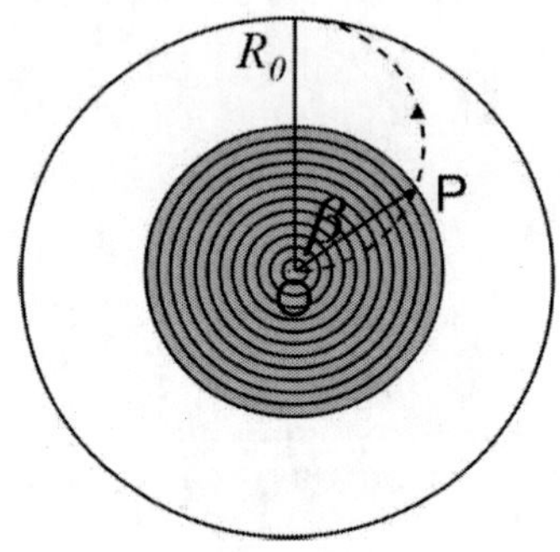

Figure 3.4. Geometric interpretation of a radial vector integral over its orthogonal spherical surface area leading to a solid sphere.

By the way, both 1s spherical quantities are represented by dynamic radii of ON and OP spheres in Figure 3.3 respectively, one decreasing while the other increasing sinusoidally. They are orthogonal at any moment and constitute two legs of a right triangle whose hypotenuse is the radius of the maximum sphere R_0. The extremity of them is that one spherical quantity Ω_2 equals $R_0 \sin\beta$ at $\beta = \pi/2$, the maximum sphere, while the other Ω_0 equals $\cos\beta$ at $\beta = \pi/2$, an infinitesimal point. According to the Pythagorean theorem, the total surface area of both electrons remains constant all the time (i.e., $4\pi R_0^2 = 4\pi r_1^2 + 4\pi r_2^2$ where r_1 represents radius OP and r_2 radius ON). This is interesting for it is the surface area that accounts for charge activity, the inner volume being shielded electromagnetically.

3.3. Vector Calculus in Quaternity Space

Four-dimensional space has been described for a neon shell in dynamic calculus, trigonometry, and geometry. Here we characterize the motion of electrons by vector calculus in Cartesian coordinates. As shown in Figure 3.5, spherical quantity Φ_0 is a zero-dimensional sphere in space at the atomic center. By increasing a space dimension, it displaces off the center and becomes a one-dimensional $2p_x$ orbital. By increasing another space dimension, the dislocated point Φ_1 becomes a semicircular arc Φ_2, which then becomes a three-dimensional hemispherical surface Φ_3, which in turn transforms into a four-dimensional solid sphere Φ_4. At this state, electronic space reaches the maximum and begins to dwindle by reducing the same space dimensions in the reverse order while the time dimension increases. Time dimension increment means electron travelling backwards in time. Here we switch the positions between spherical quantities of $\mathbf{\Phi}_7$ and $\mathbf{\Phi}_5$ compared with Figure 2.10. Please also refer to Table 10.6 for the space component expression of the polors, metors, vitors, and scalors in integral and derivative forms. Through four consecutive clockwise rotations in the negative direction of quaternity axes, four-dimensional space sphere Φ_4 gradually becomes spherical quantities of three-dimensional $\mathbf{\Phi}_7$, two-dimensional $\mathbf{\Phi}_6$, one-dimensional $\mathbf{\Phi}_5$, and zero-dimensional space sphere Φ_0, completing a full cycle. At this state, the electron Φ_0 becomes a single spherical point centered at the nucleus to initiate another cycle. We shall consider four integral operations over space on the right-hand side of Figure 3.5 and their differential counterparts on the left-hand side of the figure step by step as follows.

Because electrons undergoing clockwise rotations $\mathbf{\Phi}_7 \mapsto \mathbf{\Phi}_6 \mapsto \mathbf{\Phi}_5$ are travelling backwards in time, spherical quantity $\mathbf{\Phi}_5$ precedes spherical quantity $\mathbf{\Phi}_7$ in the time direction. Thus we put the position of $\mathbf{\Phi}_5$ before that of $\mathbf{\Phi}_7$ in the circulation to reflect the time flow in section 2.2.3 and Figure 4.3. Spherical quantity $\mathbf{\Phi}_7$ is a $2p_z$ orbital containing three-dimensional space with one-dimensional time whereas spherical quantity $\mathbf{\Phi}_5$ is a $2p_x$ orbital containing one-dimensional space with three-dimensional time. Spherical quantity $\mathbf{\Phi}_7$ is in symmetry with spherical quantity $\mathbf{\Phi}_3$ in spacetime by crossing space dimension l_1 to and fro in dynamic calculus. They belong to spherical quadrant III (Figures 2.4 and 2.10). Spherical quantities $\mathbf{\Phi}_5$ and Φ_1 are symmetry in spacetime in terms of differentiation with respect to or integration over space dimension l_3. They are two $2p_x$ orbitals colonizing the first spherical quadrant.

We shall present four pairs of electrons ($2p_x$, $2p_y$, $2p_z$, and 2s) within a neon shell in details [8]. The electron octet coexists at the same time but at different space, or at the same space but at different time, reflecting the symmetry of atomic space and time. Each pair of electrons occupy a whole space dimension (i.e., l_3, l_2, l_1, or l_0) corresponding to uniform circular motion of various geometric manifestations.

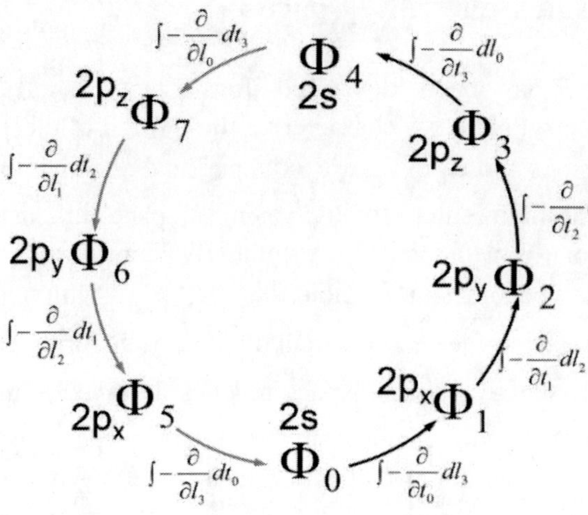

Figure 3.5. Rotation relationships between spherical quantities of the 2s2p electron octet. Solid arrows indicate electrons rotating counterclockwise. Electrons in the negative direction of quaternity axes rotate clockwise, and gray arrows indicate their travelling backwards in time.

3.3.1. The Polor in the First Dimension

Electronic transformation from 2s to $2p_x$ along the first dimension l_3 is somewhat similar to harmonic oscillation of a rigid ball under the influence of springs (Figure 1.12). As shown in Figure 3.6, charge movement from O toward A corresponds to $\int \Phi_0 dl_3$ operation in Figure 3.5. At any specific moment, the electron center D is at the vertical projection of a point C onto horizontal line segment OA. As a point C traces from O toward A along the dashed semicircle smoothly, point D gradually shifts away from nuclear center O. Given OA= R, then the distance of the electron from the nucleus is a unidirectional vector OD= $R\cos^2\psi$ with ψ rotating uniformly from π/2 to 0. If we set up an X-axis along OA with its origin at point O, then the vector is along the X-axis, measuring the displacement of the electron from the nucleus. The dynamic integral of spherical quantity Φ_0 can be expressed by the integral of vector OD with respect to x.

$$\int \Phi_0 dl_3 = \int_R \Phi_0 dx. \tag{3.19}$$

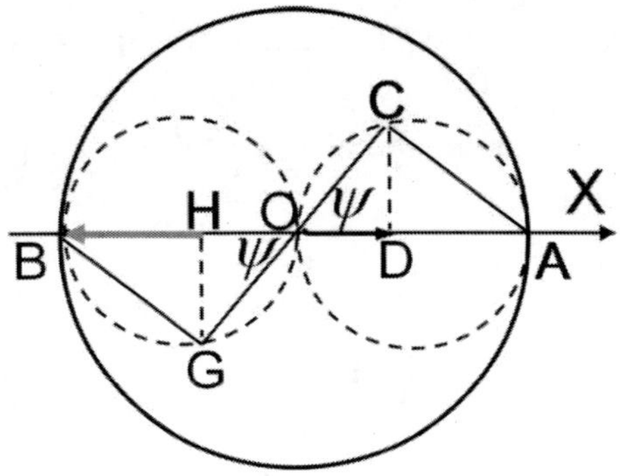

Figure 3.6. Integral of vector OD with respect to x along the path of the unidirectional radius OA, the first dimension of quaternity space. As vector OD progresses forward, vector HB for another electron retreats by differentiation.

Conversely, on the left-hand side of electron Φ_0, electronic transformation from $2p_x$ to 2s may be characterized by vector differentiation with respect to variable x:

$$-\frac{\partial \Phi_5}{\partial l_3} = \frac{\partial \Phi_5}{\partial x}, \tag{3.20}$$

which describes another $2p_x$ electron moving in spacetime symmetry with the one in the integral operation. Vector HB is equal to $-R\sin^2\psi$ with ψ rotating uniformly from $\pi/2$ to 0 and with the minus sign denoting the negative X-axis direction. In the first space dimension l_3, chord OG indicates the space component reduced, and chord GB indicates the remaining space component to be reduced. The retreating of vector HB from a full dimension OB to vanishing indicates a differential operation of equation (3.20). Since this process is against the time direction, the vanishing of vector HB actually indicates electronic displacement of 2s to $2p_x$ from O to B, spanning the space dimension l_3.

This polor dimension sets a reference for the other three dimensions of quaternity space in Cartesian coordinates. It is the displacement of the 2s electrons away from the nuclear center along the X-axis. The path of the vector integral is straight line OA along the X-axis, but follows a hypothetical semicircle OCA uniformly. Because dynamic differentiation is a process, it has a path, too. This extra feature of dynamic calculus is different from traditional infinitesimal calculus. The path of differentiation with respect to l_3 dimension is harmonic oscillation along straight line OB in the minus X-axis. As above, so below.

3.3.2. The Metor in the Second Dimension

The second dimension stretches a point into a semicircular arc. As shown in Figure 3.7, the stretching process from point A into semicircle ACB is characterized by $\int \Phi_1 dl_2$ operation that corresponds to electronic transformation of $2p_x \mapsto 2p_y$ whereas the shrinking process from the other semicircle to point A is characterized by dynamic differentiation $-\partial\Phi_6/\partial l_2$ that corresponds to electronic transformation of $2p_y \mapsto 2p_x$ in the opposite direction of the same quaternity axes. The spherical quantity in dynamic integral can be expressed trigonometrically as chord AC=AB$\sin\psi$ with ψ rotating from 0 to $\pi/2$ smoothly. Under Cartesian coordinates, it is a line integral of

$$\int \Phi_1 dl_2 = \int_C \Phi_{1x} dx + \Phi_{1y} dy \,, \tag{3.21}$$

where path C is semicircle ACB defined by $x^2 + y^2 = R^2$, R being the radius of semicircle ACB. The line integral can be derived by the traditional concept of vector integral calculus. The metamorphosis of the electron from a point to an arc increases kinetic energy equivalent to the work done by the electron during the process. The vector of chord AC can be written as $\Phi_1 = \Phi_{1x}\mathbf{i} + \Phi_{1y}\mathbf{j}$; and the unit tangent vector as $\mathbf{T} = \sin\vartheta\,\mathbf{i} + \cos\vartheta\,\mathbf{j}$ where ϑ is the angle between the positive Y-axis and the unit tangent vector in direction of increasing arc length s. Hence the component of Φ_1 in direction of the unit tangent vector $\mathbf{T}$ in direction of increasing arc length s can be expressed by

$$\phi_{\mathrm{T}} = \Phi_1 \cdot \mathbf{T} = \Phi_{1x} \sin\vartheta + \Phi_{1y} \cos\vartheta \,, \tag{3.22}$$

so that

$$\int_C \phi_{\mathrm{T}} ds = \int_C (\Phi_{1x} \sin\vartheta + \Phi_{1y} \cos\vartheta) ds = \int_C \Phi_{1x} dx + \Phi_{1y} dy \,. \tag{3.23}$$

This is a classical case of a line integral as an integral of a vector. If the vector is regarded as a force field, then the line integral represents the work done by this force in stretching the electron along the semicircle. The electronic orbital along the second dimension transforms from a point into a semicircle in geometry. We shall mark this transform by a curved arrow AC, the integral path. Spherical quantity Φ_1 is treated as a vector under Cartesian coordinates as $\Phi_1 = \Phi_{1x}\mathbf{i} + \Phi_{1y}\mathbf{j} + \Phi_{1z}\mathbf{k}$, but in this special case the Φ_{1z} component is zero so that the vector is on the X-Y plane. As a point C traces along the semicircle ACB uniformly, the direction and magnitude of the vector vary continuously.

Although equation (3.21) establishes the connection, as best as we can, between dynamic integral and vector integral with respect to arc length, the latter describes a point tracing the semicircle while the former expresses the metamorphosis of object stretching rather than the displacement of a single point.

Conversely, the differential process of $2p_y \mapsto 2p_x$ transformation can be expressed by a vector DA shrinking along the path of arc DA by differentiation

$$-\frac{\partial \Phi_6}{\partial l_2} = \frac{\partial \Phi_{6y}}{\partial x} - \frac{\partial \Phi_{6x}}{\partial y} \,, \tag{3.24}$$

which is a two-dimensional *curl* operation that measures angular motion of the electron along a semicircle. During the process, chord BD is the space component reduced, and chord DA is the space component to be reduced as expressed by equation (3.24). The electron shrinks uniformly from a semicircle BDA to arc DA and finally to a point A, which principally involves angular motion around the origin. Since this process is against the time direction, the vanishing of chord DA indicates electronic stretching from B to A along the semicircular arc BDA, spanning the space dimension l_2. Both terms of the partial differentiation are coordinated so as to ensure the semicircular path of the differentiation, i.e., $x^2 + y^2 = R^2$. This process concatenates electron Φ_6 of semicircular arc to electron Φ_5 of point B.

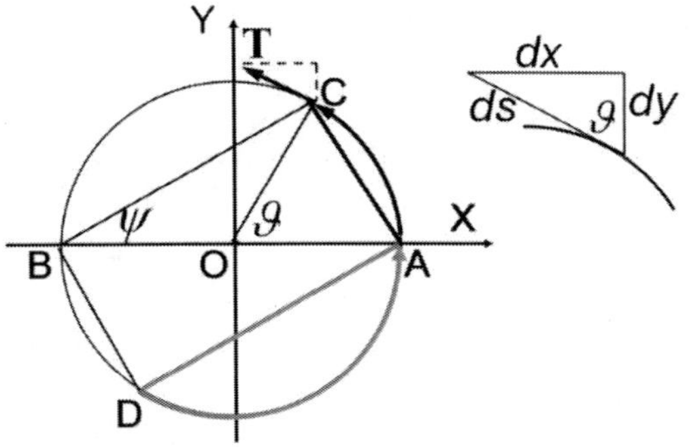

Figure 3.7. Vector integral with respect to arc length along the path of a semicircle, the second dimension of quaternity space.

3.3.3. The Vitor in the Third Dimension

Orbital $2p_z$ takes the shape of a hemispherical surface as semicircle PA'AQ rotates uniformly around axis PQ with point C sweeping through semicircle ACB, point C' through semicircle A'C'B', and so on as ψ rotates from $\pi/2$ to 0 (Figure 3.8). Every point on semicircle PA'AQ stretches from a point into a semicircle with accompanying proper vertical migration during the process to ensure a uniform surface density. In other words, semicircle PA'AQ is smashed and redistributed into hemispherical surface PAQBC uniformly. Under Cartesian coordinates, the spreading of the electron over the hemispherical surface can be expressed by a surface integral of

$$\int \Phi_2 dl_1 = \iint_H \Phi_{2x} dydz + \Phi_{2y} dzdx + \Phi_{2z} dxdy, \quad (3.25)$$

where H represents hemispherical surface PAQBC defined by $x^2 + y^2 + z^2 = R^2$, R being the radius of the hemisphere. The integral path H is the shaded region in Figure 3.8 marked by a representative curved arrow AC. The three-dimensional spherical quantity Φ_2 corresponds to an extending curved surface APCQ which is difficult to define in linear algebra.

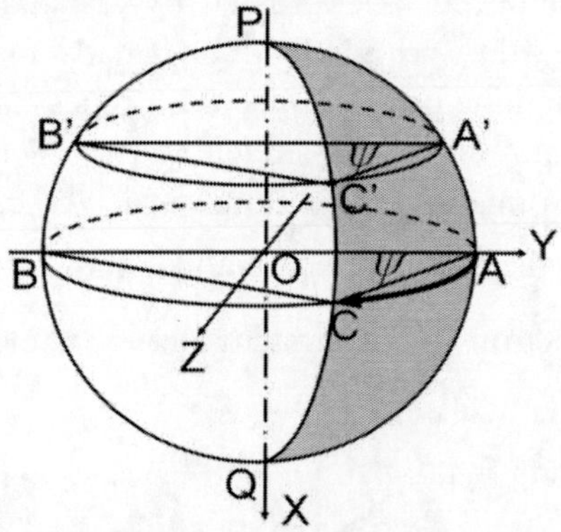

Figure 3.8. Vector integral over hemispherical surface area, the third dimension of quaternity space.

A clockwise rotation operation proceeds in the negative direction of the quaternity axis. In contrary to space growth, the differential transformation in $2p_z \mapsto 2p_y$ decreases a space dimension l_1 from a hemispherical surface Φ_7 to a semicircular arc Φ_6, which can be characterized by

$$-\frac{\partial \Phi_7}{\partial l_1} = (\frac{\partial \Phi_{7z}}{\partial y} - \frac{\partial \Phi_{7y}}{\partial z})\mathrm{i} + (\frac{\partial \Phi_{7x}}{\partial z} - \frac{\partial \Phi_{7z}}{\partial x})\mathrm{j} + (\frac{\partial \Phi_{7y}}{\partial x} - \frac{\partial \Phi_{7x}}{\partial y})\mathrm{k}, \quad (3.26)$$

where three vector components are coordinated during the *curl* of vector field Φ_7 to satisfy the hemispherical surface path of $x^2 + y^2 + z^2 = R^2$. We introduce this path for differentiation to characterize the pattern of electronic contraction. Because the differentiation is against the time direction, the actual process is the orbital spreading from $2p_y$ to $2p_z$.

3.3.4. The Scalor in the Fourth Dimension

The fourth dimension describes electronic transformation of $2p_z \mapsto 2s$ from a hemispherical surface to a hemisphere geometrically (Figure 3.9). As an electron reaches $2p_z$ of a hemispherical surface PAQ, it will wrap back into a solid hemisphere through a rotatory operation ($\Phi_3^0 = \Phi_4$ as $2p_z \mapsto 2s$). The radius AO represents the fourth space dimension for the electron to traverse harmonically as if it were moving along the dashed semicircular arc ACO uniformly. At any specific moment, chord AC is perpendicular to CO and represents the space component of the spherical quantity of the $2p_z$ electron. Radian angle ψ denotes angle $\angle COA$ that is increasing from 0 to $\pi/2$ smoothly. As ψ reaches $\pi/2$, the hemisphere becomes a solid half that satisfies equation:

$$\int \Phi_3 dl_0 = \iiint_V (\frac{\partial \Phi_{3x}}{\partial x} + \frac{\partial \Phi_{3y}}{\partial y} + \frac{\partial \Phi_{3z}}{\partial z}) dxdydz, \quad (3.27)$$

where volume V refers to solid hemisphere PAQ. Spherical quantity Φ_3 is represented by chord AC, which determines the thickness of the hollow hemisphere of vector DC.

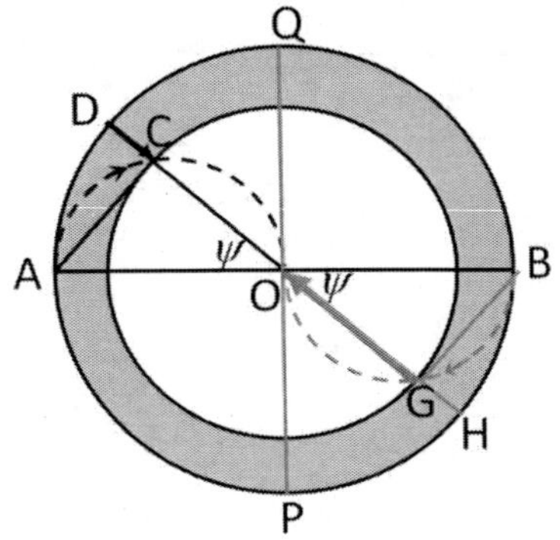

Figure 3.9. Vector integral over spherical volume, the fourth dimension of quaternity space. Both hemispheres have contrary rotatory operations: The $2p_z$ electron of PAQ hemisphere is undergoing a counterclockwise rotation ($\Phi_3^0 = \Phi_4$) whereas the other hemisphere QBP proceeds by a clockwise rotation ($\Phi_4^{-0} = \Phi_7$), both connected eventually by a spherical 2s electronic orbital. Arrows denote the time direction.

The transformation of the other hemisphere is not as straightforward as the above $2p_z$ electron. While Φ_3 undergoes a counterclockwise rotation, the clockwise rotation of Φ_4 proceeds as 2s $\mapsto$ $2p_z$ simultaneously. Radius BO represents the fourth space dimension l_0 for Φ_4 to be differentiated or reduced in the dynamic process of $-\partial\Phi_4/\partial l_0$. At any moment during the process, chord BG in the dashed semicircular arc BGO is always perpendicular to chord OG. Chord BG denotes the space component reduced and chord OG denotes the remaining space component to be reduced. As radian angle ψ (which is $\angle GOB$) increases from 0 to $\pi/2$ smoothly, the hollow hemispherical surface Φ_7 ingrows into a solid hemisphere. The differentiation is represented by a four-dimensional radial vector GO, which retreats from a full space dimension BO to center O as a state point G traces semicircular arc BGO uniformly. Vector GO is the radius of the empty sphere. Against the time direction, the differentiation experiences flattening of a solid hemisphere into a hemispherical surface sinusoidally:

$$-\frac{\partial\Phi_4}{\partial l_0} = \frac{\partial\Phi_4}{\partial x}\mathrm{i} + \frac{\partial\Phi_4}{\partial y}\mathrm{j} + \frac{\partial\Phi_4}{\partial z}\mathrm{k}, \tag{3.28}$$

where **i**, **j**, and **k** orientations are coordinated so as to ensure that the gradient field of the volume satisfies $x^2 + y^2 + z^2 = R^2$ with R reducing sinusoidally from the maximum radius to zero during the harmonic oscillation. Both counterclockwise and clockwise rotations of $2p_z \mapsto 2s \mapsto 2p_z$ are synchronized, giving the continuous shaded electronic solid as a thickening hollow sphere at any moment. Flattening backwards in space against time from Φ_4 to Φ_7 amounts to thickening forward with time from Φ_7 to Φ_4.

Because the clockwise rotation $\int(-\partial\Phi_4/\partial l)dt$ is in spacetime symmetry with the counterclockwise rotation $\int(-\partial\Phi_3/\partial t)dl$, spherical quantities Φ_3 and Φ_7 are symmetric geometrically as both hemispheres (Figure 3.9). This pair of opposite $2p_z$ orbitals combine into a whole sphere seamlessly, and the empty inner sphere can be treated as a zero-dimensional point contracting sinusoidally in complement with the implosion of both $2p_z$ orbitals. In this way, a whole sphere is divided into concentric inner and outer 2s spheres. As the front moving points C and G meet at point O in Figure 3.9, the outer sphere becomes a full homogeneous scalor, attaining maximum space and minimum time, followed by both $2p_x \mapsto 2s$ and $2s \mapsto 2p_x$ electrons emitted from the nuclear center O. The inner sphere is zero-dimensional in space while the outer sphere is four-dimensional in space, both are channeled by 1s electrons or by the $2p_x$, $2p_y$ and $2p_z$ electrons. Hence the next cycle begins.

This illustration has its advantages and disadvantages compared with that in Figure 2.12(d). Obviously, this description does not involve the asymmetrical skewing of the 2s crescent ($2p_z \mapsto 2s$ and $2s \mapsto 2p_x$ in Figure 2.14) because both symmetrical processes ($2p_z \mapsto 2s$ and $2s \mapsto 2p_z$ in Figure 3.9) proceed simultaneously. This clean model is well consistent with the geometrical shapes of four orbital types as shown in Figure 2.14 for a full neon shell. However, the beauty of orbital complementarity in section 5.2.2 seems to be diminished if we stick to this model for an argon shell. And when there are vacancies of electrons in the neon shell, such as in a carbon atom, the geometric transformation depicted in Figure 2.12 would better characterize the transformation cycle of $2s2p_x2p_y2p_z2s$. Thus this discussion reveals the subtle features of atomic structure depending on circumstances and perspectives. Nonetheless, both geometrical accounts (sections 2.3 and 3.3) for the electron octet within a neon shell could be unified in considering that they perceive the electronic orbitals in different orders (i.e., $2p_z \mapsto 2s \mapsto 2p_x$ versus $2p_z \mapsto 2s \mapsto 2p_z$) (cf. Figure 5.2 in section 5.2.2). Each model partitions a whole sphere in an elegant method according to the principle of dynamic calculus. You may choose one way or the other to explain best the orbital geometry in a particular situation. Further explanation of the 2s2p electron octet structures will be given in section 5.2.2.

The motion of electrons in four dimensions within a neon shell is summarized in Figure 3.10. Arrows are used to represent the movement of orbital geometric centers, the solid line arrows indicating four $\int \Phi dl$ operations of electrons in counterclockwise rotations, and the dotted line arrows indicating four clockwise rotations in the negative direction of quaternity axes. A differential operation with respect to a space dimension travelling backwards with time (such as equation 3.20) is complementary and symmetric to an integral operation (such as $\int \Phi_5 dt_0 = -\int \Phi_0 dl_3 = -\int_R \Phi_0 dx$) over the same space dimension travelling forwards with time. Eight calculus paths concatenate into a full cycle within the spherical layer in correspondence with Figure 3.5. Every two adjacent paths are orthogonal to each other in geometry, i.e., the downstream arrow always connects with the upstream one but turns to a direction perpendicular to it.

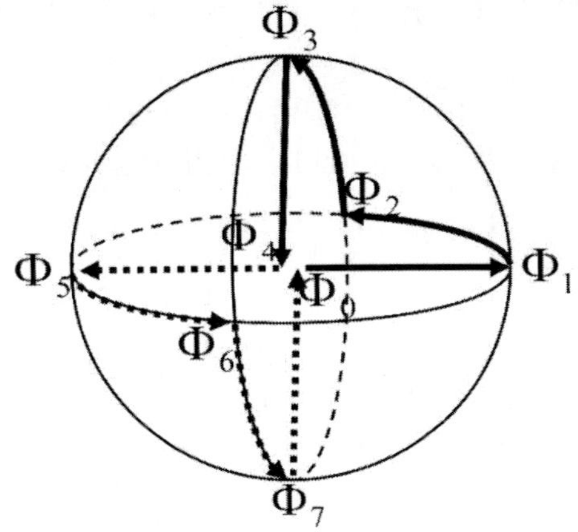

Figure 3.10. Geometric paths of vector calculus for the electron octet within a neon shell. The solid arrows represent four electrons in counterclockwise rotations whereas the dotted arrows represent another four electrons travelling forwards with time in the negative direction of quaternity axes.

A neon shell is a spacetime continuum filled by eight electrons in an oscillatory cycle. From the standpoint of a specific electron, quaternity spacetime includes not only the current electron itself but also the other seven electrons in the downstream and upstream dimensions along the rotatory operation pathway, or in the future and the past of the current electron. In the cyclic system, the future and the past are interconnected forming the whole neon shell continuum. Eight electrons represent eight spacetime dimensions, which graduate the scale of the spacetime continuum. Time exists in the rotation link in equation (2.29), each electron having a unique sinusoidal time flight in the river of rotatory operations. For example, the current time of electron Φ_2 is the future of electron Φ_1. This cyclic spacetime worldview is more fundamental than Euclidean space of the infinitely long X, Y, and Z axes with uniform time, or absolute time for all. It is particularly valuable for describing the entities such as an atom and an organism. Newtonian uniform time is useful because each electron in a dimension diagram transforms sinusoidally in its spherical quantity as if it were travelling uniformly along a semicircle like the motion of a clock.

Quaternity spacetime indicates that spherical quantities of an electron throughout all time dimensions in a spherical layer are similar to spherical quantities of various electrons in all spherical quadrants. This spacetime worldview supports the suggestion by Edward Forbes (1815-1854) that there is a striking parallelism in the laws of life throughout time and space. The laws governing the succession of forms in past time are nearly the same with those governing at the present time the differences in different areas.

3.4. Orbitals Synchronization

As described above, spherical quantities for various electrons obey dynamic calculus with various geometric manifestations. In Figure 3.6, the polor in dynamic integral is a point displacing harmonically from the origin towards the periphery along the X-axis; In Figure 3.7, the metor in dynamic integral is an arc stretching from the point into a semicircle on the X-Y plane; In Figure 3.8, the vitor in dynamic integral is a curved surface extending from the semicircle toward a hemispherical surface; and in Figure 3.9, the scalor in dynamic integral is a waxing crescent progressing from the hemispherical surface into a hemisphere. As radian angle ψ decreases from $\pi/2$ to 0 uniformly, the electrons in their various dimensions transform from one state to another. At any moment along the integral paths, their spherical quantities satisfy

$$\int \Phi dl = A\cos\psi, \tag{3.29}$$

where A is a dimension constant orthogonal to the trigonometric term and l denotes the dimension to be traversed by various electrons Φ_0, Φ_1, Φ_2, and Φ_3. For brevity, we assume that radian angle ψ is always in the scope of $[\pi/2, 0)$ in a right triangle of dimension diagram in four spherical quadrants. Radian angle ψ keeps four rotating electrons at the same pace at any moment, which automatically synchronizes the phases of four spherical quantities. We shall discuss these kinds of synchronization in detail as follows.

3.4.1. Green's Theorem

Electronic transformations of $\Phi_1^0 = \Phi_2$ and $\Phi_6^{-0} = \Phi_5$ draw two semicircles on the X-Y plane while transformations of $\Phi_0^0 = \Phi_1$ and $\Phi_5^{-0} = \Phi_0$ draw a diameter across the circle (Figure 3.11). Together they form two closed loops ACBOA and ADBOA. Because the centers of the electrons are never inside the interior of the loop on the plane during the transformations, and the surface integral of any zero vector over the interior of both loops is zero, Green's theorem indicates that the vector integral along the curve ACBOADB is zero. Because of the symmetric relationship between spherical quantities of the same spherical quadrant, we only have to consider the black arrows in integral form in Figure 3.11.

$$u\int \Phi_0 dx - \int_C \Phi_{1x} dx + \Phi_{1y} dy = 0, \tag{3.30}$$

where the first term indicates a vector integral along the radius, the second term indicates a vector integral along the semicircle, and u is a dimension compensator. Hence

$$u\int \Phi_0 dx = \int_C \Phi_{1x} dx + \Phi_{1y} dy, \tag{3.31}$$

$$u\int \Phi_0 dl_3 = \int \Phi_1 dl_2. \tag{3.32}$$

These equations indicate the synchronization of a polor ($2s \mapsto 2p_x$) and a metor ($2p_x \mapsto 2p_y$) in dynamic calculus. A pair of electrons traveling across a diameter harmonically are balanced by another pair of electrons stretching along its subtending semicircular arcs uniformly.

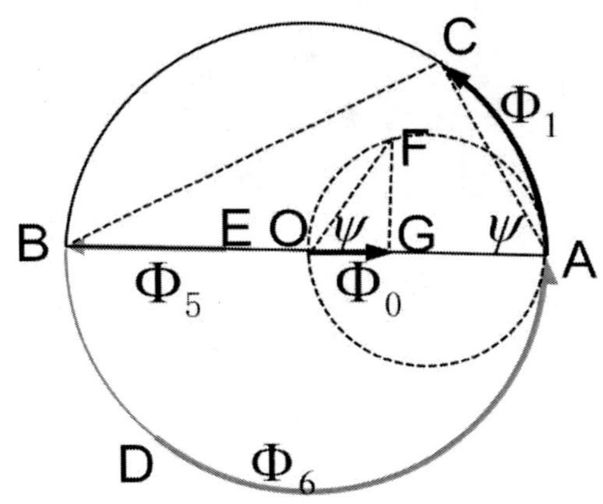

Figure 3.11. Line integral relationship between $2p_x$ and $2p_y$ electronic orbitals within a neon shell. Black arrows (OG and AC) indicate the paths of electronic orbitals during dynamic integration; and gray arrows (EB and DA) indicate the paths of electronic orbitals during dynamic differentiation. As radian angle ψ decreases from π/2 to 0 uniformly, point C traces the semicircular arc ACB while point F traces the semicircular arc OFA. Please refer to Figures 3.6 and 3.7.

3.4.2. Stokes' Theorem

Electronic transformations of $\Phi_2^0 = \Phi_3$ and $\Phi_7^{-0} = \Phi_6$ paint two hemispherical surfaces APBQD and APBQE while electronic transformations $\Phi_1^0 = \Phi_2$ and $\Phi_6^{-0} = \Phi_5$ form two semicircles ACPB and BQA (Figure 3.12). Both semicircles form the boundary of both hemispherical surfaces. Because of the complementarity between the differential and integral operations, we only have to consider the integral relationship between a $2p_y$ orbital and a $2p_z$ orbital in the positive direction of quaternity axes. Stokes' theorem governs the relationship between the surface integral and the line integral around the boundary:

$$u\int_C \Phi_{1x}dx + \Phi_{1y}dy = \iint_H \Phi_{2x}dydz + \Phi_{2y}dzdx + \Phi_{2z}dxdy, \tag{3.33}$$

$$u\int \Phi_1 dl_2 = \int \Phi_2 dl_1 , \tag{3.34}$$

where integral path C is semicircle ACPB and integral area H refers to the extending shaded hemispherical surface ACPBD. The motion of a $2p_z$ electron along the surface is counterbalanced by the flow of a $2p_y$ electron around the periphery. Stokes' theorem synchronizes the rotatory operations of a metor electron $2p_x \mapsto 2p_y$ and a vitor electron $2p_y \mapsto 2p_z$ in dynamic integral.

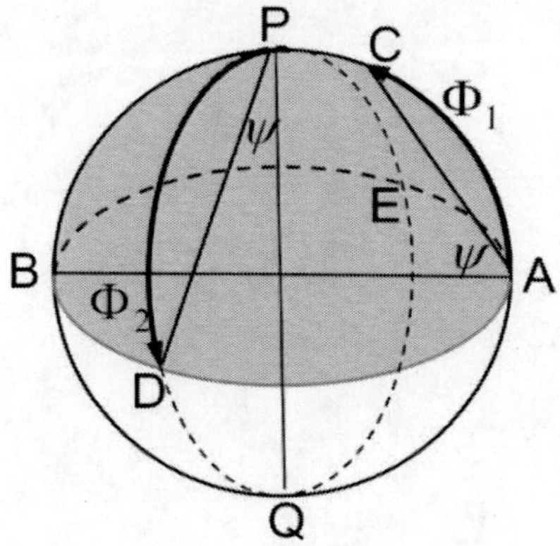

Figure 3.12. The relationship between the surface integral of a $2p_z$ electron $\int \Phi_2 dl_1$ and the line integral of a $2p_y$ electron $\int \Phi_1 dl_2$ around the surface boundary. Curved arrows (PD and AC) represent the paths of electronic orbitals during dynamic integration. As radian angle ψ decreases from $\pi/2$ to 0 uniformly, point C traces the semicircular arc ACPB, and point D traces the semicircular arc PDQ.

3.4.3. Gauss's Theorem

Electronic transformations of $\Phi_3^0 = \Phi_4$ and $\Phi_4^{-0} = \Phi_7$ forms a solid spherical 2s orbital; and electronic transformations of $\Phi_2^0 = \Phi_3$ and $\Phi_7^{-0} = \Phi_6$ are two hemispherical surfaces surrounding the solid sphere. Because of the spacetime symmetry between Φ_3 and Φ_7 and between Φ_2 and Φ_6, we only have to consider the integral relationship between $\int \Phi_3 dl_0$ and $\int \Phi_2 dl_1$, the differential relationship between $-\partial\Phi_4/\partial l_0$ and $-\partial\Phi_7/\partial l_1$ following the same pattern.

$$u\iint_H \Phi_{2x} dydz + \Phi_{2y} dzdx + \Phi_{2z} dxdy = \iiint_V \left(\frac{\partial\Phi_{3x}}{\partial x} + \frac{\partial\Phi_{3y}}{\partial y} + \frac{\partial\Phi_{3z}}{\partial z}\right) dxdydz\,, \tag{3.35}$$

$$u\int \Phi_2 dl_1 = \int \Phi_3 dl_0\,. \tag{3.36}$$

Gauss's theorem is commonly known as the divergence theorem. Physically, it implies that the density change of charges within a region of space is balanced by the electric flux into or away from that region through its spherical surface boundary. This relationship is valid if we consider the flux of electrons of $2p_x$ across the spherical boundary of 2s electrons during their transformation. On the flip side, here we give Gauss's theorem another new interpretation that the motion of an electron from $2p_z$ to 2s within the sphere V is shielded electromagnetically by the motion of another electron from $2p_y$ to $2p_z$ around the hemispherical surface H in an equilibratory manner (equation 3.35). This is perhaps the manifestation of Gauss's theorem that synchronizes the rotatory operations of a $2p_z$ electron and a 2s electron.

3.5. Electronic Interaction through Virtual Photons

We have characterized electronic orbitals in four different dimensions by vector calculus and demonstrated their tight synchronization in motion by Green's theorem, Stokes' theorem, and Gauss's theorem, but how electrons interact to achieve synchronization at the same phase of radian angle ψ ? What is the physical mechanism underlying the apparent phenomena? We shall borrow a Feynman diagram on virtual photons and give it a fresh interpretation here.

As shown in Figure 3.13, Φ_1 emits a virtual photon carrying a time dimension t_1 away while Φ_0 receives that virtual photon as a space dimension l_3. Because a virtual photon travels as an electromagnetic wave, electronic transformation through virtual photon transfer manifests sinusoidal wave behavior, consistent with the principle of dynamic calculus. Because space dimension l_3 directly comes from time dimension t_1, there is a relationship between the space and time dimensions:

$$l_3 = -ut_1 . \tag{3.37}$$

Physically, we may interpret this relation as the transfer of a virtual photon between electrons Φ_1 and Φ_0. There are virtual photons between other electrons in the same manner. Thus, electrons are synchronized by the transfer of virtual photons.

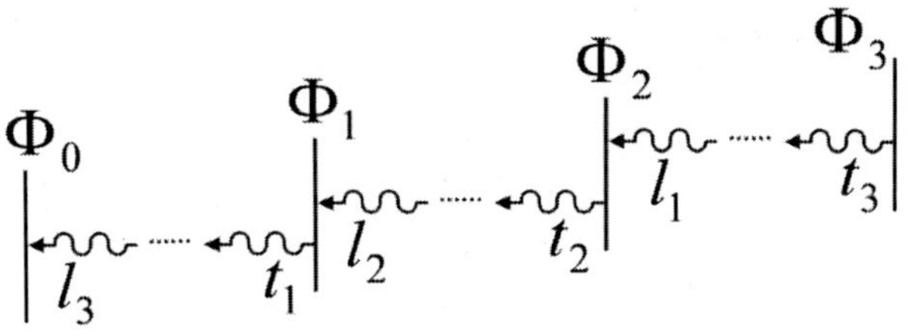

Figure 3.13. Electronic interaction through virtual photon transfer.

We draw an outgoing wave arrow as photon emission and an incoming wave arrow as photon reception. In the case of $l_3 = -ut_1$, the virtual photon wave emitted by electron Φ_1 can be expressed by $-ut_1 \cos\Psi$ carrying a time dimension away whereas the virtual photon wave received by electron Φ_0 can be expressed by $l_3 \cos\psi$ bringing a space dimension back. Therefore, the virtual photon implements in a rotatory operation in this case.

A home axis shift signifies a rotatory operation (Figure 2.4). During the exchange, radian angle ψ decreases from $\pi/2$ to 0 while radian angle Ψ increases from 0 to $\pi/2$, both being complementary. Electron Φ_1 passes a virtual photon to electron Φ_0 losing a time dimension whereas Φ_0 receives the virtual photon from electron Φ_1 gaining a space dimension as a point C traces from A toward B along semicircle ACB (Figure 3.14). Chord BC reduces while chord AC gains sinusoidally. Similarly, electron Φ_2 passes a virtual photon to electron Φ_1 while electron Φ_3 delivers a virtual photon to electron Φ_2, and so forth. Hence we have implemented dynamic calculus by virtual photon transfer physically. As long as virtual photon waves between the electrons maintain the same phase, all electrons in the transformation cycle are in good synchrony with each other within the neon shell.

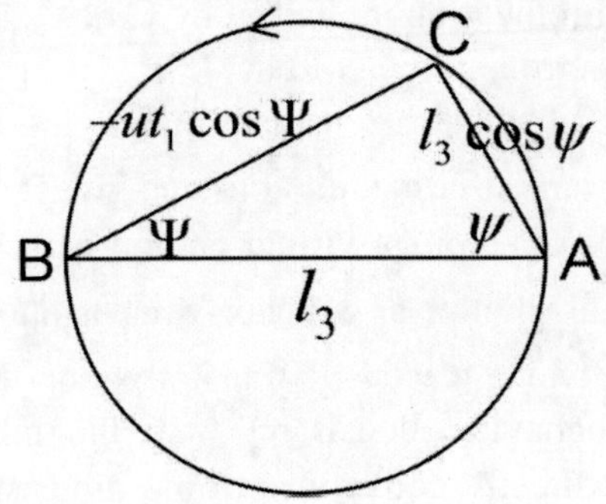

Figure 3.14. Dimension diagram of a trigonometric relationship between waves of virtual photon emission and reception.

3.6. About Quaternity Space

I was asked by a mathematician about how to define the distance between any two points in quaternity coordinates. Once the question was raised, the answer was inescapable. First, the quaternity coordinate system is a four-dimensional spacetime system as shown in Figure 2.4. There are four different spherical quadrants. A point in the first spherical quadrant is a point geometrically; a point in the second spherical quadrant represents an arc; a point in the third spherical quadrant represents a surface; and a point in the fourth spherical quadrant represents a solid (sections 2.3 and 3.3). Therefore, instead of a homogenous medium, the quaternity space are sinusoidal dimension gradients formed by various electrons in dynamic calculus (cf. Figure 2.13).

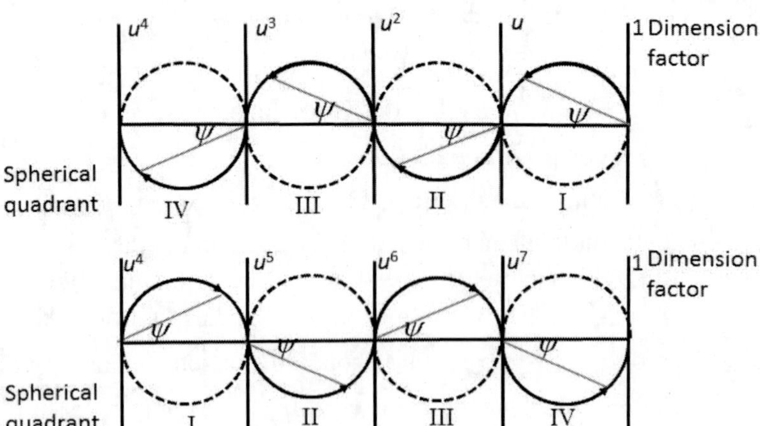

Figure 3.15. Quaternity coordinates in planar form with reference to Figure 2.10 showing eight electrons are transforming in eight different dimensions with a similar phase ψ in a neon shell. Arrows indicate the positions and directions of the rotating electrons.

Second, quaternity coordinates are a simple way to illustrate the dimensional identity of a spherical quantity. All spherical quantities in dynamic calculus are rotating along the semicircular arcs (Figure 2.10) but measured by the subtended chords at any moment. They represent harmonic oscillation of electrons within an atom. Therefore, not every point in the spherical quadrants is meaningful, but only points on the semicircular arcs make sense to represent spherical quantities within an atomic shell. As shown in Figure 3.15, the solid arcs represent the paths of the electron octet in counterclockwise rotations in a neon shell, each semicircular arc being a dimension; and the dashed arcs represent the paths of the electron octet with clockwise rotations in an argon shell as will be elaborated in Chapter 5.

Third, the distance between any two points within the same spherical quadrant can be measured by the difference between their circumferential angles. It represents the space interval between electrons in both phases. A radian angle of $\pi/2$ covers a whole space dimension. The distance between two points within the different spherical quadrants can be calculated by accumulating their dimension difference and circumferential angle difference. For example, a $2p_x$ electron and a $2p_z$ electron of a similar phase within a neon shell have a distance of two space dimensions. Analogously, one would say that the interval between you and your grandfather is two generations apart. This explanation should clarify the property of quaternity coordinate system in distance measurement.

Conclusion

We started this chapter by expressing a spherical quantity in terms of a rotatory vector under quaternity coordinates and Cartesian coordinates. In a dimension diagram, a spherical quantity in dynamic calculus represents the transformation of an electron along a semicircle that corresponds to a unique geometric path for each type of electrons. In this chapter, we have tried to establish the relationship between spherical quantities in dynamic calculus and vector calculus with the instantiation of both 1s electrons in the two-dimensional helium shell and the instantiation of the 2s2p electron octet in the four-dimensional neon shell. The path of vector calculus could be a line segment, a semicircular arc, a hemispherical surface, or a spherical volume. We were able to specify the paths, but left the vectors undefined algebraically. We portrayed the vectors in geometry (see Figure 2.12) instead of algebra because there are no simple expressions for such compressible fluids geometries in algebra. Because electrons are not rigid bodies in kinematic movement, it is difficult to characterize their trajectories by vectors in the Cartesian coordinate system.

In quaternity space, we have defined four spherical quantities in dynamic calculus in terms of vector calculus. The polor, metor, vitor, and scalor characterize the motion of electrons in point displacement, arc elongation, surface extension, and crescent solid expansion over various space dimensions, respectively. The integral calculi in equation (3.38) in one, two, three, and four dimensions represent the gradual and continuous unfolding of an electron within the neon shell. The differential calculi in equation (3.39) are the *curl* operations ($\nabla\times$) in one, two, three, and four dimensions, respectively, because each electron is transforming along a hypothetical semicircle uniformly, even though there are not *curl* operations in one and four dimensions in current mathematical texts. These *curl* operations describe the direction and strength of vectors in rotational motion in four various dimensions.

A *curl* operation can be expressed by a circular function with a uniformly varying angle subtended by the arc along the hypothetical semicircle in a dimension diagram, recalling, for example, the radian angle of α in $C_1 \cos\alpha$ in Figure 1.7 and equation 2.30. Faraday's law with a *curl* operation describes an electron in harmonic oscillation. Please refer to section 2.2.2 for the equivalence of Faraday's law and the principle of a rotatory operation for an electron in harmonic oscillation. Electrons in the negative direction of quaternity axes are transforming in clockwise rotations according to Figure 3.5 by gradually decreasing a space dimension while increasing a time dimension. They are traveling against the time direction and may alternatively be arranged into equation (2.29) to follow the time flow.

$$\begin{cases} \int \Phi_0 dl_3 = \int\limits_R \Phi_{0x} dx, \\ \int \Phi_1 dl_2 = \int\limits_C \Phi_{1x} dx + \Phi_{1y} dy, \\ \int \Phi_2 dl_1 = \iint\limits_H \Phi_{2x} dydz + \Phi_{2y} dzdx + \Phi_{2z} dxdy, \\ \int \Phi_3 dl_0 = \iiint\limits_V (\frac{\partial \Phi_{3x}}{\partial x} + \frac{\partial \Phi_{3y}}{\partial y} + \frac{\partial \Phi_{3z}}{\partial z}) dxdydz. \end{cases} \tag{3.38}$$

$$\begin{cases} -\frac{\partial \Phi_5}{\partial l_3} = \frac{\partial \Phi_5}{\partial x}, \\ -\frac{\partial \Phi_6}{\partial l_2} = \frac{\partial \Phi_{6y}}{\partial x} - \frac{\partial \Phi_{6x}}{\partial y}, \\ -\frac{\partial \Phi_7}{\partial l_1} = (\frac{\partial \Phi_{7z}}{\partial y} - \frac{\partial \Phi_{7y}}{\partial z})\mathrm{i} + (\frac{\partial \Phi_{7x}}{\partial z} - \frac{\partial \Phi_{7z}}{\partial x})\mathrm{j} + (\frac{\partial \Phi_{7y}}{\partial x} - \frac{\partial \Phi_{7x}}{\partial y})\mathbf{k}, \\ -\frac{\partial \Phi_4}{\partial l_0} = \frac{\partial \Phi_4}{\partial x}\mathrm{i} + \frac{\partial \Phi_4}{\partial y}\mathrm{j} + \frac{\partial \Phi_4}{\partial z}\mathrm{k}. \end{cases} \tag{3.39}$$

The equation quartets represent dynamic transformations of four types of electrons: $2p_x$, $2p_y$, $2p_z$, and 2s. The integral forms are performed in the time direction; but the differential forms are performed against the time direction. The space components of these electrons are synchronized to the same radian phase ψ at any moment within the neon shell. Each electronic orbital is in dynamic equilibria with its adjacent ones during a rotatory operation. We have hypothesized that the synchronization for all electrons should be physically realized through virtual photon transfer between adjacent electrons, i.e., one electron emits a photon wave as a time dimension while another electron receives the photon wave as a space dimension. The coupling of the electrons manifests as Green's theorem, Stokes' theorem, or Gauss's theorem in vector calculus. These theorems synchronize various electrons in rotatory operations. They can be recapitulated by the general Stokes' theorem: The integral of a differential form Φ over the boundary of some orientable manifold M is equal to the integral of its exterior derivative $d\Phi$ over the whole of M as

$$\int_{\partial M} \Phi = \int_M d\Phi . \tag{3.40}$$

A pair of $2p_x$ electrons are traversing across a diameter, which is the dimensional boundary of both $2p_y$ electrons in the shape of open semicircular arcs; the circle of $2p_y$ electrons forms the dimensional boundary of both $2p_z$ electrons in the shape of open hemispherical surfaces; the spherical surface of $2p_z$ electrons constitutes the dimensional boundary of both 2s electrons in the shape of open spheres (cf. Review question 2.7); and the two concentric spheres in turn delimit both $2p_x$ electrons polarizing within the diametrical segment. As an electron transforms from $2p_x$ to $2p_y$, to $2p_z$, and to 2s orbital, each step reduces a dimension in density or in time dimension, i.e., from Φ to $d\Phi$, while increasing a space dimension from the boundary ∂M to the whole M. These boundary relationships and their rotation relationships are essential for the general Stokes' theorem to hold. The general Stokes' theorem is indeed another expression of a rotatory operation.

It is important to elucidate the physical correspondence of the general Stokes' theorem in atomic spacetime. Microscopically, we have put forward that electrons are synchronized through emitting and receiving virtual photons. Macroscopically, we still don't know exactly whether it is force on an electric charge in an electric field, or force on an electric current in a magnetic field, or whatever electromagnetic force in action, but it is certain that an electron continuously unfolds itself over four dimensions in quaternity space as a flower blooms gradually. An electron may exist as a point, an arc, a surface, or a solid. Electrons in diverse forms are interactive entities in synchrony, not windowless monads. They are governed by the general Stokes' theorem. This is a work in progress.

Moreover, because the general Stokes' theorem has broad physical implications in dynamics, fluid dynamics, electromagnetism, heat conduction, thermodynamics, etc., one may explain electronic behavior in diverse physical meanings. For example, $\int \Phi_1 dl_2$ is an integral around a semicircle describing the arc stretching of a $2p_y$ electron and can be regarded as the circulation of a velocity field in fluid dynamics. Thus, the general Stokes' theorem opens up prospects for unifying classical physics with particle physics, and electrodynamics with the theory of quaternity. This chapter gives you a fresh taste and feel of the general Stokes' theorem for synchronizing various electrons within a neon atom.

Chapter 4

Quaternity, Relativity, and Quantum Mechanics

After formulating wave functions for electrons within a neon atom, we have to stop to make comments on quantum mechanics because there are claims that quantum mechanics has already given a complete description of electronic behavior, and the motion of particles cannot be described precisely. How can the quaternity theory characterize electrons in more detail and better precision? Besides, what is the relationship between quaternity and relativity when both are dealing with spacetime? Does quaternity predict the same consequences, such as length contraction and time dilation, as special relativity does or even more? In this chapter, we shall answer these questions unambiguously from the quaternity perspective with an effort to unify both relativity and quantum mechanics, two active fields of modern physics. We shall prove that the duality and quaternity equations essentially conform to one-dimensional and three-dimensional Schrödinger's equations respectively through the interpretation of spherical quantities in dynamic calculus.

4.1. Introduction

Relativity and quantum mechanics excel at different arenas. The special and general theories of relativity mainly deal with space and time in the universe where astronomical distance is normally involved. For example, we commonly adopt light-year as a distance unit and millions of years as a time unit. These magnitudes are much larger than human's accustomed rulers, say, meter and hour. On the other hand, quantum mechanics is struggling to describe particles on the atomic scale. For instance, an electron has a radius of 2.818×10^{-15} m and rest mass of 9.109×10^{-31} kg. The observation of the tiny size is beyond human naked eye capability. The difference in the yardsticks of relativity and quantum mechanics relative to humans gives rise to serious deviations from each other not only in experimental approaches but also in conceptual formations. The discovery of this issue is mine.

4.2. Scope of Views

Man is the intrepid observer in both fields of relativity and quantum mechanics. This is the basic requirement of experimental physics, in which human sensible evidence must be gathered to prove or disprove a proposition. Here human sensible evidence includes, of course, data that are collected by devices made by humans. In studying the behavior of electrons, man is a monster standing far away of the atomic sphere. He probes the electrons by inducing them to detectors for measuring parameters, such as angular and spin momentums. Such a perturbation is remarkable from the viewpoint of an electron, which normally results in irreversible changes of electronic states after measurement. This situation is created because human's sensible scales are so macro to the electrons. This means that electrons are violently captured out of their primitive caves for the purpose of experiments. Man intrudes abruptly. Though the proper role of physical science is to find out what the world would be like unobserved, this goal has not been achieved for particles so far. To date, our knowledge of particles is mostly obtained by bombarding atoms with high energy protons or other charges.

In contrast, when examining celestial bodies, man is a tiny point inside the universe. With all instruments and apparatuses he invented, he has only penetrated a small fraction of the universe simply because his rulers are dwarfed by the vast space. He never masters the universe as a whole entity like an atomic sphere, nor does he know whether or not there are multifarious spacetime structures within it.

In both relativity and quantum mechanics, we view entities from the human perspective disregarding what they are, and consequently we are actually examining them from very different standpoints and angles relative to the objects. Specifically, we adopt a huge ruler relative to a particle to measure the particle in quantum mechanics; and we open our near-sighted eyes to stare at the vast universe in relativity. This may lead to different conclusions. In the former case, we get a bird's-eye view of the entire particle with large uncertainty; and in the latter case, we see only a fraction of the universe like a blind man touching a part of an elephant by chance. An entity as a whole and a small part of an entity are of course different. Even though both views are real, they cannot be mentioned in the same breath due to the perspective difference.

The fact that man is the center of the world in experimental physics, objects being his surroundings only, is a dilemma of science. We have to gain our knowledge based on human perception and judgments, yet this man-centered view actually biases and limits scientific progress. It wasn't until the proposition of the heliocentric theory by Copernicus (1473-1543) that we realized human races are not at the center of the universe. It wasn't until the release of "*On the Origin of Species*" by Darwin (1809-1882) that we realized human races are not angels. Instead, humans and monkeys share a common ancestor. We have spent so long getting away the general human conceit, but we haven't quite escaped this trap yet. To tackle this problem, here we bring out the quaternity perspective.

The quaternity perspective is the perspective of observing an entity from outside its activity sphere and at a proper distance and time interval that match the size and oscillating rhythm of the entity. The imaginary observer uses a proper measuring rod that matches the size of the observed object (Figure 4.1). For example, an ant carries a clay-sized particle; a man rides a bicycle; and the earth interacts with the moon. There are comparable scales between the actors and the objects in these examples. From such a proper perspective, Newton observed an apple falling down from a tree that sparked a brilliant thought in his mind about gravity. When applied to our living surroundings, the quaternity perspective is the perspective of classical mechanics. However, the term "the quaternity perspective" is especially reserved for examining electrons within inert atoms on a proper scale. The theory of quaternity is essentially founded on new space and time concepts that extend our human senses in a logical way. Because the proper observer is hypothetical, the system of the truths has to be derived from our rational capacities. A rigorous mathematical derivation is the key to establishing the theory.

The quaternity perspective leads to the spherical view. After characterizing electronic orbitals within atoms by the spherical view successfully, we may apply the same methodology to characterizing the structures of cosmic space on a proportional scale. The entire universe, as observed from a hypothetical standpoint outside the universe, is comparable to an atomic sphere under the quaternity perspective. However, we do not suggest that the whole universe is similar to an atom with regard to spacetime dimensions.

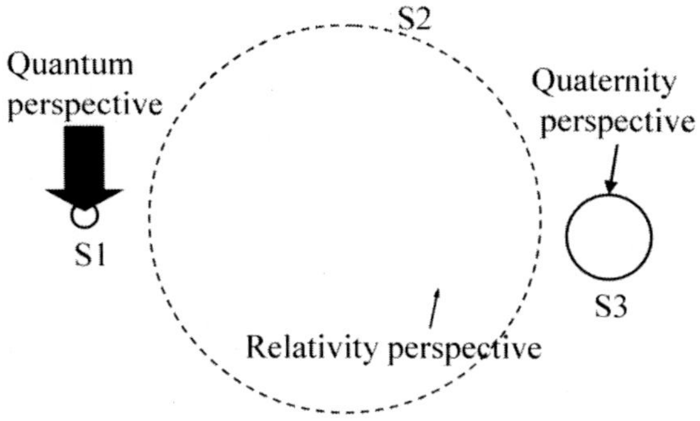

Figure 4.1. Comparison of the quantum perspective, the relativity perspective, and the quaternity perspective where S1, S2, and S3 refer to the spheres of the concerned objects and the arrows mean the observing perspectives. There are certain mismatches between the arrows and spheres under the quantum perspective (too large an arrow for a tiny sphere) and the relativity perspective (too small an arrow for an enormous sphere).

The spacetime structures of the whole universe might be more complex than those of a single atom. The reason that we have not experienced spacetime structures other than three-dimensional space with one-dimensional time is because of our limited reach in the universe. Or we might have already detected strange media such as black holes that are radically different from other regions, but we still don't regard them as distinct spacetime dimensions. Quaternity space is not homogeneous everywhere as Cartesian coordinates define. It is fair to say that we still lack the general picture of entire universe.

On the other hand, relativity may be applied to quantum mechanics for studying particles. We shall express special relativity in terms of dynamic calculus in section 4.4. However, dynamic calculus is the proper way to characterize electrons within inert atoms.

Granting that quaternity perceives the world from a wider spectrum than human direct experience does not warrant it to escape the confirmation by man-centered experimental physics. On the contrary, the results that quaternity yields must be able to withstand human scrutiny. For instance, looking at a tree from different angles, one sees various shapes and shadows. The observer could be an ant that resides at the tree trunk, or it could be a bird flying over the tree. Both animals would certainly provide different pictures about the tree, but there is still a way to unify them if we take their perspectives into proper consideration. To gain more information on the tree, a good photographer should try to capture the snapshots of it from various angles, such as from inside and outside, and from far and near. The best photo to describe the whole tree is perhaps one taken at a distance that matches the tallness of the tree. However, a photo taken from a new angle, though having different looks and feels, must be able to justify its silhouette of the tree in consistent with the previous photos. Specifically, quaternity must be compatible with relativity and quantum mechanics in a certain way and should not contradict with established experimental evidence. Thus, the readers can't help asking whether quaternity agrees with quantum mechanics in the description of electrons. We shall proceed to answer this question through the discussion of EPR paradox.

4.3. Quaternity on EPR Paradox

Albert Einstein, Boris Podolsky, and Nathan Rosen (EPR) published a paper in 1935 to question the completeness of quantum mechanical description of particle behavior [9]. They said: "every element of the physical reality must have a counterpart in the physical theory... If, without in any way disturbing a system, we can predict with certainty the value of a physical quantity, then there exists an element of physical reality corresponding to this physical quantity." This statement formed the theme of the EPR paradox. However, when applied to the behavior of an electron, this statement made a big assumption that the electron is an unchanging particle, i.e., it possesses a certain physical quantity at all space and time. This is flaw by the judgement of the spherical view.

As was discussed in Chapter 2, an electron must be described by a series of wave functions rather than of a single one. For example, within a neon shell, each of eight electrons is circulating and switching among eight states, obeying the quaternity equation. Each state has its peculiar physical quantities associated with its space and time dimensions. Since each electron in its unperturbed atomic shell is changing between states continuously, it cannot be described by a fixed physical quantity. The concept of spherical quantity was introduced to account for electrons under cyclic rotatory operations in Chapters 1 and 2.

EPR's statement, of course, has something to do with the timescale. Within a picosecond, an electron might change states thousands of times according to the principle of rotatory operations. You cannot associate a specific physical quantity with it due to its dynamic nature. However, if you were an observer with a time ruler of 10^{-100} second, then the slowly changing rate of electrons relative to the observer could be negligible. In the latter case, Einstein was correct that an element of physical reality corresponds to a physical quantity. This problem is caused by adopting an improper scale inadvertently. For example, we cannot use a time scale of millennium to measure the growth rate of a baby. Nor should we try to characterize electronic behavior using a measuring rod that is intended for a child. Can you see an ant clearly from a spacecraft five hundred miles away?!

Early in the 20th century, physicists found that the actions of elementary particles could not be absolutely predicted, but their actions could only be forecast to a certain level of inexactness through the use of probability functions. This is not only because the measurement techniques are too heavy-handed, but also because particles are changing states intrinsically along their tracks unknown to quantum mechanics. When we use our human standard, such as position and momentum, to measure an electron, the electron is forced to stop its normal track of rotation and response to the measuring agency. At that moment, the electron transforms into a positive or negative property of demand from its current state. This gives two impressions: a) the property of an electron is not real until observed; and b) the property of an electron is not permanently associated with it. These two points formed the main idea of the Copenhagen interpretation of quantum mechanics [10].

The spherical view provides deep insight into the Copenhagen interpretation. The measurement process is really an event. If it were not under the interference of the measuring agency, an electron would behave as it normally does by dynamic calculus, i.e., it would not detour its track unless forced to do so. An electron within atoms must be described by spherical quantities instead of physical quantities. A spherical quantity represents dimension transformation of an electron from one physical quantity to another. An electron within an atom does not possess the property of a certain fixed physical quantity. Any physical quantities of an electron exist in the form of continuously changing spherical quantities until we measure them. The point here is that we are not taking proper approaches to describing electrons. We don't know the language of the particle world at all when imposing the concept of physical quantity to electrons. And as a result, we only got statistical data in most cases in quantum mechanics.

Like integers on a real number axis, physical quantities are discrete points on the curve of spherical quantities (Figure 2.13). How continuous spherical quantities collapse into discrete physical quantities has something to do with experimental system to test the electrons. To explain this, suppose that we, humans, are at the mercy of an extraterrestrial being who wants to test a property of us, and the test requires us to stand on our head at a testing machine. We shall hypothetically call the demanded property "QRate." Assuming that those who are at sleep fail at the test while the others pass, thus the number of people who pass the test has to do with whether they are at sleep or not. However, because the extraterrestrial being does not know the concept of "asleep," he only adopts probability to analyze the results. If on average, people sleep eight hours in a day, then the "QRate" obtained by the extraterrestrial being would be close to 8/24 in the failure probability after a large sample is tested, but the problem is that humans sleep and wake up every day whereas an individual is sampled every thousand years in the test. Because of the large timescale of the extraterrestrial being, the test is about the stability of a human society over a long history rather than the property of a person. Thus the extraterrestrial being has confused the property of a person with the property of a population in the probability test. "QRate" is the property of a person but it is statistically obtained from the population. This situation is analogous to the case of our examining electrons. The timescale of human is too large to measure the cycling property of an electron. As a result, we are testing the stability of an experimental system rather than the property of an electron in motion.

When a pair of electrons in the singlet state with total spin 0 are separated, if one electron has spin up in the X-component, the other must have spin down in the X-component. Actual measurements did show the expected approximately 100% correlation. In this case, the electronic states are predetermined by the separation process because they are no longer characterized by spherical quantities in dynamic calculus after being separated. Each electron is in a definite physical quantity state after separation and remains constant until detection. It would be totally nonsense to talk about superluminal communication between electrons. That quantum entanglement remains valid across a long distance is a fake proposition. The undiscriminating application of the Copenhagen interpretation leads to a misconception.

Concerning with the second impression, whether the property of an electron is permanently associated with it or not depends on the interpretation of the words "property" and "permanently." If "permanently" refers to a timescale of a thousand years, then a baby has an uncertain growth rate because her life span is only at best a hundred year. If "permanently" refers to an interval of 10^{-100} second, then an electron has a fixed set of physical quantities even though it is changing states continuously. One might argue that "permanently" refers to the whole lifetime of the object, but is "asleep" a human property?

Humans change states between "asleep" and "awake" every day, and hence "asleep" should be considered to be a property inherently associated with humans. There must be some physical quantities associated with the human circadian rhythm, but the extraterrestrial being can only grasp the probability of "QRate," not the "asleep" property. Whether the property is existed or not for an individual remains undecided until tested by the extraterrestrial being. Although the "QRate" probability happens to correspond to the "asleep" property assuming population uniformity over history in this case, not every probability tested on populations in other cases may reflect a certain property of an individual! Thus the probability methodology by quantum mechanics is either incomplete in describing electrons or simply beside the point.

The theory of quaternity has clearly demonstrated that there are precise ways to describe electronic orbitals, and hence that the description of atomic structures by quantum mechanics is incomplete. Does this violate Heisenberg's principle of uncertainty? Not at all. Being hampered by the principle of uncertainty does not preclude us from characterizing electronic behavior accurately in alternative ways. Heisenberg's principle of uncertainty ($\Delta p \cdot \Delta \lambda \geq \hbar/2$) is based upon the assumptions of discrete physical quantities with linear algebra. Note that those assumptions are often unnoticed. Modern physicists even claim that there are not any assumptions underlying the principle of uncertainty and regard it as an absolute law for particles, not realizing that the principle is based on faith (of physical quantities in Euclidean space) rather than reason. This is a terrible mistake. The history of science tells us an appalling fact. After an assumption is repeated a million times, it becomes a law because the confidence of the speakers is assured; and after a law is repeated a million times, it becomes an obstacle to scientific progress for the confidence has developed into prejudice and faddism. The principle of uncertainty belongs to this case. Extending an assumption beyond its valid zone leads to a bad doctrine.

The spherical view is not limited by those assumptions. It is based upon spherical quantities in dynamic calculus and hence may describe electronic behavior in an alternative manner. When we adopt a proper timescale to study human behavior, the human circadian rhythm is easily understandable. Spherical quantities in circular functions indicate that the behavior of electrons is entirely deterministic, i.e., each electron follows its own track traveling between space and time within the atomic shell, just as a man sleeps at night and works at daytime having a daily living pattern, which is cognizable. There is not a so-called principle of uncertainty in dynamic calculus.

4.4. Special Relativity in Quaternity

Under the principle of rotatory operations as was introduced in Chapters 1 and 2, a physical entity gaining a dimension in space is accompanied by losing a time dimension, and vice versa. For example, an electron might switch its state from a polor to a metor, i.e., changing its state from one-dimensional space with three-dimensional time to two-dimensional space with two-dimensional time. Such a switch is so rapidly within a neon atom that we have only considered its full consequence. However, in the case of a slow switch, we have to consider its partial effect during the transformation process. If a spherical quantity is undergoing a clockwise rotation from states M to N, then contraction in space dimension and dilation in time dimension proceed simultaneously. We may express such a process as

$$\int -\frac{\partial M}{\partial l}\, dt = N \,, \tag{4.1}$$

where the differential and integral operations are carried out simultaneously during the tradeoff between space and time dimensions. The consequence of the process is

$$-\frac{\partial M}{\partial l} = \frac{\partial N}{\partial t} \,. \tag{4.2}$$

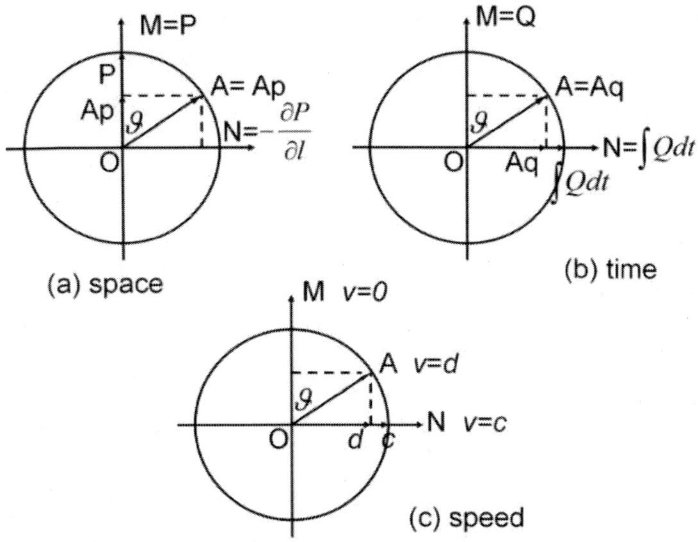

Figure 4.2. Illustration of the transformation of a spherical quantity in relation to velocity d in a partial clockwise rotation.

How can two operations be carried out simultaneously? Dynamic calculus can be implemented by trigonometric functions. When a spherical quantity M with space component P and time component Q has undergone a clockwise rotatory operation, the resultant spherical quantity N is composed of space component $-\partial P/\partial l$ and time component $\int Qdt$ (Figure 4.2). The clockwise rotatory operation can be interpreted by the rotation of radian angle ϑ. As ϑ increases from 0 to $\pi/2$ continuously and smoothly, the space component of M decreases a full dimension from P to $-\partial P/\partial l$ while the time component of M increases a full dimension from Q to $\int Qdt$.

As shown in Figure 4.2, if the velocity of a physical entity is zero ($v=0$), then ϑ equals zero and no rotation occurs; and if the physical entity travels at the speed of light ($v=c$), then ϑ equals $\pi/2$ indicating that a clockwise rotation has proceeded completely, i.e., spherical quantity M transforms into N. Light represents a clockwise rotation or space to time dimension conversion, contrary to the counterclockwise rotation of a virtual photon (section 3.5). If the velocity of the physical entity equals d in the range of $0<d<c$, then the value of ϑ is between 0 and $\pi/2$. We shall call the process a partial clockwise rotatory operation, indicating that M partially transforms toward N with a speed d, less than the speed of light c. If A is a middle state in the clockwise rotation pathway, then from Figure 4.2(a), the radius of

the circle is P and the space component of A is A_p; and from Figure 4.2(b), the radius of the circle is $\int Qdt$ and the time component of A is A_q.

$$A_p = P\cos\vartheta, \tag{4.3}$$

$$A_q = \int Qdt \cdot \sin\vartheta. \tag{4.4}$$

These are so because a spherical quantity in quaternity coordinates coincides with a rotatory vector around a central point on the Cartesian plane (see section 1.5.1). From Figure 4.2(c), the radius of the circle is c, which is also the hypotenuse of a right triangle.

$$\sin\vartheta = \frac{d}{c}, \tag{4.5}$$

whence

$$A_p = P\sqrt{1-\left(\frac{d}{c}\right)^2}. \tag{4.6}$$

It should be understood that when $\sin\vartheta = 0$, A_q declines into Q, reducing a time dimension from $\int Qdt$, and when $\cos\vartheta = 0$, A_p declines into $-\partial P/\partial l$, reducing a space dimension from P. Because A_p is less than P and A_q is larger than Q, the state point A experiences space contraction and time dilation during the process of a partial clockwise rotation. The larger the velocity d is, the larger the radian angle ϑ is, and the more severe the transformation is. Thus special relativity boils down to a problem of a partial clockwise rotation in quaternity. Equation (4.6) is actually Lorentz's transformation with regard to length contraction where P is the length of an object at rest and A_p is the length of the object at velocity d. The relationship between spherical quantities in full and partial clockwise rotatory operations is summarized in Table 4.1.

Table 4.1. The transformation of a spherical quantity under a partial clockwise rotation and a full clockwise rotation with reference to Figure 4.2

Clockwise rotation	Rotating angle	Speed	Space component	Time component
Original state	$\vartheta=0$	0	P	Q
Partial clockwise rotation	$0<\vartheta<\pi/2$	d	A_p	A_q
Full clockwise rotation	$\vartheta=\pi/2$	c	$-\partial P/\partial l$	$\int Qdt$

4.5. Quaternity Solutions to Schrödinger's Equations

The duality and quaternity equations govern spherical quantities in dynamic calculus whereas Schrödinger's equations govern physical quantities in infinitesimal calculus. Physical quantities correspond to spherical quantities at the full dimension positions. In this section, we shall establish that the description of duality in Chapter 1 constitutes the solution to one-dimensional Schrödinger's equation; and the description of quaternity in Chapter 2 constitutes the solution to three-dimensional Schrödinger's equation. Duality and quaternity are spherical quantity interpretations of Schrödinger's equations. A spherical quantity in duality or quaternity equation corresponds to a wave function in Schrödinger's equation. These alternative explanations of electronic orbitals in continuous atomic spacetime complement the quantum mechanical description of discrete electronic orbitals within atoms and carry a significant step towards the unification of the world order by dynamic calculus.

4.5.1. Duality as a Solution to One-dimensional Schrödinger's Equation

Although the duality equation (equation 1.13) defines spherical quantities in transformation by dynamic calculus, it is in conformity with one-dimensional Schrödinger's equation for physical quantities:

$$i\hbar \frac{\partial \Omega}{\partial t} = -\frac{\hbar^2}{2m}\frac{\partial^2 \Omega}{\partial x^2} + V\Omega, \tag{4.7}$$

where Ω is the wave function and V is the potential energy of the particle. To prove their equivalence, we shall derive a common formula from both equations. First, we don't think it is necessary to adopt X, Y, and Z components to formulate the Schrödinger's equation for a helium atom because 1s electronic orbitals are spherical and isotropic, each of which can sufficiently be explained by a radial quantity. Because l is one-dimensional in the radial direction, we may use l instead of x. However, the second derivative and the first derivative are antiparallel with respect to dimension l so that $\partial x^2 = -\partial l^2$. Second, realizing the following basic relationships:

$$\omega = \frac{-E}{\hbar}, \quad \frac{1}{r} = \frac{p}{\hbar}, \tag{4.8}$$

$$-E = \frac{p^2}{2m}, \tag{4.9}$$

where E and p are energy and momentum for a 1s electron within the helium shell, we have

$$-\frac{\hbar^2}{2m} = -\frac{r^2 p^2}{2m} = r^2 E. \tag{4.10}$$

Thus, after both sides are divided by Ω, equation (4.7) becomes

$$i\hbar \frac{1}{\Psi}\frac{d\Psi}{dt} = -r^2 E \frac{1}{\psi}\frac{d^2\psi}{dl^2} + V\,. \tag{4.11}$$

Furthermore, from equation (1.2), we may derive

$$i\Psi = C_1(i\cos\alpha + \sin\alpha) = -\frac{d\Psi}{d\alpha}, \tag{4.12}$$

so that

$$\omega i\Psi = -\frac{d\Psi}{d\alpha}\omega = -\frac{d\Psi}{d\alpha}\cdot\left(-\frac{d\alpha}{dt}\right) = \frac{d\Psi}{dt}\,. \tag{4.13}$$

The left-hand side of equation (4.11) becomes

$$i\hbar \frac{1}{\Psi}\frac{d\Psi}{dt} = i\frac{-E}{\omega}\frac{1}{\Psi}\frac{d\Psi}{dt} = -E\frac{\omega i}{\omega^2}\frac{1}{\Psi}\frac{d\Psi}{dt} = \frac{-E}{\omega^2}\frac{1}{\Psi}\frac{d^2\Psi}{dt^2}\,. \tag{4.14}$$

Thus one-dimensional Schrödinger's equation can be rewritten as

$$-\frac{E}{\omega^2\Psi}\frac{d^2\Psi}{dt^2} = -\frac{Er^2}{\psi}\frac{d^2\psi}{dl^2} + V\,. \tag{4.15}$$

Because of the isotropy of 1s electrons, the atomic sphere may be regarded as one-dimensional along the radial direction. Upon separating the time and space variables, both sides of the equation are assigned to an energy constant *E*. The left-hand side becomes a harmonic oscillator as expressed by equation (1.1), and the right-hand side becomes

$$-\frac{Er^2}{\psi}\frac{d^2\psi}{dl^2} + V = E\,, \tag{4.16}$$

or

$$-r^2\frac{d^2\psi}{dl^2} + \frac{V}{E}\psi = \psi\,. \tag{4.17}$$

On the other hand, the duality equation as separable time and space variables reads

$$-\frac{1}{\Psi\omega^2}\frac{d^2\Psi}{dt^2} = -\frac{r^2}{\psi}\frac{d^2\psi}{dl^2}\,. \tag{4.18}$$

To separate the space and time components of the duality equation, both sides of the equation are assigned to a whole dimension constant 1. As before, the left-hand side becomes a harmonic oscillator, and the right-hand side becomes

$$-\frac{r^2}{\psi}\frac{d^2\psi}{dl^2} = 1. \tag{4.19}$$

However, in the process of dynamic calculus, the second derivative of the space component contains the first derivative that partially shields the second derivative and corresponds to the existence of a potential energy function in one-dimensional Schrödinger's equation. To include the rotation path or operational history of the second derivative, equation (4.19) becomes

$$-\frac{r^2}{\psi}\left(\frac{d^2\psi}{dl^2} + \frac{d\psi}{dl}\right) = 1. \tag{4.20}$$

The addition of the first derivative to the second derivative is somewhat in symmetry to the addition of a constant term to an infinite integral for generality. Because r is a unit imaginary number (r^2 =-1), this equation takes the form of

$$-r^2\frac{d^2\psi}{dl^2} + \frac{d\psi}{dl} = \psi. \tag{4.21}$$

Comparing equations (4.21) with (4.17), the following formula must hold if the duality equation would conform to Schrödinger's equation:

$$\frac{V}{E}\psi = \frac{d\psi}{dl}. \tag{4.22}$$

Because the potential energy function V and the energy E have the same units, the quotient V/E in front of ψ can be omitted in terms of a sinusoidal spherical quantity for ψ. Hence

$$\psi = \frac{d\psi}{dl}, \tag{4.23}$$

So that

$$\int\frac{d\psi}{\psi} = \int dl = \int r d\beta. \tag{4.24}$$

Integrating both sides of the equation leads to

$$\ln\psi = r\beta + C_3, \tag{4.25}$$

$$\psi = e^{j\beta} \cdot e^{C_3} = C_2(\cos\beta + j\sin\beta), \tag{4.26}$$

where $r = j$, C_3 is a constant, and C_2 is e to the power of C_3. Spherical quantity ψ in equation (4.26) does agree with that in equation (1.5). Thus, despite their different perceptions, one-dimensional Schrödinger's equation and the duality equation are equivalent in defining the wave functions for 1s electrons. The evidence that supports the Schrödinger's equation also agrees with the duality equation.

4.5.2. Quaternity as a Solution to Three-dimensional Schrödinger's Equation

We have formulated eight spherical quantities for eight spacetime dimensions in a sphere. They are solutions to the quaternity equation, describing eight 2s2p electrons in harmonic oscillations along four various dimensions in a neon shell. Although the quaternity equation governs spherical quantities instead of physical quantities, it will be shown that the quaternity equation is stricter than three-dimensional Schrödinger's equation in Cartesian coordinates in the definition of wave functions. Three-dimensional Schrödinger's equation is defined by

$$i\hbar\frac{\partial\Phi}{\partial t} = -\frac{\hbar^2}{2m}\nabla^2\Phi + V\Phi, \tag{4.27}$$

where Φ is the wave function and V is the potential energy of the particle. The strictness arises from the fact that quaternity distinguishes the levels of space dimensions among X, Y, and Z directions and regards time as a spherical quantity containing four levels of dimensions complementary to space rather than a one-dimensional quantity only. It is rational to assume that this anisotropy of 2p orbitals may be applied to the Schrödinger's equation as an implicit constraint (Figure 2.14) in the following discussion, as evidence of this constraint will be provided from the organic chemistry perspective (Figure 6.2).

From the quaternity equation (equation 2.7), we may separate the time variable Θ and the space variables θ of Φ as follows:

$$\frac{1}{\dot{\Psi}^4\Theta}\frac{d^4\Theta}{dt^4} = \frac{1}{\psi'^4\theta}\frac{d^4\theta}{dl^4}. \tag{4.28}$$

Assigning both sides of the equation to dimension 1 yields two linear differential equations:

$$\frac{d^4\Theta}{dt^4} = \dot{\Psi}^4\Theta\,, \tag{4.29}$$

$$\frac{1}{\psi'^4}\frac{d^4\theta}{dl^4} = \theta\,. \tag{4.30}$$

These are the four-dimensional harmonic oscillators in time and space as characterized in the dimensional array of spherical quantities Φ_0, Φ_1, Φ_2, Φ_3, and Φ_4, Φ_5, Φ_6, Φ_7 in equations (2.18) and (2.25). Because the fourth derivative contains the third, second, and first derivatives, we may include these differential steps in equation (4.30) to reflect the electron in various stages of oscillations (i.e., Φ_1, Φ_2, Φ_3), as if the identity of a person would include his yesterday and the day before yesterday, etc., because this is the path he has walked along. This expansion is also like the data structure in the C++ programming language where an array name can be thought of as a constant pointer, which takes the address of the first element followed by those of the other elements sequentially.

$$\frac{1}{\psi'^4}\left(\frac{d^4\theta}{dl^4} + \frac{d^3\theta}{dl^3} + \frac{d^2\theta}{dl^2} + \frac{d\theta}{dl}\right) = \theta\,, \tag{4.31}$$

which can be rewritten as

$$\frac{1}{\psi'^4}\frac{d^4\theta}{dl^4} + \frac{1}{\psi'^2}\frac{d^2}{dl^2}\left(\frac{1}{\psi'^2}\frac{d\theta}{dl} + \frac{1}{\psi'^2}\theta + \frac{1}{\psi'^2}\int \theta dl\right) = \theta\,. \tag{4.32}$$

Inside the parentheses, the factor $1/\psi'^2$ placed directly before a spherical quantity can be treated as a double integral operator over two space dimensions (see equation 5.24 in section 5.2.3) so that

$$\frac{1}{\psi'^2}\frac{d^2}{dl^2}\left(\int \theta dl + \iint \theta dl^2 + \iiint \theta dl^3\right) + \frac{1}{\psi'^4}\frac{d^4\theta}{dl^4} = \theta\,. \tag{4.33}$$

It becomes clear that the three terms inside the parentheses represent the space components of 2p electrons along the X, Y, and Z directions. They correspond to $2p_x$, $2p_y$, and $2p_z$ orbitals in a neon shell which have different levels of space dimensions as illustrated in Figure 2.14.

$$\frac{1}{\psi'^2}\frac{d^2}{dl^2}(\theta_x + \theta_y + \theta_z) + \frac{1}{\psi'^4}\frac{d^4\theta}{dl^4} = \theta. \tag{4.34}$$

Performing the second derivative upon the space component of a particular orientation only affects that space orientation and hence the double integrals are equivalent to a Laplacian operator.

$$\frac{1}{\psi'^2}\nabla^2\theta + \frac{1}{\psi'^4}\frac{d^4\theta}{dl^4} = \theta. \tag{4.35}$$

Moreover, angular velocity $\dot{\Psi}$ and orbital radius $-1/\psi'$ have some basic relationships with energy E_2 and momentum p_2 for a 2s2p electron:

$$\dot{\Psi} = \frac{-E_2}{\hbar}, \; -\frac{1}{\psi'} = \frac{\hbar}{p_2}; \tag{4.36}$$

$$-E_2 = \frac{p_2^2}{2m}. \tag{4.37}$$

Therefore, we have

$$\frac{1}{\psi'^2} = \frac{\hbar^2}{p_2^2} = -\frac{\hbar^2}{2mE_2}. \tag{4.38}$$

Substituting this equation into equation (4.35) yields

$$-\frac{\hbar^2}{2m}\nabla^2\theta + E_2\frac{1}{\psi'^4}\frac{d^4\theta}{dl^4} = E_2\theta, \tag{4.39}$$

which conforms to the time-independent Schrödinger equation with a 2s spherical quantity multiplied by E_2 serving as the potential energy term $V\theta$ (the second term on the left-hand side of the equation). Energy E_2 is the separation constant of the Schrödinger's equation. Both $V\theta$ and $E_2\theta$ are functions only of the distance from the origin, corresponding to the isotropic spheres of both 2s electrons. Thus three 2p orbitals (the first term on the left-hand side of the equation) are delimited by both 2s electronic orbitals (i.e., $V\theta$ and $E_2\theta$) in the atomic shell. This unique interpretation of the time-independent Schrödinger equation provides fresh insight into the structure of a four-dimensional atomic sphere where physical quantities correspond to spherical quantities.

On the other hand, three-dimensional Schrödinger's equation for separable solutions reads

$$i\hbar \frac{1}{\Theta}\frac{d\Theta}{dt} = -\frac{\hbar^2}{2m}\frac{1}{\theta}\nabla^2\theta + V. \tag{4.40}$$

Both sides of this Schrödinger's equation are assigned to the separation constant E_2 so that the space variable on the right-hand side conforms to equation (4.39) while the time variable on the left-hand side becomes

$$i\hbar \frac{1}{\Theta}\frac{d\Theta}{dt} = E_2, \tag{4.41}$$

Because the imaginary unit i is normally applied to a two-dimensional plane, this equation may be degraded from the neon shell to the helium shell into

$$i\hbar \frac{1}{\Psi}\frac{d\Psi}{dt} = E_1, \tag{4.42}$$

provided the connection shortcut between $\Phi_0 = \Omega_0$ and $\Phi_4 = \Omega_2$ in section 2.3 and Figure 4.3. This shortcut is implemented by the chain rules of dynamic calculus inwards and dimension encapsulation mechanism outwards between the helium and neon shells. E_1 is energy for a 1s electron. An electron enters from the neon shell to the helium shell along the shortcut at the 2s and 1s junction.

$$i\frac{-E_1}{\omega}\frac{1}{\Psi}\frac{d\Psi}{dt} = E_1, \tag{4.43}$$

or

$$i\frac{1}{\omega}\frac{1}{\Psi}\frac{d\Psi}{dt} = -1. \tag{4.44}$$

Because the imaginary unit i shifts its physical quantity operand toward a perpendicular orientation, it may be regarded as a differential operator. From equation (1.2), we obtain

$$i\Psi = C_1(i\cos\alpha + \sin\alpha) = -\frac{d\Psi}{d\alpha}, \tag{4.45}$$

so that

$$\omega i\Psi = \omega \cdot (-\frac{d\Psi}{d\alpha}) = -\frac{d\alpha}{dt} \cdot (-\frac{d\Psi}{d\alpha}) = \frac{d\Psi}{dt}. \tag{4.46}$$

Substituting this into equation (4.44) yields

$$\frac{1}{\omega^2\Psi}\frac{d^2\Psi}{dt^2} = -1, \tag{4.47}$$

or

$$\frac{d^2\Psi}{dt^2} = -\omega^2\Psi, \tag{4.48}$$

which is a one-dimensional harmonic oscillator that proceeds in the same phase as the four-dimensional harmonic oscillator in equation (4.29) because of dimension encapsulation mechanism of the helium shell within the neon shell. The four-dimensional harmonic oscillation is channeled by the shortcut of the one-dimensional harmonic oscillator (equation 4.48) because of the spacetime continuity between both s-orbitals in helium and neon shells (Figure 4.3).

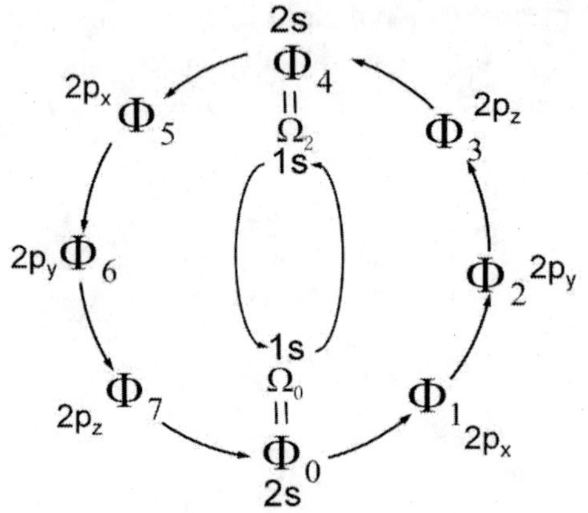

Figure 4.3. Rotation relationships between ten electrons within a neon atom in spherical quantities and orbital types. Each arrow indicates a counterclockwise rotation of an electron. Both 1s electrons are spacetime continuous with both 2s electrons respectively at the whole dimension positions, which are indicated by vertical equal signs between them.

Moreover, the duality equation for electrons in the helium shell separated into time and space variables reads:

$$-\frac{1}{\omega^2\Psi}\frac{d^2\Psi}{dt^2} = -\frac{r^2}{\psi}\frac{d^2\psi}{dl^2}, \tag{4.49}$$

where time and space components of the 1s electrons transform harmonically and synchronously, and equation (4.48) describes the time component of the 1s electrons in harmonic oscillation. In Figure 2.10, the quaternity equation defines the 2s2p electron transformation cycle of $\Phi_0 \mapsto \Phi_1 \mapsto \Phi_2 \mapsto \Phi_3 \mapsto \Phi_4 \mapsto \Phi_5 \mapsto \Phi_6 \mapsto \Phi_7 \mapsto \Phi_0$ while the duality equation defines the 1s electron transformation cycle of $\Omega_0 \mapsto \Omega_2 \mapsto \Omega_0$ with the spacetime continuity of 1s and 2s electrons at $\Phi_0 = \Omega_0$ and $\Phi_4 = \Omega_2$. These electrons transform by various harmonic oscillations characterized in geometry and in counterclockwise rotations (Figure 2.12). On the other hand, according to the above deduction, three-dimensional Schrödinger's equation actually corresponds to the following harmonic oscillator:

$$-\frac{1}{\omega^2\Psi}\frac{d^2\Psi}{dt^2} = \frac{1}{\psi'^4\theta}\frac{d^4\theta}{dl^4}, \tag{4.50}$$

which establishes the relationship between the cycles of electrons within the helium and neon shells. The right-hand side is a four-dimensional harmonic oscillator while the left-hand side is a one-dimensional harmonic oscillator when both sides are assigned to a whole dimension 1. The expansion of the four-dimensional harmonic oscillation has been demonstrated to be equivalent to the time-independent Schrödinger's equation in equations (4.31) to (4.39). Therefore, the corrected Schrödinger's equation in spherical quantities is as follows:

$$-\frac{1}{\omega^2\Psi}\frac{d^2\Psi}{dt^2} = \frac{1}{\psi'^4\theta}\left(\frac{d\theta}{dl} + \frac{d^2\theta}{dl^2} + \frac{d^3\theta}{dl^3}\right) + \frac{1}{\psi'^4\theta}\frac{d^4\theta}{dl^4}, \tag{4.51}$$

Because of the spacetime continuity of 1s and 2s electrons in Figure 4.3, equation (4.50) defines the electron transformation cycle of $\Phi_0 \mapsto \Phi_1 \mapsto \Phi_2 \mapsto \Phi_3 \mapsto \Phi_4 \mapsto \Omega_0$ in the positive direction of the quaternity axes while the electron transformation cycle in the negative direction of the quaternity axes would be $\Phi_4 \mapsto \Phi_5 \mapsto \Phi_6 \mapsto \Phi_7 \mapsto \Phi_0 \mapsto \Omega_2$. Thus the electron cycles defined by the quaternity equation, with the encapsulation of the duality equation, are similar to those defined by the corrected Schrödinger's equation (Figure 4.4). In this sense, a solution to the quaternity equation is also a solution to the Schrödinger's equation, and it is also true conversely if we assume the anisotropy of 2p orbitals (Figure 2.14) in the X, Y, and Z directions for the Schrödinger's equation. Therefore the quaternity equation is equivalent to the Schrödinger's equation even though their approaches to the differential equations are remarkably different. With dynamic calculus implemented by circular functions, geometrical shapes in accordance with quaternity space, and the quaternity equation proven to be compatible with the Schrödinger's equation, we have succeeded in delineating a vivid picture of the atomic structure within a neon atom by a coherent theory. Because the quaternity equation governs multiple electrons in a neon atom and is in conformity with three-dimensional Schrödinger's equation, the solutions to three-dimensional Schrödinger's equation for a hydrogen atom can also be used (with appropriate modifications) to give states for the electrons of higher elements. This fresh exploration undoubtedly constitutes a fruitful complement to quantum mechanics. A more elaborate solution in circular complex functions to equation (4.50) can be found in section 5.3.1 with Figure 5.9. Because we can derive ten spherical quantities for 1s2s2p electrons from the duality equation and the quaternity equation (equation 4.50), the solutions to the Schrödinger's equation (4.40) for a hydrogen atom can also be used to give states for the electrons of higher elements (i.e., 2s2p electrons).

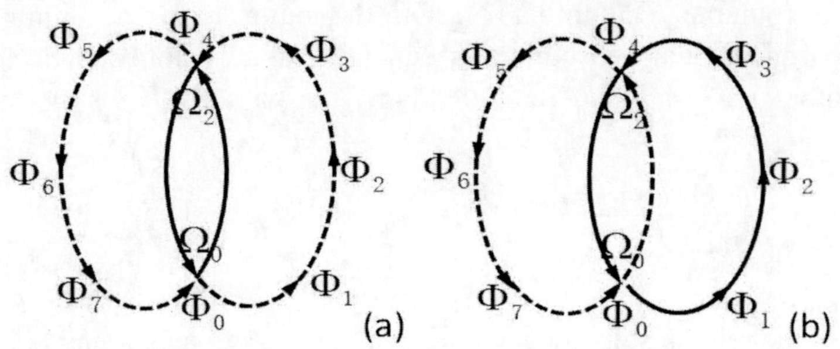

Figure 4.4. The electronic cycle defined by the quaternity equation (a) in dashed lines with connection to the electronic cycle defined by the duality equation (a) in solid lines; and the electronic cycles defined by the Schrödinger's equation (b) in dashed and solid lines. Plots (a) and (b) are physically equivalent in characterizing the transformation of ten electrons within a neon atom.

Concerning about the anisotropy of 2p electronic orbitals, we have a substantial amount of evidence from organic chemistry. In traditional organic chemistry, 2p orbitals are known to be dumbbell-shaped, isotropically aligned along the X, Y, and Z axes. So they are assigned $2p_x$, $2p_y$, and $2p_z$ orbitals. In contrast to this traditional view, the quaternity theory characterizes 2p orbitals as two points, two semicircles, and two hemispheres, which possess different levels of space dimensions (Table 2.2). Attracted by the nucleus, a $2p_x$ orbital is a line segment pointing from the nucleus to either pole; a $2p_y$ orbital is a planar sector spreading out from the nucleus to a semicircular arc; and a $2p_z$ orbital is a three-dimensional hemispherical shape. These fresh perceptions of anisotropic 2p orbitals in a carbon atom lead to an alternative and parallel account to orbital hybridization theory on the nature of covalent bonds in molecules as basic as methane, ethene, and ethyne. The suitability of quaternity space and dynamic calculus for explaining the geometries and chemical bonds of organic molecules is the basic evidence of 2p orbitals anisotropy, as will be elaborated in Chapter 6.

Conclusion

Quantum mechanics and relativity are two cornerstones of modern physics, yet as they are applied to the atomic sphere and the universe, both have generated a lot of controversies that testify their limitations in theory and applications. These limitations are not due to the construction defect of these theories but due to the imperfection of their undergirding foundation that we have taken for granted. This foundation is physical quantities in Euclidean space. The limitations of quantum mechanics and relativity echo its incompetence in dealing with space and time within the atoms and the universe. When the basement of space and time deviates from the reality, the entire establishment built upon it is flaw. However, since physical quantities and Euclidean space are deeply ingrained in people's mind and imagination and have dominated the knowledge base for thousands of years, changing them seems out of the question. This makes it difficult to introduce the spherical view, especially when it is not full-fledged. The scientific communities are reluctant to accept such a fresh proposal that does not fall into any existing categories of research fields. Thus, it is important for scientists to keep an open mind on fundamental questions, not to be carried away by the fashion of the day.

Relativity and quantum mechanics describe the world from two extreme perspectives and in very different scope relative to the studied objects. In relativity, man observes the universe in a way similar to a tiny ant, who resides at the trunk of a big tree, seeing a small portion of the tree from inside the tree. In quantum mechanics, man studies the atoms from outside its sphere. He gets a vague bird's-eye view of the object from a relatively far away distance. Thus the unification of relativity and quantum mechanics lies in correctly evaluating their perspectives and kens of vision. By restoring their pictures to their proper perspectives, we may reconcile these two seemingly inconsistent accounts that have confused so many physicists to date.

In the quaternity perspective, we see that a geodesic line on the earth corresponds to a straight line and the sea surface corresponds to a flat plane. There is not an absolute straight line or flat plane in a real entity but only conditional lines and planes in a certain region from a certain observer. This may apply to all entities ranging from the atoms, the earth, and the whole universe.

Besides the difference in the scope of views, one significant property of a spherical quantity is that it describes an entity as a dynamic oscillating one. Within an atomic shell, electrons are changing space and time states thousands of times within a picosecond continuously and harmonically. The principle of rotatory operations is valid for describing electronic behavior. As for the entire universe, the space and time oscillation is going on at an agonizingly slow pace due to its enormous scale, and even the partial rotation rule seems negligible from the perspective of an earthly life. Nonetheless, the red shift of peculiar light spectra from distant galaxies has confirmed this dynamic nature of the universe. It is believed that the universe is expanding and becoming less dense since its beginning; and we think this situation is like the counterclockwise rotation of a 1s electron Ω_0 within a helium shell. The space and time in both the atoms and the universe are oscillating harmonically by the same principle though on very dissimilar scales. Hence we unify the microcosm and macrocosm from the quaternity perspective.

We suggest that the dimensions of space and time should be defined in relation to the entity of concern, contrary to infinite Euclidean space and absolute Newtonian time that are independent of real objects (cf. Section 1.5.2). The measuring rod and the recording clock for measurement must match the dimension of the real entity to be meaningful. This is why we choose the quaternity perspective. In a dimension diagram for an entity (such as Figure 2.6), circular Newtonian time is represented by arc length along the semicircle, varying uniformly under a constant angular velocity whereas spherical quantity time is represented by the chord subtended by the arc. Like sinusoidal time pegged to a specific dimension, spherical quantity space is true physical existence that transforms in that dimension sinusoidally relative to linear uniform space. Space denotes size while time denotes density. Both are complementary in a certain dimension and can be expressed by sine and cosine functions with a constantly varying angle, respectively. This establishes the connection between space and time in spherical quantities and those in physical quantities.

We have demonstrated in rigorous mathematics that the duality equation in spherical quantities is equivalent to one-dimensional Schrödinger's equation in physical quantities; and the quaternity equation in spherical quantities is comparable to three-dimensional Schrödinger's equation in physical quantities. A spherical quantity corresponds to a wave function. This effectively integrates the quaternity theory with quantum mechanics. While quantum mechanics describes the electronic orbitals in a discrete manner, the quaternity theory provides further information on the atomic structure in a continuous manner. Discrete physical quantities sit at the whole dimension positions of continuous spherical quantities like integers at the real number axis. Ten electrons within a neon atom actually form two multiple electron circulations (Figures 4.3 and 4.4). They naturally constitute ten continuous spacetime dimensions in the atomic sphere. The quaternity perception of the world is based on a new spacetime perspective and has been shown to be advantageous over quantum mechanics in explaining the motion of electrons within a neon atom. Although the validity of quaternity remains to be tested, experimental evidence that has supported quantum mechanics and relativity so far is fully compatible with quaternity. The theory of quaternity is built upon an alternative spacetime concept and is a significant stride forward in physics. It is expected that quaternity will become the mainstream science in decades and centuries yet to come.

With regard to space dimensions, we are still in the Stone Age for we are able to count up to three only, namely X dimension, Y dimension, and Z dimension, as in the ancient Chinese history three was the maximum number. Where is the fourth space dimension? It is a riddle that no one really knows. Although we are taught in linear algebra about n-dimensional vector space, geometrical shapes of more than three dimensions have never been delineated satisfactorily. As I work on this treatise, I convince myself that we are walking out of the Stone Age as the shining light of quaternity gradually pierces the darkness on the eastern horizon – the age of quaternity is coming! After relativity and quantum mechanics boldly cause a century of philosophical perplexities, here comes quaternity quietly at last with a fresh spacetime worldview that is destined to change the landscape of science.

Chapter 5

The Structure of Atomic Shells

We described electronic orbitals within both helium and neon shells in Chapters 1 and 2. After integrating them into modern physics in Chapters 3 and 4, we resume our discussion on electronic orbitals within inert atoms. Let us plow ahead. The discovery of dimension encapsulation mechanism allows us to further construct spherical quantities for electrons beyond the electron octet in a neon shell. Step by step, we shall build mathematical structures to account for various electrons in heavy inert atoms such as radon and oganesson. Oganesson is a synthetic chemical element with symbol Og and atomic number 118. The elucidation of 3d- and 4f-orbitals is one of the most exciting and enthralling parts of the story. The splendid beauty of dynamic calculus convinces us that it describes nature squarely. Also of particular importance in this chapter is the recognition of orbital complements in geometry between 2s2p and 3s3p electrons due to a clockwise direction shift and between 1s and $3d^{1s}$ electrons due to a standpoint shift in rotatory operations, which will provide vital clues to cellular models in Chapter 8.

5.1. Introduction

The first spherical layer is the helium shell where both 1s electrons constitute a pair. The second spherical layer is comprised of sixteen electrons, among which 2s2p electrons are the counterpart of 3s3p electrons. The third spherical layer contains thirty-six electrons in two groups, 4s3d4p electrons and 5s4d5p electrons, that form an electronic rope. The fourth spherical layer includes thirty-two 6s4f5d6p electrons and an equal number of 7s5f6d7p electrons that are twisted into a multiple electron rope. Even though the four-dimensional structure paradigm is scalable, the outer spherical layers are always more complex than the inner ones. With electrons in the helium and neon shells having been described in the previous chapters, this chapter will characterize all the other electrons in the argon, krypton, xenon, argon, and oganesson atomic shells step by step. The construction of spherical quantities for these electrons will shed light on the unique property of each characteristic electron in the periodic table of elements.

5.2. Electronic Orbitals in an Argon Shell

An atomic shell refers to the continuous spacetime occupied by the characteristic electrons in the same period of the periodic table of elements, corresponding to an inert atom. The argon and neon shells share the same spherical layer within an argon atom as both 1s electrons share a helium shell. We have characterized the 2s2p electron octet in reasonable detail under quaternity spacetime. Here we present its counterpart, the 3s3p electron octet, within the same framework. Space and time symmetry and orbital complementarity are the key to understanding both 2s2p and 3s3p octets in the spherical world. As you will see, dimension factors as calculus operators comply with the rule of spherical quantities in dynamic calculus; and electronic ropes from conceptual abstractions demonstrate the symmetry of spherical quantities about space and time. Counterclockwise and clockwise rotations produce pairs of complementary orbitals in geometry. The order of electrons within an argon atom is by the rule of dynamic calculus, never chaotic.

5.2.1. Spherical Quantities for 3s3p Electrons

Spherical quantities for electrons within an argon shell still rely upon both handles, Ψ and ψ , which are spherical quantities of 1s electrons in time and space respectively but are radian angles when viewed from the argon shell. Basically spherical quantities for 2s2p electrons have been constructed by eight consecutive counterclockwise rotations from $A_1A_2\cos\Psi\cos\psi$; and spherical quantities for 3s3p electrons will be built by clockwise rotations from it instead. A counterclockwise rotation transforms a spherical quantity in the direction of reducing a time dimension while increasing a space dimension; and a clockwise rotation transforms a spherical quantity in the direction of increasing a time dimension while reducing a space dimension. Both are spacetime symmetric in dynamic calculus. All rotatory operations that we have encountered in Chapters 1 and 2 are counterclockwise. A rotation normally refers to a counterclockwise rotation unless declared as clockwise. Since 3s3p electrons are built by clockwise rotations, they are in symmetry with 2s2p electrons in spacetime.

Upon an initial spherical quantity $\varphi_{00} = A_1A_2\cos\psi\cos\Psi$, we may perform a clockwise rotation as follows:

$$\begin{aligned}\varphi_{00}^{-0} &= -\frac{dA_2\cos\psi}{dl}\int A_1\cos\Psi dt \\ &= (A_2\frac{d\psi}{dl}\sin\psi)\cdot(-A_1\frac{1}{\dot{\Psi}}\int\cos\Psi d\Psi) \\ &= -A_1A_2(-\frac{1}{\dot{\Psi}}\psi')\sin\psi\sin\Psi \\ &= -A_1A_2 n\sin\psi\sin\Psi,\end{aligned} \tag{5.1}$$

where

$$\psi' = -\frac{d\psi}{dl}, \tag{5.2}$$

$$n = \frac{1}{u} = -\frac{1}{\dot{\Psi}}\psi' = \Omega_2^{-0}. \tag{5.3}$$

Variables Ψ and ψ are time and space radian angles in the argon shell, respectively. Through consecutive clockwise rotations, we obtain four spherical quantities for $3s3p_z3p_y3p_x$ electrons in the negative direction of quaternity axes:

$$\begin{pmatrix} \varphi_0 \\ \varphi_1 \\ \varphi_2 \\ \varphi_3 \end{pmatrix} = \begin{pmatrix} \varphi_{00} - \dfrac{\partial \varphi_{00}}{\partial l} \\ \varphi_{00}^{-0} - \dfrac{\partial \varphi_{00}^{-0}}{\partial l} \\ \varphi_{00}^{-02} - \dfrac{\partial \varphi_{00}^{-02}}{\partial l} \\ \varphi_{00}^{-03} - \dfrac{\partial \varphi_{00}^{-03}}{\partial l} \end{pmatrix}, \tag{5.4}$$

where φ_{00}^{-02} and φ_{00}^{-03} denote performing clockwise rotations twice and three times upon φ_{00} respectively. In terms of trigonometric functions, the matrix becomes

$$\begin{pmatrix} \varphi_0 \\ \varphi_1 \\ \varphi_2 \\ \varphi_3 \end{pmatrix} = A_1 A_2 \begin{pmatrix} \cos\psi\cos\Psi - \psi'\sin\psi\cos\Psi \\ -n(\sin\psi\sin\Psi + \psi'\cos\psi\sin\Psi) \\ n^2(\cos\psi\cos\Psi - \psi'\sin\psi\cos\Psi) \\ -n^3(\sin\psi\sin\Psi + \psi'\cos\psi\sin\Psi) \end{pmatrix}. \tag{5.5}$$

After carrying out four consecutive clockwise rotations upon φ_{00}, we arrive at function

$$\varphi_{40} = A_1 A_2 n^4 \cos\psi\cos\Psi. \tag{5.6}$$

At this moment, time and space radian angles switch their roles, i.e., Ψ represents a space radian angle while ψ denotes a time radian angle for subsequent clockwise rotations. The calculus operators exchange their operands so that

$$\begin{aligned} \varphi_{40}^{-0} &= -n^4 \frac{dA_1\cos\Psi}{dl} \int A_2 \cos\psi \, dt \\ &= A_1 A_2 n^4 \left(\frac{d\Psi}{dl}\sin\Psi\right) \cdot \left(-\frac{1}{\dot{\psi}} \int \cos\psi \, d\psi\right) \\ &= -A_1 A_2 n^4 \left(-\frac{1}{\dot{\psi}}\Psi'\right)\sin\Psi\sin\psi \\ &= -A_1 A_2 n^5 \sin\Psi\sin\psi, \end{aligned} \tag{5.7}$$

where

$$\Psi' = -\frac{d\Psi}{dl}, \tag{5.8}$$

$$-\frac{1}{\dot{\psi}}\Psi' = \frac{1}{u} = n. \tag{5.9}$$

In the same manner, we may derive spherical quantities for another electron quartet $3s3p_z3p_y3p_x$ in the positive direction of quaternity axes from

$$\begin{pmatrix} \varphi_4 \\ \varphi_5 \\ \varphi_6 \\ \varphi_7 \end{pmatrix} = \begin{pmatrix} \varphi_{40} - \frac{\partial \varphi_{40}}{\partial l} \\ \varphi_{40}^{-0} - \frac{\partial \varphi_{40}^{-0}}{\partial l} \\ \varphi_{40}^{-02} - \frac{\partial \varphi_{40}^{-02}}{\partial l} \\ \varphi_{40}^{-03} - \frac{\partial \varphi_{40}^{-03}}{\partial l} \end{pmatrix}, \tag{5.10}$$

which expands into trigonometric terms as

$$\begin{pmatrix} \varphi_4 \\ \varphi_5 \\ \varphi_6 \\ \varphi_7 \end{pmatrix} = A_1 A_2 n^4 \begin{pmatrix} \cos\Psi\cos\psi - \Psi'\sin\Psi\cos\psi \\ -n(\sin\Psi\sin\psi + \Psi'\cos\Psi\sin\psi) \\ n^2(\cos\Psi\cos\psi - \Psi'\sin\Psi\cos\psi) \\ -n^3(\sin\Psi\sin\psi + \Psi'\cos\Psi\sin\psi) \end{pmatrix}. \tag{5.11}$$

Like electrons in a neon shell, there are clockwise rotation relationships between every two adjacent spherical quantities, e.g., between φ_0 and φ_1, φ_1 and φ_2, φ_2 and φ_3, φ_7 and φ_0, and so on. Each electron is transforming through eight states continuously. After eight consecutive clockwise rotations, each electron returns to its original state. The octet cycle gives remarkable stability for the argon shell. Spherical quantities φ_0 and φ_4 in the octet are 3s electrons at the quaternity dimensions of *1* and n^4; and the other six spherical quantities make three pairs of 3p-orbitals. Spacetime dimensions of 3s3p electrons are reciprocal to those of 2s2p electrons, one by one.

$$un = 1. \tag{5.12}$$

By the way, we have assumed that electrons within a neon shell rotate counterclockwise whereas those within an argon shell rotate clockwise invariably. Each octet alignment constitutes a conservative cycle so that an argon atom is inert. As a matter of fact, both neon and argon atoms are noble gases. However, within a fluorine or chlorine atom, the order of electrons is not so neat. Depending on the position of the unfilled void in the atomic sphere, an electron may rotate either counterclockwise or clockwise in order to fill the vacancy. Thus each electron has the potential of rotating back and forth. For example, if Φ represents a metor at a dimension of two-dimensional space with two-dimensional time, then it may transform into either a polor or a vitor depending on the position of the vacancy. One question arises as to whether or not two electrons collide at a particular dimension when both are moving toward the same vacancy at the same moment. The existence of a vacancy is the source of instability of an unsaturated atom such as fluorine or chlorine.

5.2.2. Complementary Orbitals Due to a Clockwise Direction Shift

The geometric relationship between 2s2p and 3s3p electrons deserves further scrutiny. Here we first present a typical case of orbital complement and then apply it to more cases. As shown in Figure 5.1, we compare the process of $2s \mapsto 2p_x$ transformation with that of $3s \mapsto 3p_z$ transformation. Transformation of $2s \mapsto 2p_x$ involves the sinusoidal contraction of a solid sphere and the rightwards displacement of the spherical center. The whole sphere eventually converges into a point at the periphery. In contrast, transformation of $3s \mapsto 3p_z$ has a reverse time direction, which involves the waxing of a hemispherical surface at the left hemisphere into a crescent (shaded solid). The entire empty sphere is eventually compressed rightwards until it disappears at the periphery. Both electronic orbitals under transformations are complementary in geometrical shapes, i.e., the solid sphere in the former exactly coincides with the empty sphere in the latter. Both orbitals share the same spatial sphere but their electron cloud distributions at any moment are complementary like a pair of plug and socket. The geometric complement is at the level of radian angles complement, by which two spherical quantities observe the Pythagorean theorem, having a constant hypotenuse dimension, e.g., $(A_2 \cos\psi)^2 + (A_2 \sin\psi)^2 = A_2^2$ in space for both Φ_0 and φ_0 electrons with $\psi \in (\pi/2, 0)$. The rightwards cosinusoidal displacement of the solid sphere indicates the space increment of the $2s \mapsto 2p_x$ process while the sinusoidal waxing of the crescent rightwards indicates reversely the space decrement of the $3s \mapsto 3p_z$ process. The space components of both spherical quantities, in a sense, constitute a pair of conjugate complex numbers, so do their time components $A_1(\cos\Psi - \dot{\Psi}\sin\Psi)$ and $A_1(\cos\Psi - 1/\dot{\Psi}\sin\Psi)$ with $\dot{\Psi}$ as the imaginary unit i.

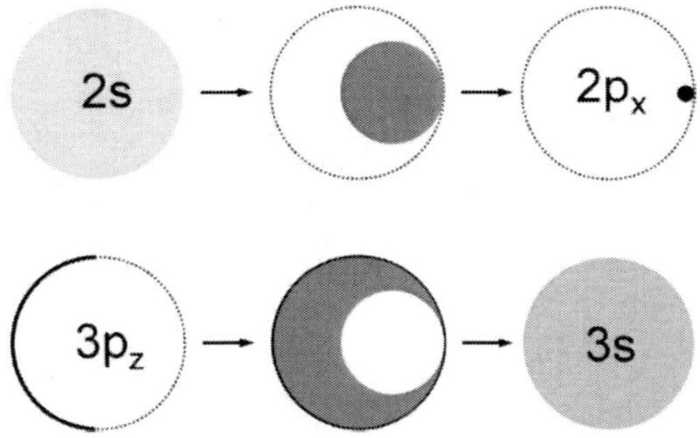

Figure 5.1. Profile diagrams of two complementary orbitals in transformations.

The complementarity is achieved because $3p_z$ orbitals are oriented along the opposite direction of $2p_x$ orbitals. Within the neon shell, eight electrons are built by consecutive counterclockwise rotations from the dimensionless *1* axis; and within the argon shell, eight electrons are constructed by consecutive clockwise rotations from the same notch when zero-dimensional space of a 2s orbital switches into four-dimensional space of a 3s orbital. The clockwise direction shift results in anti-parallel dimensional matches between both octets. As shown in Figure 5.2, the space dimensions of every orbital pair in the same row are complementary in considering that the time flow direction of 3s3p orbitals is contrary to that of 2s2p.

We have illustrated complementary processes of $1 \mapsto u$ and $n \mapsto 1$ orbital transformations in Figure 5.1. From Figure 5.2, we may find other complementary pairs such as $u^3 \mapsto u^4$ and $n^4 \mapsto n^3$, $u^4 \mapsto u^5$ and $n^5 \mapsto n^4$, and $u^7 \mapsto 1$ and $1 \mapsto n^7$. Each pair are complementary in a unique dimensional match. Note that Figure 5.1 is a profile diagram where $3p_z$ orbital is drawn as a semicircle; but Figure 5.2 is a stereo sketch where p_z-orbitals are drawn as hemispheres. Although u^7 dimension is equivalent to n dimension according to equations (2.26) and (5.12), the $2p_z$ orbital is different from the $3p_z$ orbital in orientation. The $2p_z$ orbital is a hemispherical surface with a mouth opening upwards whereas the $3p_z$ orbital is a hemispherical surface with a mouth facing right. The series of u^i dimensions define the orientations of 2s2p electrons while the series of n^i dimensions indicate the orientations of 3s3p electrons.

By the way, the relationship between $2p_y$ and $3p_y$ orbitals is also complementary in geometry (Figure 5.3). In these ways, both 2s2p and 3s3p shells share the same spherical layer and complement each other to a maximum extent at the scope of the sphere in contrast with those complements at the spherical quadrant level (Figure 2.10).

It is also noteworthy that a clockwise shift in rotations from 2s2p electrons to 3s3p electrons is different from time and space roles switch between radian angles within the neon shell. As radian angles Ψ and ψ switch their time and space roles, the differential and integral operations of a rotation switch their operands without changing the operator. A clockwise direction shift changes the operator rather than its operands, from $\int -(\partial/\partial t)dl$ to $\int -(\partial/\partial l)dt$ for the counterclockwise to clockwise shift. Roles switch does not involve changes between u^i and n^i dimension series for any spherical quantities, but clockwise rotations produce spherical quantities in the n^i dimension series instead of the u^i dimension series, which implicitly alter the geometrical orientations of the p-orbitals. However, both 2s2p and 3s3p octets are exchanging their shells in an argon atom every cycle as both 1s electrons would exchange their dimensions in a helium shell every cycle. These reflect the symmetry and complementarity of electronic orbitals under the control of dynamic calculus. An alternative perception of 2s2p electron octet involves the rotation cycle along dimensions $1 \mapsto u \mapsto u^2 \mapsto u^3 \mapsto u^4 \mapsto n^3 \mapsto n^2 \mapsto n \mapsto 1$ (Figure 5.2).

It should be recognized that the complement between space and time components of each electron is at the dimensional level (Figure 2.11); the complement between each pair of similar orbital type electrons is at the spherical quadrant level (Figure 2.10); and the complement between 2s2p electrons and 3s3p electrons is at the spherical shell level (Figure 5.2). These relationships reflect the beauty of natural law in regulating various electrons through good coordination within an atom. The alienation and interplay between space and time are the eternal theme of nature.

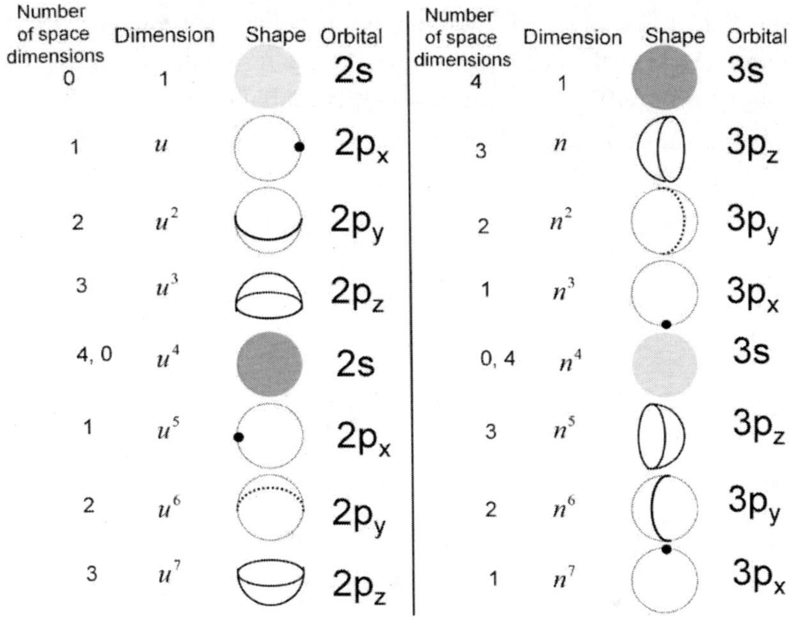

Figure 5.2. Complementary dimensions and orbital shapes of 2s2p and 3s3p electrons due to the clockwise direction shift in rotations. For each electronic orbital, the number of time dimensions plus the number of space dimensions equals four.

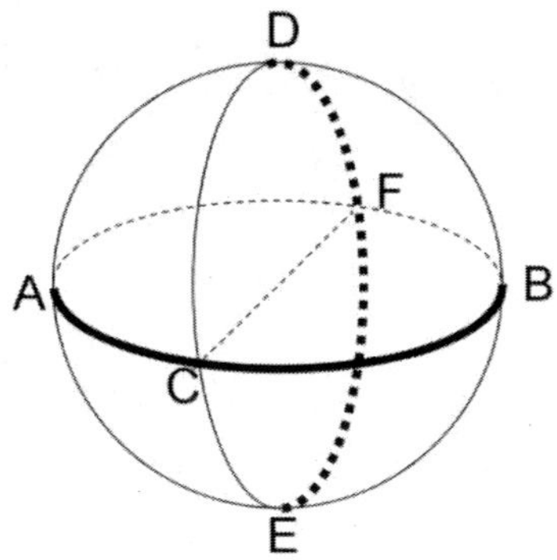

Figure 5.3. Spatial complementary relationship between $2p_y$ and $3p_y$ orbitals represented by two interlaced arcs ACB and DFE on a sphere.

5.2.3. Dimension Factors as Operators

In agreement with equations (1.1) and (1.4), we may write

$$\omega\Omega = -\frac{\partial\Omega}{\partial t}, \tag{5.13}$$

$$\omega^2\Omega = \frac{\partial}{\partial t}\left(-\frac{\partial\Omega}{\partial t}\right), \tag{5.14}$$

where ω is an angular velocity dimension in the function expressions of $\omega\Omega$ and $\omega^2\Omega$. This dimension factor may be regarded as a differential operator with respect to a time dimension. The first derivative is negative while the second differential operator is positive $\partial/\partial t$ because there is only one dimension of time within a helium shell, the second time dimension being the first one wrapped back in the opposite direction. By similar logic, orbital radius r may be treated as an integral operator when placed before a proper spherical quantity directly.

$$r\Omega = \int \Omega dl , \tag{5.15}$$

$$r^2\Omega = -\iint \Omega dl^2 . \tag{5.16}$$

Combining operators ω and r yields

$$\omega r\Omega = v\Omega = \int -\frac{\partial\Omega}{\partial t} dl , \tag{5.17}$$

where a velocity dimension factor serves as a rotatory operator. Thus the duality equation may be expressed concisely by

$$v^2\Omega = \Omega , \tag{5.18}$$

which indicates that rotating a spherical quantity twice returns it to its original state since $v^2 = 1$ in the two-dimensional helium shell.

The interpretation of dimension factors as operators may be extended to the four-dimensional case. Given $\dot{\Psi}$ as the angular velocity in a neon shell, we may treat it as a negative differential operator:

$$\dot{\Psi}\Phi = -\frac{\partial\Phi}{\partial t} , \tag{5.19}$$

$$\dot{\Psi}^2\Phi = -\frac{\partial}{\partial t}\left(-\frac{\partial\Phi}{\partial t}\right) = \frac{\partial^2\Phi}{\partial t^2} , \tag{5.20}$$

$$\dot{\Psi}^3\Phi = -\frac{\partial^3\Phi}{\partial t^3} , \tag{5.21}$$

$$\dot{\Psi}^4\Phi = -\frac{\partial}{\partial t}(\dot{\Psi}^3\Phi) = \frac{\partial^4\Phi}{\partial t^4} . \tag{5.22}$$

Given ψ' as a dimension factor in equation (5.2), we may also write

$$-\frac{1}{\psi'}\Phi = \int \Phi dl , \tag{5.23}$$

$$\frac{1}{\psi'^2}\Phi = \iint \Phi dl^2, \tag{5.24}$$

$$-\frac{1}{\psi'^3}\Phi = \iiint \Phi dl^3 , \tag{5.25}$$

$$\frac{1}{\psi'^4}\Phi = \int\iiint \Phi dl^4 . \tag{5.26}$$

Combining the differential and integral operators yields

$$-\frac{1}{\psi'}\dot{\Psi}\Phi = u\Phi = \int -\frac{\partial\Phi}{\partial t}dl , \tag{5.27}$$

where u is a velocity dimension in the four-dimensional spacetime serving as a rotatory operator upon function Φ . From equations (2.16) and (2.23), we have $u = v\Omega$. Since both helium and neon shells are coupled together, the rotation of a 1s electron is a rotatory operator upon a 2s2p electron. In other words, the cycling of 1s electrons is coupled with the cycling of the 2s2p electron octet. After radian angles Ψ and ψ switch their time and space roles, we have another pair of differential and integral operators instead:

$$-\frac{1}{\Psi'}\dot{\psi} = u . \tag{5.28}$$

Thus the quaternity equation may be expressed in a concise way as

$$u^4\Phi = \Phi , \tag{5.29}$$

where u is a counterclockwise rotatory operator upon a spherical quantity. In this way, the 2s2p octet may be compactly expressed as $\sum_{i=0}^{7} u^i\Phi_0$ where index i indicates the number of rotatory operations to perform upon the spherical quantity and plus signs indicate continuity of orbitals in spacetime. Likewise, because n is a clockwise rotatory operator, the 3s3p electron octet may be expressed as $\sum_{i=0}^{7} n^i\varphi_0$ as well where index i indicates the number of clockwise rotations to perform upon the spherical quantity and Σ indicates the spacetime continuity.

Treating dimension factors as operators is neat. However, we must still treat dimension factors associated with circular functions as dimension identifiers and unit imaginary numbers to avoid any confusion. These are the cases as were expressed in equations (2.18), (2.25), (5.5), and (5.11). Only when a dimension factor is placed before a spherical quantity directly should we regard it as a calculus operator.

5.2.4. Electronic Ropes

A counterclockwise rotation transforms a spherical quantity in the direction of reducing a time dimension while increasing a space dimension. In the two-dimensional helium shell, we have

$$\Omega_0^0 = \Omega_2 \,; \Omega_2^0 = \Omega_0 \,. \tag{5.30}$$

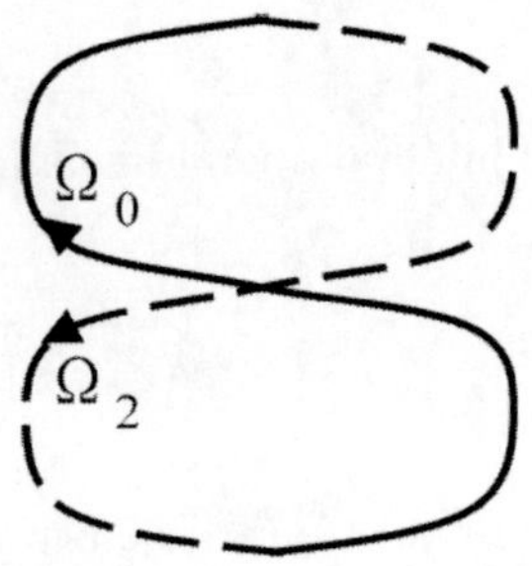

Figure 5.4. Conceptual illustration of both 1s electrons forming a rope structure.

The rotation of one electron results in another electron, and rotating an electron twice results in itself after a transformation cycle. Mechanically, when you twist a strand of fiber or hemp in a fixed direction and then let it fold back, a rope is formed. If we regard a rotatory operator $\int(-\partial/\partial t)dl$ upon an electron as a kind of twisting tension, then electronic transformation of equation (5.30) constitutes the art of making a rope with Ω_0 and Ω_2 as two partaking strands (Figure 5.4). If one strand represents a time component, the other strand is a space component, both being intertwined together. If one strand represents momentum p, then the other strand denotes energy E so that the rotation relationship indicates the balance of forces binding both strands.

$$-\frac{\partial p}{\partial t} = \frac{\partial E}{\partial l} \,. \tag{5.31}$$

In formulating spherical quantities for 2s2p electrons, we performed eight consecutive counterclockwise rotations upon function $A_1 A_2 \cos\Psi \cos\psi$. The head and tail of 2s2p electrons are interconnected so that the octet is a two-strand rope. One strand is composed of four electrons in the positive direction of quaternity axes, the other strand is composed of another four electrons in the negative direction of quaternity axes. Four pairs of electrons (Φ_0 and Φ_4, Φ_1 and Φ_5, Φ_2 and Φ_6, and Φ_3 and Φ_7) are located at opposite directions of the same quaternity axes with anti-parallel alignments so as to form a rope properly (see Figure 2.10). Likewise, spherical quantities for 3s3p were obtained through eight clockwise rotations from function $A_1 A_2 \cos\psi \cos\Psi$ so that the 3s3p octet constitutes a rope as well.

Electronic orbital Φ_0 is constructed by dimension encapsulation through the counterclockwise rotation of Ω_2 whereas electronic orbital φ_0 is constructed by dimension encapsulation through the clockwise rotation of Ω_0. The counterclockwise rotations of 2s2p electrons and clockwise rotations of 3s3p electrons within an argon atom do not counteract each other, but rather twist the 1s electrons in a fixed direction because they are performed upon 1s sphere in the opposite directions. To illustrate, if we stand at a certain point and look right and forward at a helical cord (Figure 5.5), the helix is wound counterclockwise; and if we face left but look backward from the same standpoint, the helix is wound clockwise. The helix is actually coiled in a fixed pattern. Dimension factor n in 3s3p electrons indicates a clockwise rotation upon Ω_2 (equation 5.3) while dimension factor u in 2s2p electrons indicates a counterclockwise rotation upon Ω_0 (equation 2.16). Because Ω_0^0 and Ω_2^{-0} indicate the same transformation, the calculus strain of 2s2p and 3s3p octets from the outer spherical layer to the inner spherical layer reinforces 1s sphere in a rotation cycle.

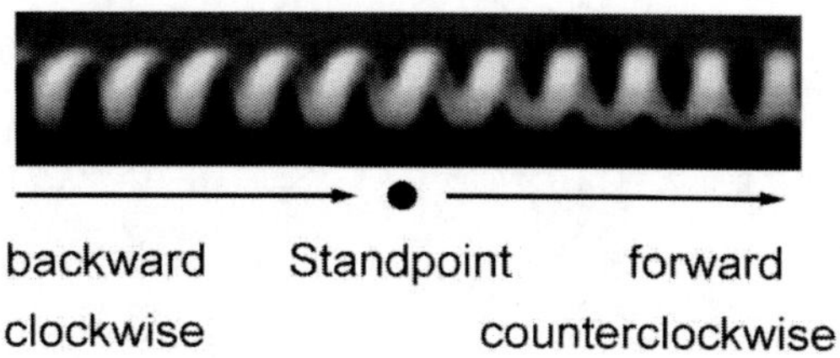

Figure 5.5. The perspective direction shift accounting for a clockwise shift of a helical structure. Arrows represent the direction of time flow and hence the observation direction.

There are two junctions between the 2s2p octet and 3s3p octet connecting the inner 1s shell (Figure 5.6). Both Φ_0 and φ_0 electrons are spacetime reversal, so are φ_4 and Φ_4. Each pair have a rotation relationship through the inner 1s electrons. At the junctions, we have

$$\Phi_0 = \Omega_0, \quad (5.32)$$

$$\varphi_0 = \Omega_2, \quad (5.33)$$

$$\Omega_0^0 = \Omega_2, \quad (5.34)$$

and

$$\Phi_4 = v^2\Omega_2 = \Omega_2, \quad (5.35)$$

$$\varphi_4 = v^2\Omega_0 = \Omega_0, \quad (5.36)$$

$$\Omega_2^{-0} = \Omega_0. \quad (5.37)$$

Thus both 2s2p and 3s3p octets constitute an electronic rope comprising four strands, each strand having four electrons in unique alignments. They are intertwined and complementary within an argon atom. Electrons at the same quaternity axis such as Φ_1 and φ_7 (with $u=n^7$ from equation 5.12) rotate in phase and perpendicular to each other in space (Figure 5.2). The rope of 2s2p and 3s3p octets actually behave like two pairs of inductive magnetic and electric waves confined with the argon sphere.

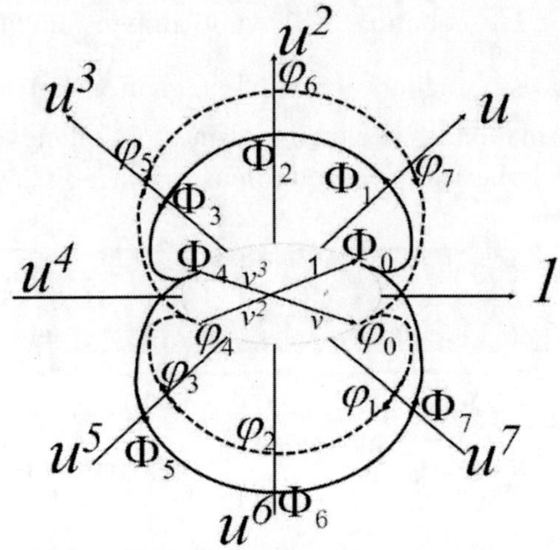

Figure 5.6. Hypothetical model of an argon atom indicating the dimensional relationship between 2s2p and 3s3p electrons in quaternity coordinates. Dimensions 1, u, u^2, ⋯, u^7 represent quaternity axes in the neon and argon shells and dimensions 1, v, v^2 and v^3 represent duality axes in the helium shell. The solid arrows indicate counterclockwise rotations of 2s2p electrons and the dashed arrows indicate clockwise rotations of 3s3p electrons.

5.3. Electronic Orbitals in a Krypton Shell

Because electronic orbitals within a krypton shell occupy a spherical layer outside the spherical layer of the neon and argon shells, spherical quantities for 4s4p may be constructed by rotatory operations upon an initial function encapsulating the 2s2p shell. This dimension encapsulation involves casting two circular complex functions into complementary radian angles by cosine operators. Both circular complex functions are obtained by dividing spherical quantities for 2s2p electrons into time and space components. Before building any electronic orbitals, we have to define circular complex functions and learn the properties of circular complex functions as radian angles and as exponential functions.

5.3.1. Coordination of Circular Complex Functions

Because sine and cosine functions characterize one-dimensional projections of uniform circular motion, they are called circular functions in mathematics. Here we introduce the concept of circular complex function. A circular complex function is a complex function composed of circular functions. A circular complex function represents a complete set of time or space dimensions of spherical quantities within an inert atomic shell. For example, both Ψ and ψ functions are circular complex functions within the helium shell. Function Ψ contains a time dimension composed of both terms $C_1 \cos\alpha$ and $-C_1\omega\sin\alpha$; and function ψ contains a space dimension composed of both terms $C_2 \cos\beta$ and $C_2 r \sin\beta$ in the two-dimensional spacetime system. Each function expresses a full cycle of a spherical quantity in time or space. Because of this, performing a differential or integral operation upon a circular complex function shifts the sequence of its trigonometric terms, but does not change the function itself. Dimension wrapping around in a circular complex function determines its invariant property upon a calculus operation.

$$\Psi = C_1(\cos\alpha - \omega\sin\alpha), \tag{5.38}$$

$$\omega\Psi = -\frac{d\Psi}{dt} = C_1(-\omega\sin\alpha + \cos\alpha) = \Psi. \tag{5.39}$$

$$\psi = C_2(\cos\beta + r\sin\beta), \tag{5.40}$$

$$r\psi = \int \psi \, dl = C_2(r\sin\beta + \cos\beta) = \psi. \tag{5.41}$$

Strictly speaking, $\omega\Psi$ is different from Ψ because of the sequence change in its two terms, so is $r\psi$ from ψ. From equations (1.8) and (1.9), we obtain a complete set of dimensions in two-dimensional spacetime by combining time and space components:

$$\Omega = C_1(\cos\alpha - \omega\sin\alpha)\cdot C_2(\cos\beta + r\sin\beta), \tag{5.42}$$

where the dot operation should be regarded as coordination between time and space components instead of conventional multiplication. The rule of a coordination operation is that only adjacent terms between both functions are associated (Figure 5.7). The coordination of time and space components in equation (5.42) results in four terms as expressed in equation (1.9). These four terms correspond to four pairs of associations indicated by zigzag lines. Although the result in this case happens to coincide with the consequence of conventional multiplication, coordination is not multiplication.

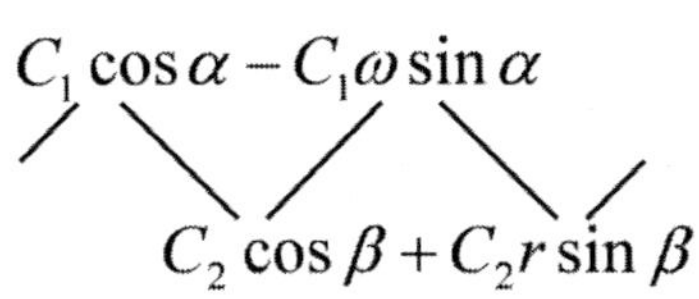

Figure 5.7. Association pairs related by lines between adjacent terms of circular complex functions in coordination. The last half line connects with the first half to form an association pair.

In quaternity spacetime, we may partition spherical quantities of 2s2p electrons into time and space components as follows:

$$\Theta = A_1(\cos\Psi - \dot{\Psi}\sin\Psi - \dot{\Psi}^2\cos\Psi + \dot{\Psi}^3\sin\Psi) + A_1\dot{\Psi}^4(\cos\Psi - \frac{1}{\Psi'}\sin\Psi - \frac{1}{\Psi'^2}\cos\Psi + \frac{1}{\Psi'^3}\sin\Psi), \tag{5.43}$$

$$\theta = A_2(\cos\psi - \frac{1}{\psi'}\sin\psi - \frac{1}{\psi'^2}\cos\psi + \frac{1}{\psi'^3}\sin\psi) + A_2\frac{1}{\psi'^4}(\cos\psi - \dot{\psi}\sin\psi - \dot{\psi}^2\cos\psi + \dot{\psi}^3\sin\psi), \tag{5.44}$$

Eight terms in function Θ are obtained by performing four consecutive differential operations upon $A_1\cos\Psi$ with respect to time dimensions followed by four consecutive integral operations over space dimensions; and eight terms in function θ are similarly obtained by performing four consecutive integral operations upon $A_2\cos\psi$ over space dimensions followed by four consecutive differential operations with respect to time dimensions. Because four space dimensions in function Θ are actually four time dimensions in sequence upon switching space and time roles, there are

$$\dot{\Psi}^4\frac{1}{\Psi'^4} = 1, \tag{5.45}$$

and similarly,

$$\frac{1}{\psi'^4}\dot{\psi}^4 = 1. \tag{5.46}$$

Functions Θ and θ are four-dimensional versions of circular complex functions because each of them comprises eight trigonometric terms in the spherical layer. Both are complementary to maintain a constant total number of space and time dimensions. The coordination of them yields eight electrons of 2s2p.

$$\Phi = \sum_{i=0}^{7}\Phi_i = \Theta\cdot\theta. \tag{5.47}$$

The association pairs of the coordination are related by 16 lines in Figure 5.8, and the calculation result can be found in equations (2.18) and (2.25). It becomes evident that coordination is different from multiplication in four-dimensional spacetime.

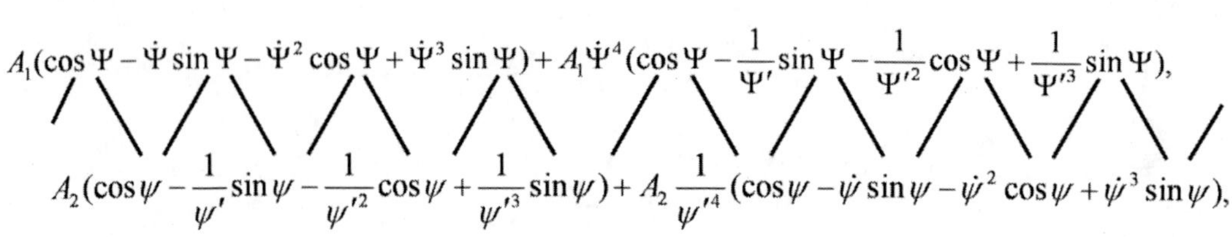

Figure 5.8. The coordination of both circular complex functions resulting in 16 pairs of associations in zigzag lines.

Eight terms of circular complex function θ in equation (5.44) are shown in solid curves as radian angle ψ varies from 2π to 0, and from 0 to 2π again (Figure 5.9). During the process, the value of θ undergoes four consecutive integral operations, followed by four consecutive differential operations, completing a full cycle from point A to point B and return to point A again. Two terms of circular complex function Ψ in equation (5.39) are shown in dashed curves as radian angle α varies from $\pi/2$ to 0, and from 0 to $\pi/2$ again. During the process, the value of Ψ undergoes an integral operation from point A to point B and then a differential operation from point B to point A. Both functions θ and Ψ are continuous, representing the solutions to equation (4.50), the modified form of three-dimensional Schrödinger's equation as was discussed in section 4.5.2.

Because of the imaginary number i on the left-hand side of equation (4.40), circular complex function Ψ shifts from equation (5.38) to equation (5.39) so that both terms in Ψ are actually equivalent to both terms in circular complex function ψ in equation (5.40) one by one. Thus, eight electrons in the neon shell are continuous with two electrons in the helium shell in space through the connections of $\Phi_0 = \Omega_0$ at point A and $\Phi_4 = \Omega_2$ at point B (cf. Figures 2.13 and 4.3). Space components in both shells are in continuity by dimension encapsulation mechanism and the chain rules of dynamic calculus.

Electrons within a neon shell and a helium shell are not only continuous in waveforms, but also rotating at the same phase with an overall angular velocity relationship of equation (2.35). Both circular complex functions Ψ and Θ represent time components of the helium and neon shells respectively. They are geared together as was illustrated in Figure 2.9 by cogwheels in mesh. Specifically, from the helium shell to the neon shell, the coupling of spherical quantities is by dimension encapsulation mechanism (cf. Section 2.2.4); and from the outer shell to the inner one, the coupling of spherical quantities is by the chain rules of dynamic calculus. The chain rules of dynamic calculus from an outer spherical layer to an inner spherical layer are the reverse mechanisms of dimension encapsulation upon perspective shift.

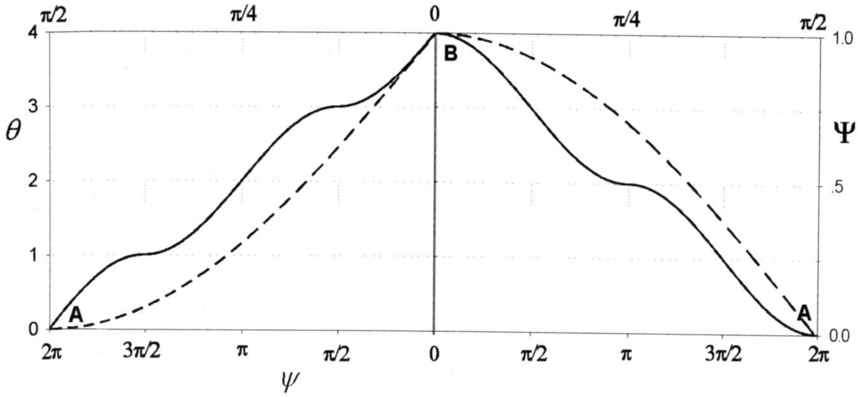

Figure 5.9. Circular complex functions as solutions to equation (4.50). Eight terms in θ (equation 5.44) in solid curves gear with two terms in Ψ (equation 5.39) in dashed curves exactly at state point A and point B in the graph. Both points A are one upon folding the two plots together and completing a cycle.

Given the differential chain of time component $-d\Theta/dt = -d\Theta/d\Psi \cdot d\Psi/dt$ at spherical quantity Φ_0, the left-hand side can be written as $-d\Theta/dt = A_1(\cos\Psi - \dot{\Psi}\sin\Psi)$ with $\dot{\Psi} = -d\Psi/dt = C_1(\cos\alpha - \omega\sin\alpha)$. Because functions $C_1 \cos\alpha$ and $\dot{\Psi}$ are time continuous whereas functions $\dot{\Psi}$ and $A_1 \cos\Psi$ are time continuous, spherical quantities Φ_0 and Ω_0 are time continuous, so are their space components (Figure 4.3). The dynamic transition of derivatives from $-d\Theta/dt$ to $-d\Psi/dt$ observes the differential chain rule strictly. Both differential steps are geared together like two cogwheels in mesh. Circular complex functions of space components θ and ψ are coupled together likewise by dimension encapsulation mechanism and the integral chain rule (Figure 5.9). At the negative direction of quaternity axes, spherical quantities Φ_4 and Ω_2 are in spacetime continuity, too (equations 2.13 and 2.16). These handshakes (dimension encapsulation mechanism and calculus chain rules) between 1s and 2s electrons represent bidirectional matter and energy exchanges that may balance the tension and synchronize the waveforms between the helium and neon shells. Figure 5.9 is equivalent to Figure 4.3 in expressing the fundamental relationships and interactions between ten electrons within a neon atom. They are graphical solutions to three-dimensional Schrödinger's equation in spherical quantities.

5.3.2. Circular Complex Functions as Exponential Functions

In line with their calculus invariant property, functions Ψ and ψ can be expressed in exponential form:

$$\Psi = C_1 e^{-\omega\alpha}, \tag{5.48}$$

$$\psi = C_2 e^{r\beta}, \tag{5.49}$$

where α and β are radian angles, and ω and r are the imaginary units. Performing a differential or integral operation upon each exponential function does not change its expression. This is an important property of an exponential function and a circular complex function.

$$\begin{aligned} -\frac{d\Psi}{dt} &= -C_1 \frac{de^{-\omega\alpha}}{dt} \\ &= -C_1 e^{-\omega\alpha} \cdot (-\omega) \cdot \frac{d\alpha}{dt} \\ &= -\Psi \cdot \omega^2 \\ &= \Psi. \end{aligned} \tag{5.50}$$

and similarly,

$$\begin{aligned}\int \psi dl &= \int C_2 e^{r\beta} dl \\ &= \int C_2 e^{r\beta} r d\beta \\ &= C_2 e^{r\beta} \\ &= \psi .\end{aligned} \tag{5.51}$$

We applied equations (1.3) and (1.6) to the above calculations. Equations (5.50) and (5.51) agree with equations (5.39) and (5.41) in dynamic calculus, respectively.

Upon the four-dimensional circular complex functions, we may also perform a calculus operation as follows:

$$\begin{aligned}\dot{\Psi}\Theta &= -A_1 d(\cos\Psi - \dot{\Psi}\sin\Psi - \dot{\Psi}^2\cos\Psi + \dot{\Psi}^3\sin\Psi)/dt \\ &+ A_1\dot{\Psi}^4\int(\cos\Psi - \frac{1}{\Psi'}\sin\Psi - \frac{1}{\Psi'^2}\cos\Psi + \frac{1}{\Psi'^3}\sin\Psi)dl \\ &= A_1(-\dot{\Psi}\sin\Psi - \dot{\Psi}^2\cos\Psi + \dot{\Psi}^3\sin\Psi + \dot{\Psi}^4\cos\Psi) \\ &+ A_1\dot{\Psi}^4(-\frac{1}{\Psi'}\sin\Psi - \frac{1}{\Psi'^2}\cos\Psi + \frac{1}{\Psi'^3}\sin\Psi + \frac{1}{\Psi'^4}\cos\Psi) \\ &= \Theta .\end{aligned} \tag{5.52}$$

The above calculations disregard the sequence advance (or queue iteration) of the trigonometric terms. Strictly speaking, a sequence advance in a circular complex function indicates a home axis shift in quaternity coordinates. Function Θ is circular because its head and tail are continuous by dynamic calculus, i.e., the last trigonometric term of the function is continuous with the first term in equation (5.52). A circular complex function represents a continuum in space or time. In the same manner, we also have

$$-\frac{1}{\psi'}\theta = \theta . \tag{5.53}$$

Because of their invariant property in calculus, these circular complex functions may be expressed in exponential form as well:

$$\Theta = A_1 e^{\frac{1}{\dot{\Psi}}\Psi}, \theta = A_2 e^{-\frac{1}{\psi'}\psi}, \tag{5.54}$$

where Ψ and ψ are radian angles, and $\dot{\Psi}$ and $-1/\psi'$ are the imaginary units. Both exponential equations in four-dimensional spacetime parallel Euler's formula (equations 5.48 and 5.49) in the two-dimensional setting. Performing differentiation upon Θ with respect to a time dimension leads to the function itself.

$$
\begin{aligned}
-\frac{d\Theta}{dt} &= -A_1 de^{\frac{1}{\Psi}\Psi} / dt \\
&= -A_1 e^{\frac{1}{\Psi}\Psi} \cdot \frac{1}{\dot{\Psi}} \cdot \frac{d\Psi}{dt} \\
&= A_1 e^{\frac{1}{\Psi}\Psi} \cdot \frac{\dot{\Psi}}{\dot{\Psi}} \\
&= \Theta.
\end{aligned} \tag{5.55}
$$

This indicates that performing a minus differential operation with respect to a time dimension does not change the exponential function. In the same manner, we may also write

$$
\int \theta dl = \int A_2 e^{-\frac{1}{\psi'}\psi} dl = \int A_2 e^{-\frac{1}{\psi'}\psi} \cdot (-\frac{1}{\psi'}) d\psi = A_2 e^{-\frac{1}{\psi'}\psi} = \theta. \tag{5.56}
$$

Hence we have set the rule for dynamic calculus upon exponential functions with imaginary exponents and the rule for dynamic calculus upon circular complex functions. They are not only self-consistent, but also consistent with infinitesimal calculus. For example,

$$
e^x = 1 + x + \frac{x^2}{2!} + \frac{x^3}{3!} + \ldots, \tag{5.57}
$$

where the right-hand side can be regarded as a special circular complex function whose second term is the integral of the first term, $\int 1dx$, and the third term is the integral of the second term, $\int xdx$, and so forth. This is in good agreement with the property of a circular complex function that every two adjacent terms have a calculus relationship. Equation (5.57) is special because it is a unidirectional open ray. We shall suggest in section 8.8 that biological evolution should follow the unidirectional ray from organisms of less dimensions to organisms of more dimensions.

5.3.3. Circular Complex Functions as Radian Angles

Of great interest is that a circular complex function can be treated as a radian angle, which permits the expression of a nested exponential function such as:

$$
\Theta = A_1 e^{\frac{1}{\Psi} \cdot C_1 e^{-\alpha x}}, \tag{5.58}
$$

where Θ itself is a radian angle, too. Within the scope of a helium shell, both Ψ and ψ are circular complex functions; but within the scope of a neon shell, they are radian angles whose values have been assigned in Figure 2.6. Within the scope of a neon shell, Θ and θ are circular complex functions; but within the scope of a krypton shell, they are radian angles.

Circular complex functions are chords transforming sinusoidally by dynamic calculus while radian angles are physical quantities changing uniformly. As shown in a dimension diagram (Figure 5.10), the magnitudes of radian angles Θ and θ are measured by arcs AC and BC, respectively. They are equivalent to radian angles Ψ and ψ respectively and are complementary at any moment as a point C traces semicircle ACB uniformly. Because the cosine of an angle equals the sine of its complement, sine and cosine are cofunctions of one another.

$$\cos\Theta = \sin\theta . \tag{5.59}$$

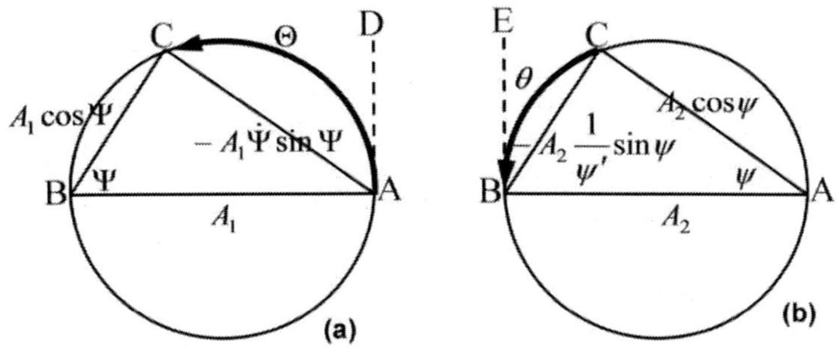

Figure 5.10. Dimension diagram showing the complement of radian angles Θ and θ in semicircle ACB. Note that $\Theta = \angle DAC$ and $\theta = \angle EBC$.

In Figure 5.10, the state point C is codetermined by chords $A_1\cos\Psi$ and $-A_1\dot{\Psi}\sin\Psi$ in time as well as by chords $A_2\cos\psi$ and $-A_2\dfrac{1}{\psi'}\sin\psi$ in space. This indicates the transformation of a spherical quantity from $A_1A_2\cos\Psi\cos\psi$ to $-A_1A_2u\sin\Psi\sin\psi$.

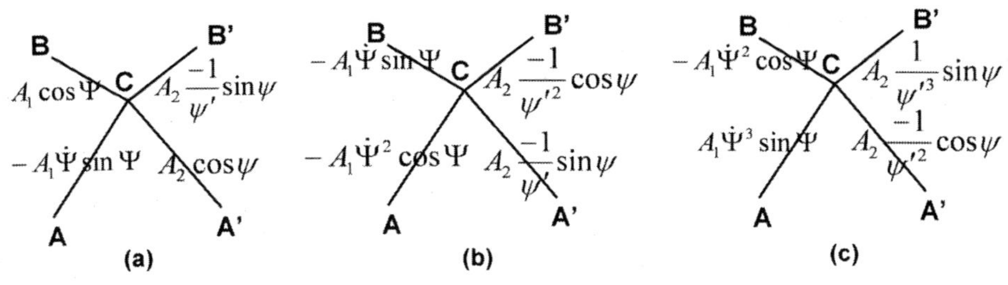

Figure 5.11. Conceptual illustration of four chords connected at a centromere in each of three consecutive rotatory operations.

For subsequent rotations, each state point C is codetermined by four similar chords (Figure 5.11). We shall call the state point C a centromere because it connects with four chords and moves from one end of the semicircle to the other during a rotation. The segregation time and space components of spherical quantity Φ into circular complex functions Θ and θ is somewhat analogous to the meiosis of cells because variables Ψ and ψ are separated during the process, which reduces the number of variables by half. In the same manner, spherical quantity φ within an argon shell may be segregated into two circular complex functions $\hat{\Theta}$ and $\hat{\theta}$ as well.

$$\varphi = \sum_{i=0}^{7} \varphi_i = \hat{\theta} \cdot \hat{\Theta}, \tag{5.60}$$

$$\begin{aligned} \hat{\theta} &= A_2 e^{\frac{1}{\psi'}\psi} = A_2(\cos\psi - \psi'\sin\psi - \psi'^2\cos\psi + \psi'^3\sin\psi) \\ &+ A_2\psi'^4(\cos\psi - \frac{1}{\dot{\psi}}\sin\psi - \frac{1}{\dot{\psi}^2}\cos\psi + \frac{1}{\dot{\psi}^3}\sin\psi), \end{aligned} \tag{5.61}$$

$$\begin{aligned} \hat{\Theta} &= A_1 e^{-\frac{1}{\dot{\Psi}}\Psi} = A_1(\cos\Psi - \frac{1}{\dot{\Psi}}\sin\Psi - \frac{1}{\dot{\Psi}^2}\cos\Psi + \frac{1}{\dot{\Psi}^3}\sin\Psi) \\ &+ A_1\frac{1}{\dot{\Psi}^4}(\cos\Psi - \Psi'\sin\Psi - \Psi'^2\cos\Psi + \Psi'^3\sin\Psi). \end{aligned} \tag{5.62}$$

Although $\hat{\Theta}$ and $\hat{\theta}$ as circular complex functions are different from functions θ and Θ, they take the same values as radian angles:

$$\Theta = \hat{\theta}; \ \theta = \hat{\Theta}. \tag{5.63}$$

The equivalence in terms of radian angles can be explained by the synchronization of counterclockwise rotation Φ_{00}^{0} and clockwise rotation φ_{00}^{-0} within a krypton atom. The former has been illustrated in Figure 5.10 and the latter can be found in Figure 5.12. Radian angles $\hat{\Theta}$ and $\hat{\theta}$ are measured by arcs BC and AC respectively. By comparing both figures, it is easy for us to derive equation (5.63) when the counterclockwise and clockwise rotations proceed at the same angular velocity.

How to explain the dimension diagrams? For a counterclockwise rotation, in the differentiation of Figure 5.10(a), arc BC decreases while arc AC increases. Spherical quantity $A_1\cos\Psi$ gradually transforms into spherical quantity $-A_1\dot{\Psi}\sin\Psi$. And in the integration of Figure 5.10(b), arc BC decreases while arc AC increases, which indicates a contrary effect that chord $A_2\cos\psi$ gives way to chord $-A_2\frac{1}{\psi'}\sin\psi$.

For a clockwise rotation, the dimension diagram shows a reverse trend. In the integration of Figure 5.12(b), arc BC decreases while arc AC increases. Spherical quantity $A_1\cos\Psi$ gradually transforms into spherical quantity $-A_1\frac{1}{\dot{\Psi}}\sin\Psi$. And in the differentiation of Figure 5.12(a), arc BC decreases while arc AC increases, which indicates a contrary effect that chord $A_2\cos\psi$ gives way to chord $-A_2\psi'\sin\psi$. In all cases, the state point C determines the values of the radian angles so that equation (5.63) is confirmed when both counterclockwise and clockwise rotations are synchronized within the krypton atom.

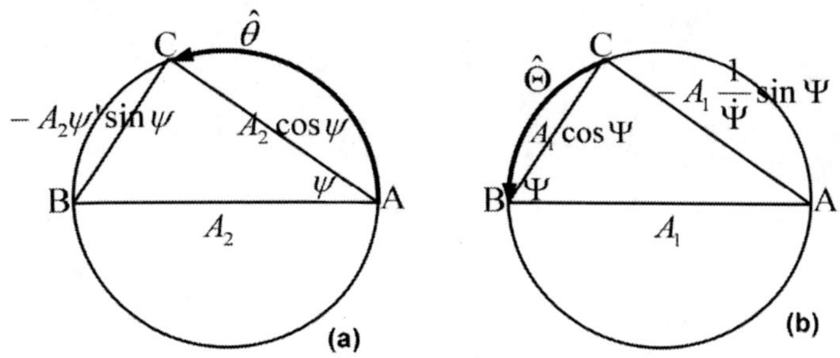

Figure 5.12. Radian angles $\hat{\Theta}$ and $\hat{\theta}$ in the dimension diagram of clockwise rotation φ_{00}^{-0} in correspondence with radian angles θ and Θ in Figure 5.10.

5.3.4. Spherical Quantities for 3d Electrons

After dividing spherical quantity Φ into two radian angles Θ and θ, the spacetime dimensions of the 2s2p shell may be encapsulated by a function:

$$\Gamma_{00} = A_3 A_4 \cos\Theta\cos\theta, \tag{5.64}$$

which serves as the starting point for building a krypton shell. We may perform a rotatory operation upon Γ_{00} as follows:

$$\begin{aligned}\Gamma_{00}^{0} &= -\frac{d(A_3\cos\Theta)}{dt}\int A_4\cos\theta dl \\ &= A_3\sin\Theta\frac{d\Theta}{dt}\cdot(-A_4\frac{1}{\theta'}\int\cos\theta d\theta) \\ &= A_3A_4(-\dot{\Theta}\cdot\frac{1}{\theta'})\sin\Theta\sin\theta \\ &= A_3A_4 w\sin\Theta\sin\theta,\end{aligned} \tag{5.65}$$

where $\dot{\Theta}$ and $-\frac{1}{\theta'}$ are the angular velocity and orbital radius, both being constants in the krypton shell.

$$\dot{\Theta} = -\frac{d\Theta}{dt}, \tag{5.66}$$

$$\theta' = -\frac{d\theta}{dl}, \tag{5.67}$$

The product of $\dot{\Theta}$ and $-\frac{1}{\theta'}$ is a velocity dimension w that initiates the branch of 3d electrons via calculus chain rules.

$$w = -\frac{1}{\theta'} \cdot \dot{\Theta} = \frac{dl}{d\theta} \cdot \left(-\frac{d\Theta}{dt}\right)$$
$$= \int A_2 \cos\psi \; dl \cdot \left(-\frac{dA_1 \cos\Psi}{dt}\right) \tag{5.68}$$
$$= \overline{\Phi}_0^0 = u\overline{\Phi}_0,$$

where $\overline{\Phi}_0$ is a 3d orbital in the krypton shell. It is similar to Φ_0 in trigonometric form but has a complementary shape in geometry. Instead of building upon 1s electrons, 3d orbitals attach to 4s electrons. The dimension encapsulation upon the neon shell leading to Γ_{00} and the calculus chains following the rotation of Γ_{00}^0 take different routes. In the krypton shell, the dimension encapsulation upon 2s2p electrons leads to 4s electrons; and the calculus chains of 4s electrons results in 3d electrons. On ten consecutive rotations of Γ_{00}, the power of w is consumed to the construction of 3d electrons. Not until 3d electrons are filled does the power of w raise up to two and more to produce 4p electrons. To express spherical quantities for 3d electrons clearly, we divide a rotatory operation into differential and integral steps artificially. Under the platform of

$$\Gamma_0 = A_3 A_4 (\cos\Theta\cos\theta - \dot{\Theta}\sin\Theta\cos\theta), \tag{5.69}$$

eight 3d electrons take the values of $\overline{\Phi}_i$ as gateway function $\dot{\Theta}$ initiates rotatory operations upon $\overline{\Phi}_0$ as follows:

$$\dot{\Theta} \cdot \frac{-1}{\theta'} = w = u\overline{\Phi}_i = \begin{pmatrix} \overline{\Phi}_0 & \overline{\Phi}_4 \\ \overline{\Phi}_1 & \overline{\Phi}_5 \\ \overline{\Phi}_2 & \overline{\Phi}_6 \\ \overline{\Phi}_3 & \overline{\Phi}_7 \end{pmatrix}, \tag{5.70}$$

which means rotations iterating through eight electronic states. Spherical quantities $\overline{\Phi}_i$ are 3d electrons similar to Φ_i in mathematical formula, so they are labeled as $3d^{2s2p}$ electrons. Another two 3d electrons are closely coupled with 1s electrons and are labeled as $3d^{1s}$. Both $3d^{1s}$ electrons attach to the $\overline{\Phi}_0$ orbital as gateway function $\dot{\Psi}$ propagates into $\overline{\Omega}_0$ and $\overline{\Omega}_2$ functions in sequence via calculus chain rules (see equations 1.15 to 1.18). The rotation of Γ_{00} drives the rotation of $\overline{\Phi}_i$, which in turn drives the rotation of $\overline{\Omega}_0$ and $\overline{\Omega}_2$.

$$\overline{\Phi}_0 = A_1 A_2 (\cos\Psi\cos\psi - \dot{\Psi}\sin\Psi\cos\psi), \tag{5.71}$$

$$\dot{\Psi} \cdot \frac{-1}{\psi} = u = v\overline{\Omega}_i = \begin{pmatrix} \overline{\Omega}_0 & \overline{\Omega}_2 \end{pmatrix}. \tag{5.72}$$

From the construction order, it is easy to see that 3d electrons are somewhat similar to $2s^2 2p^6$ and $1s^2$ electrons combined. They are similar in spherical quantity expressions, the difference being that 1s2s2p electrons wrap around the nucleus whereas the $3d^{2s2p}$ electron octet attaches to the outmost 4s sphere and $3d^{1s}$ electrons in turn attach to the octet (Figure 5.13). Both 3d and 1s2s2p electrons are supposed to be complementary in spacetime. Eight $3d^{2s2p}$ electrons pass a gateway while two $3d^{1s}$ electrons have to pass two gateways in connection with the 4s electrons. They are long-range passengers that fill the niches of the krypton sphere. The flexible mobility of 3d orbitals contributes to the good electric conductivity of their corresponding transition metals especially copper and zinc with $3d^{1s}$ electrons.

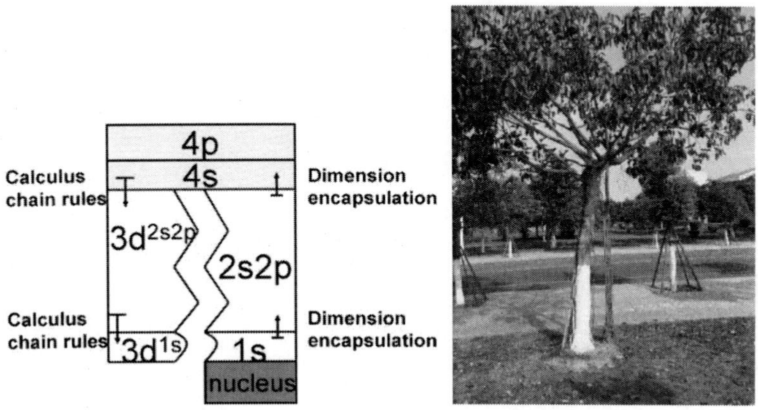

Figure 5.13. Dimension encapsulation mechanism of 2s2p building upon 1s electrons and of 4s building upon 2s2p electrons in contrast with calculus chain rules of $3d^{2s2p}$ hanging down from 4s electrons and of $3d^{1s}$ hanging down from $3d^{2s2p}$ electrons, disregarding 3s3p electrons here. Is this comparable to an Asian council tree (Ficus altissima) which grows roots hanging down from the branches to soil?

After ten 3d electrons have been instantiated, 4s4p electrons are generated by the formulae of

$$\begin{pmatrix} \Gamma_0 \\ \Gamma_1 \\ \Gamma_2 \\ \Gamma_3 \\ \Gamma_4 \\ \Gamma_5 \\ \Gamma_6 \\ \Gamma_7 \end{pmatrix} = \begin{pmatrix} \Gamma_{00} - \dfrac{\partial \Gamma_{00}}{\partial t} \\ \Gamma_{00}^{\cdot 0} - \dfrac{\partial \Gamma_{00}^{\cdot 0}}{\partial t} \\ \Gamma_{00}^{\cdot 02} - \dfrac{\partial \Gamma_{00}^{\cdot 02}}{\partial t} \\ \Gamma_{00}^{\cdot 03} - \dfrac{\partial \Gamma_{00}^{\cdot 03}}{\partial t} \\ \Gamma_{00}^{\cdot 04} - \dfrac{\partial \Gamma_{00}^{\cdot 04}}{\partial t} \\ \Gamma_{00}^{\cdot 05} - \dfrac{\partial \Gamma_{00}^{\cdot 05}}{\partial t} \\ \Gamma_{00}^{\cdot 06} - \dfrac{\partial \Gamma_{00}^{\cdot 06}}{\partial t} \\ \Gamma_{00}^{\cdot 07} - \dfrac{\partial \Gamma_{00}^{\cdot 07}}{\partial t} \end{pmatrix}. \tag{5.73}$$

In trigonometric functions, the matrix takes the form of

$$\begin{pmatrix} \Gamma_0 \\ \Gamma_1 \\ \Gamma_2 \\ \Gamma_3 \\ \Gamma_4 \\ \Gamma_5 \\ \Gamma_6 \\ \Gamma_7 \end{pmatrix} = A_3 A_4 \begin{pmatrix} \cos\Theta\cos\theta - \dot{\Theta}\sin\Theta\cos\theta \\ -w(\sin\Theta\sin\theta + \dot{\Theta}\cos\Theta\sin\theta) \\ w^2(\cos\Theta\cos\theta - \dot{\Theta}\sin\Theta\cos\theta) \\ -w^3(\sin\Theta\sin\theta + \dot{\Theta}\cos\Theta\sin\theta) \\ w^4(\cos\theta\cos\Theta - \dot{\theta}\sin\theta\cos\Theta) \\ -w^5(\sin\theta\sin\Theta + \dot{\theta}\cos\theta\sin\Theta) \\ w^6(\cos\theta\cos\Theta - \dot{\theta}\sin\theta\cos\Theta) \\ -w^7(\sin\theta\sin\Theta + \dot{\theta}\cos\theta\sin\Theta) \end{pmatrix}, \tag{5.74}$$

$$\dot{\theta} = -\frac{d\theta}{dt}. \tag{5.75}$$

The order of spherical quantity generations reflects the order of energy levels of electronic orbitals. Electrons within the krypton shell corresponding to chemical elements in the fourth period have energy levels in the sequence of 4s3d4p. A dimension encapsulation upon the 2s2p shell leads to the formation of a 4s scalor in a larger spherical layer. Then there are two branches of rotatory operations from the established 4s platform. One branch is the upward rotatory operations according to equation (5.73), which gives 4s4p electrons; the other branch is the downward rotatory operations stemming from gateway function $\dot{\Theta}$ via calculus chain rules, which results in ten 3d electrons. These two branches constitute a clear contrast between going around and entering into the inner spherical layer.

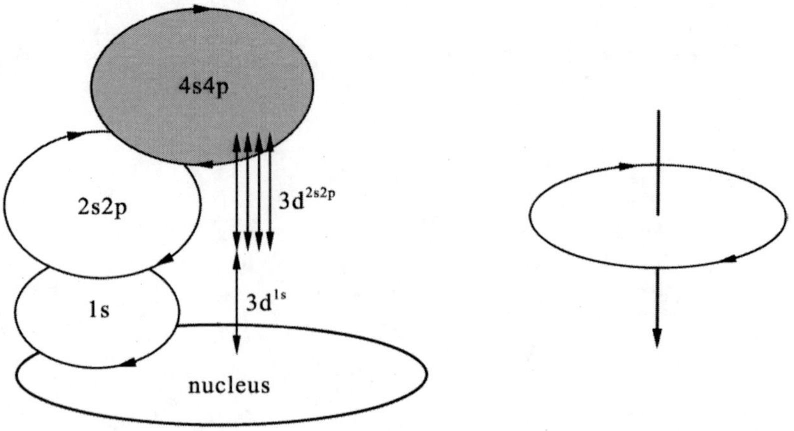

Figure 5.14. An electronic rope with two strands in a krypton atom, one being spherical shape of 2s2p1s electrons, and the other being linear shape of 3d electrons. They are intertwined together inductively albeit separate in the diagram, which is reminiscent of circular magnetic lines induced by an electric current flowing along a straight wire, disregarding 3s3p electrons here.

The downward branch has a lower energy and hence shows up before the upward branch in the periodic table of elements. Because 3d electrons attach to the platform of equation (5.69), 4s electrons must be established preceding any 3d electrons. This is the priority order of spherical quantities or wave functions in the krypton shell.

In general, after filling all available dimensions in a certain spherical layer, we roll up to a larger spherical layer by a dimension encapsulation. We built 2s2p electrons from 1s sphere and built 4s electrons from the 2s2p shell in this way. On the other hand, the calculus chains hanging down from 4s to $3d^{2s2p}$ and to $3d^{1s}$ electrons constitute a symmetrical branch with 1s2s2p electrons. Electrons $3d^{2s2p1s}$ and 2s2p1s are two anti-parallel strands that form a rope as they pair together (Figure 5.14). During the rolling up, we treated spherical quantities as radian angles using cosine operators (see equations 2.12 and 5.64) whereas in the hanging down we converted radian angles back into spherical quantities using rotation operators u and v (see equations 5.70 and 5.72). Such bidirectional conversions reflect the scope shift of the observer from an inner spherical layer to an outer spherical layer, or the other way around.

While the 4s4p octet represents the primary scale of spacetime in the krypton shell, the propagation of gateway function $\dot{\Theta}$ into various $\overline{\Phi}_i$ (see equation 5.70) is secondary in spacetime scale and has a minor distinction or effect on the overall atomic properties. This accounts for the more physical similarities among transition metal elements from Sc to Zn than among elements from Ga to Kr. However, because the mathematical formulae of $3d^{2s2p1s}$ electrons coincide with those of 2s2p1s electrons, there are somewhat similar properties between 3d electrons and 2s2p1s electrons. For instance, Cu and Zn ($3d^{1s}$) correspond to 1s electrons and tend to take valences of +1 and +2, respectively, during chemical reactions. Their chemical properties are somewhat similar to those of alkaline and alkaline earth elements, respectively.

So far we have refrained from pointing out an important rule in wave function expression because we do not want to confuse the readers from the beginning. When we said that the Γ_1 electron had a wave function designated by equation (5.74) where dimension factor w took the value of equation (5.68), it was different from

$$\begin{aligned}\Gamma_1 &= -A_3A_4w(\sin\Theta\sin\theta + \dot{\Theta}\cos\Theta\sin\theta)\\ &= -A_1A_2u(\cos\Psi\cos\psi - \dot{\Psi}\sin\Psi\cos\psi)A_3A_4(\sin\Theta\sin\theta + \dot{\Theta}\cos\Theta\sin\theta).\end{aligned} \tag{5.76}$$

Even though the mathematical substitution is correct, equation (5.76) suggests more waveforms for an electron than Γ_1 in equation (5.74). A 4s4p electron treats w as a dimension identifier and as a whole unit that stays at that level only. The oscillation of the 4s4p octet does not involve the details of w expression, i.e., the physical correspondence for equation (5.68) is not a part of 4s4p electrons. Waveforms contained in w represent 3d electrons. In other words, equation (5.76) indicates two electrons (Γ_1 and $\overline{\Phi}_0$) rather than one. As we formulate wave functions, we must consider such nuances in their physical implication. That a complex function corresponds to a physical electron is the basic criterion for writing a good wave function.

5.3.5. Complementary Orbitals Due to a Standpoint Shift

The geometries of $3d^{2s2p}$ orbitals challenge your wildest space imagination, notwithstanding the clue that $3d^{2s2p}$ orbitals must somehow complement the 2s2p octet. Unless computer aided modeling is established properly to help in this regard, we will not make any intellectual speculation this far on their geometrical shapes. However, we would like to touch on the geometry of $3d^{1s}$ orbitals that are complementary to 1s orbitals in spacetime. As shown in Figure 5.15, both 1s electrons are solid spheres that center at nucleus O. As radian angle α increases from 0 to $\pi/2$, radian angle β reduces from $\pi/2$ to 0. One 1s electron reduces its spherical radius from QO to NO (shaded area in Figure 5.15a) and to vanishing while the other 1s electron increases its spherical radius from naught to OP (shaded area in Figure 5.15b) and to OQ. Both 1s electrons oscillate cyclically, but their geometric centers are always at point O.

In contrast, both $3d^{1s}$ electrons are hollow spheres that attach to spherical surface Q. Spherical surface Q is also the boundary of 1s sphere at maximum space. As radian angle α increases from 0 to $\pi/2$, one $3d^{1s}$ electron in grows from a spherical surface to a thick hollow (shaded ring in Figure 5.15c) sphere and eventually to a solid sphere while the other $3d^{1s}$ electron changes from a solid sphere into a hollow one (the shaded ring in Figure 5.15d) and eventually to a spherical surface. Spherical surface Q is the region with which both $3d^{1s}$ electrons always connect during harmonic oscillations.

Electrons in Figure 5.15(a) and 5.15(c) are always complementary sharing the whole sphere O together whereas electrons in Figure 5.15(b) and 5.15(d) are complementary in geometry similarly. The spacetime complements arise from the fact that 1s electrons spin around the nucleus (point O) whereas $3d^{1s}$ electrons always stick to the $3d^{2s2p}$ octet (spherical surface Q). The standpoint shift determines their complements in geometry. The geometric complements are at the level of radian angles with spherical quantities obeying the Pythagorean theorem, as was the geometric complement shown in Figure 5.1. It is interesting to note that 1s electrons are introverted whereas $3d^{1s}$ are extroverted in behavior. Their complementarity ensures the harmony and integrity of the entire krypton atom.

We shall further explain the orbital complementarity by set theory here. A set is a collection of objects or a container of objects, but here a set refers to a part of spacetime continuum within an atom. Because of the standpoint shift, both $3d^{1s}$ electrons $\overline{\Omega}_0$ and $\overline{\Omega}_2$ complement both 1s electrons Ω_0 and Ω_2 in spacetime, respectively. The union of each pair represents both dimensions in a two-dimensional sphere; and the intersection of each pair is a null set.

$$\Omega_0 \cup \overline{\Omega}_0 = \Omega,\ \Omega_0 \cap \overline{\Omega}_0 = \emptyset, \tag{5.77}$$

$$\Omega_2 \cup \overline{\Omega}_2 = \Omega,\ \Omega_2 \cap \overline{\Omega}_2 = \emptyset, \tag{5.78}$$

where Ø is an empty set, and Ω denotes the universal set of both dimensions in the two-dimensional sphere. For example, electrons of Figure 5.15(a) and (c) have the same wave function expression but different geometric significances, so do those of Figure 5.15(b) and (d). Equation (5.77) indicates that $\overline{\Omega}_0$ is a complement of Ω_0 in spacetime; and equation (5.78) indicates that $\overline{\Omega}_2$ is a complement of Ω_2 in spacetime. The complementarity is an alternative expression to the Pythagorean theorem for right triangles of Figure 1.7. Because of the standpoint shift, as Ω_0 is gaining a space dimension, $\overline{\Omega}_0$ is actually reducing a space dimension from the standpoint of the central nucleus.

The counterclockwise rotation of $\overline{\Omega}_0$ has a reverse effect when viewed from the standpoint of the central nucleus so that the following two transformations proceed at the same pace. Time and space dimensions of $3d^{1s}$ electrons transform in the opposite directions to those of 1s electrons.

$$\int -\frac{\partial \Omega_0}{\partial t} dl = \Omega_2 \ , \tag{5.79}$$

$$\int -\frac{\partial \overline{\Omega}_0}{\partial l} dt = \overline{\Omega}_2 \ . \tag{5.80}$$

Both processes are synchronized automatically by the same radian angle β. The complementary orbitals are electromagnetically inductive because their spherical quantities observe Faraday's law and the Pythagorean theorem. This indicates that the electrons in the inner shell of an atom can be induced by the electrons in the outer shell. This induction effect has far-reaching significance in biology. For example, it indicates that the environment may induce the evolution direction of species.

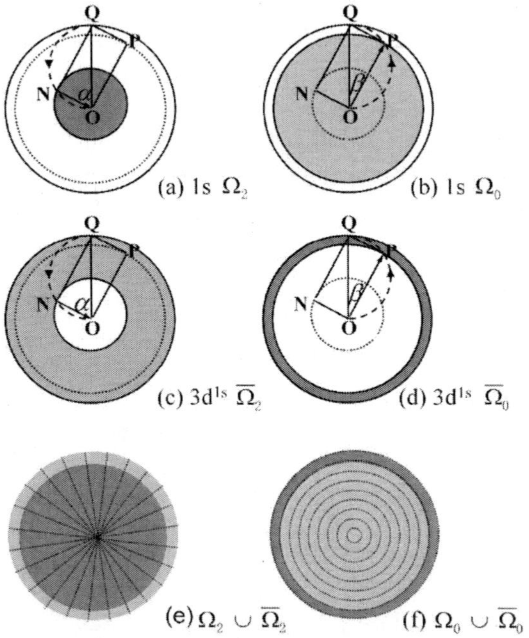

Figure 5.15. Geometric complements between $3d^{1s}$ and 1s orbitals due to their different geometrical orientations and the electromagnetic field continuum of them with reference to Figure 1.16.

It is noteworthy that Figures 5.13, 5.14, and 5.15 actually express the same atomic structure from various perspectives. Figure 5.13 draws the architecture of wave function construction in the krypton shell; Figure 5.14 exaggerates the intertwining relation of 3d electrons with 1s2s2p electrons; and Figure 5.15 focuses on the complementary geometries of $3d^{1s}$ and 1s electrons. In particular, Figure 5.15(e) and (f) give a visual aid of the electromagnetic field continuum of Ω_0 and $\overline{\Omega}_0$ electrons and of Ω_2 and $\overline{\Omega}_2$ electrons. Figure 5.15(e) is a density (time) expression whereas Figure 5.15(f) is a space expression. We ignored 3s3p electrons in these graphics for brevity.

By the way, the order of electrons within a krypton or zinc atom provides a functional model for a eukaryotic cell. According to an endosymbiotic hypothesis, a eukaryotic cell began as a large anaerobic organism. A mitochondrion arose when a small aerobic bacterium was ingested by the larger cell. Here $3d^{1s}$ electrons hang down from the outer layer of $3d^{2s2p}$ and 4s electrons like a mitochondrion coming from outside the cell. And the aerobic and anaerobic complementary roles of the mitochondrion and the larger cell are analogous to spacetime complements of $3d^{1s}$ and 1s electrons. We shall revert to such an interesting biological subject in Chapter 8.

5.4. Electronic Orbitals in a Radon Shell and Beyond

We proceed to formulate spherical quantities for electrons in an oganesson atom. Oganesson is the largest element with 118 protons. We shall arrange all elements into a quaternity periodic table of elements based on the property of their characteristic electrons. To make such an arrangement, 3d-orbitals are divided into $3d^{2s2p}$ and $3d^{1s}$ because their waveforms are similar to those of 2p2s1s electrons, and 4f-orbitals are divided into $4f^{4p}4f^{4s}4f^{2p}$ in correspondence with 4p4s2p electrons as well. Under the framework of quaternity spacetime, this alternative periodic table provides sharp insight into the structure of various atoms. Moreover, we consider the atomic shells of hydrogen, helium, neon, argon, krypton, xenon, radon, and oganesson to be an octet configuration. The atomic shell octet within an oganesson atom is a stable system in a certain way similar to the electron octet within a neon shell. This insightful perception allows us to jump to the conclusion that there are exactly 118 elements existing in the world.

5.4.1. Spherical Quantities for 4f Electrons

Spherical quantities for 5s4d5p can be constructed through consecutive clockwise rotations on the initial function of

$$\gamma_{00} = A_3 A_4 \cos\theta \cos\Theta \tag{5.81}$$

where Θ and θ are a pair of radian angles equivalent to $\hat{\theta}$ and $\hat{\Theta}$ that are circular complex functions derived from spherical quantities φ. Because $\Theta = \hat{\theta}$ and $\theta = \hat{\Theta}$ as expressed in equation (5.63), we simply adopt Θ and θ to get rid of the cumbersome caret notations.

Performing clockwise rotatory operations similar to equations (5.4) and (5.10) with substitutions of φ_{00} with γ_{00} yields the 5s5p electron octet:

$$\begin{pmatrix}\gamma_0\\ \gamma_1\\ \gamma_2\\ \gamma_3\\ \gamma_4\\ \gamma_5\\ \gamma_6\\ \gamma_7\end{pmatrix}=A_3A_4\begin{pmatrix}\cos\theta\cos\Theta-\theta'\sin\theta\cos\Theta\\ -m(\sin\theta\sin\Theta+\theta'\cos\theta\sin\Theta)\\ m^2(\cos\theta\cos\Theta-\theta'\sin\theta\cos\Theta)\\ -m^3(\sin\theta\sin\Theta+\theta'\cos\theta\sin\Theta)\\ m^4(\cos\Theta\cos\theta-\Theta'\sin\Theta\cos\theta)\\ -m^5(\sin\Theta\sin\theta+\Theta'\cos\Theta\sin\theta)\\ m^6(\cos\Theta\cos\theta-\Theta'\sin\Theta\cos\theta)\\ -m^7(\sin\Theta\sin\theta+\Theta'\cos\Theta\sin\theta)\end{pmatrix}, \tag{5.82}$$

where

$$m=-\frac{1}{\dot{\Theta}}\theta'=\frac{1}{w}, \tag{5.83}$$

or after the role exchange between time and space radian angles Θ and θ, we write

$$\Theta'=-\frac{d\Theta}{dl}, \tag{5.84}$$

$$m=-\frac{1}{\dot{\theta}}\Theta'=\frac{1}{w}. \tag{5.85}$$

Factor m is a reciprocal velocity dimension in the xenon shell that initiates the branch of 4d electrons. 4d-orbitals may be divided into $4d^{3s3p}$ and $4d^{1s}$ according to their waveforms. The first octet, $4d^{3s3p}$, bears close relationships to φ_i and the last two $4d^{1s}$ electrons are somehow complementary to both 1s electrons. The $4d^{3s3p}$ electron octet attaches to 5s electrons in the outer spherical layer and $4d^{1s}$ electrons in turn attach to the octet. These transition electrons are inductive with 3s3p1s electrons. Within the krypton and xenon spherical layer, there are two electronic strands, one composed of 4s3d4p twisted counterclockwise, and the other composed of 5s4d5p strained clockwise, both forming a rope.

The scalability of quaternity spacetime makes it possible to build spherical quantities for various electrons up to four spherical layers. Electrons are packed in each layer by similar orders even though the structural complexity of an outer spherical layer tends to increase as a result of more dimensions involved. Spherical quantities for 6s4f5d6p electrons in the sixth period can be derived by encapsulating dimensions within a krypton shell. The encapsulation regards time and space components of spherical quantities Γ and γ as radian angles. To save letters, we still use variables Γ and γ to represent those radian angles derived from the meiosis of spherical quantities Γ and γ. This is possible because it is reasonable to use a male to denote a sperm and a female to denote an ovum analogously. Thus the dimension encapsulation upon the krypton shell leads to a membrane function:

$$\mathrm{H}_{00} = A_5 A_6 \cos\Gamma \cos\gamma\ , \tag{5.86}$$

where A_5 and A_6 are dimension constants, and Γ and γ serve as two complementary radian angles in the radon shell. The function is called a membrane function because it encapsulates an atomic shell as plasma membrane encloses the cytoplasm in a cell. We construct electrons 6s4f5d6p starting from this membrane function. Let

$$\begin{pmatrix} 6s & 6s \\ 6p_x & 6p_x \\ 6p_y & 6p_y \\ 6p_z & 6p_z \end{pmatrix} = \begin{pmatrix} \mathrm{H}_0 & \mathrm{H}_4 \\ \mathrm{H}_1 & \mathrm{H}_5 \\ \mathrm{H}_2 & \mathrm{H}_6 \\ \mathrm{H}_3 & \mathrm{H}_7 \end{pmatrix}, \tag{5.87}$$

where spherical quantities in the right matrix correspond to orbital types in the left matrix. We may assign the values of H_i through consecutive counterclockwise rotations by replacing Γ_{00} with H_{00} in equation (5.73) so that

$$\begin{pmatrix} \mathrm{H}_0 \\ \mathrm{H}_1 \\ \mathrm{H}_2 \\ \mathrm{H}_3 \\ \mathrm{H}_4 \\ \mathrm{H}_5 \\ \mathrm{H}_6 \\ \mathrm{H}_7 \end{pmatrix} = A_5 A_6 \begin{pmatrix} \cos\Gamma\cos\gamma - \dot{\Gamma}\sin\Gamma\cos\gamma \\ -p(\sin\Gamma\sin\gamma + \dot{\Gamma}\cos\Gamma\sin\gamma) \\ p^2(\cos\Gamma\cos\gamma - \dot{\Gamma}\sin\Gamma\cos\gamma) \\ -p^3(\sin\Gamma\sin\gamma + \dot{\Gamma}\cos\Gamma\sin\gamma) \\ p^4(\cos\gamma\cos\Gamma - \dot{\gamma}\sin\gamma\cos\Gamma) \\ -p^5(\sin\gamma\sin\Gamma + \dot{\gamma}\cos\gamma\sin\Gamma) \\ p^6(\cos\gamma\cos\Gamma - \dot{\gamma}\sin\gamma\cos\Gamma) \\ -p^7(\sin\gamma\sin\Gamma + \dot{\gamma}\cos\gamma\sin\Gamma) \end{pmatrix}, \tag{5.88}$$

where

$$\dot{\Gamma} = -\frac{d\Gamma}{dt}, \tag{5.89}$$

$$\gamma' = -\frac{d\gamma}{dl}, \tag{5.90}$$

$$p = \dot{\Gamma} \cdot \frac{-1}{\gamma'}. \tag{5.91}$$

Factor p is a velocity dimension in the radon shell. Or after the role exchange between time and space radian angles Γ and γ, there are

$$\dot{\gamma} = -\frac{d\gamma}{dt}, \tag{5.92}$$

$$\Gamma' = -\frac{d\Gamma}{dl}, \tag{5.93}$$

$$p = -\dot{\gamma} \cdot \frac{1}{\Gamma'}. \tag{5.94}$$

Under the platform of spherical quantity H_0, gateway function $\dot{\Gamma}$ initiates the branch of 5d electrons:

$$H_0 = A_5 A_6 (\cos\Gamma \cos\gamma - \dot{\Gamma} \sin\Gamma \cos\gamma), \tag{5.95}$$

$$\dot{\Gamma} \cdot \frac{-1}{\gamma'} = p = w\overline{\Gamma} = \begin{pmatrix} \overline{\Gamma}_0 & \overline{\Gamma}_4 \\ \overline{\Gamma}_1 & \overline{\Gamma}_5 \\ \overline{\Gamma}_2 & \overline{\Gamma}_6 \\ \overline{\Gamma}_3 & \overline{\Gamma}_7 \end{pmatrix} = \begin{pmatrix} 5d^{4s} & 5d^{4s} \\ 5d^{4px} & 5d^{4px} \\ 5d^{4py} & 5d^{4py} \\ 5d^{4pz} & 5d^{4pz} \end{pmatrix}, \tag{5.96}$$

where spherical quantities in the left matrix correspond to orbital types in the right matrix. Functions Γ_i were spherical quantities of 4s4p electrons, and functions $\overline{\Gamma}_i$ refer to $5d^{4s4p}$ electrons attaching to spherical quantity H_0. They have similar mathematical formulae but complementary geometries. Hanging down from the $5d^{4s4p}$ octet is another two spherical quantities for $5d^{2s}$.

$$\overline{\Gamma}_0 = A_3 A_4 (\cos\Theta \cos\theta - \dot{\Theta} \sin\Theta \cos\theta), \tag{5.97}$$

$$\dot{\Theta} \cdot \frac{-1}{\theta'} = w = u\overline{\Phi} = \begin{pmatrix} \overline{\Phi}_0 \\ \overline{\Phi}_4 \end{pmatrix} = \begin{pmatrix} 5d^{2s} \\ 5d^{2s} \end{pmatrix}. \tag{5.98}$$

The expression means that both functions $\overline{\Phi}_0$ and $\overline{\Phi}_4$ are similar to 2s electrons in waveforms, but they attach to the $5d^{4s4p}$ octet instead of building upon 1s electrons. And the $5d^{4s4p}$ octet attaches to 6s electrons. We have labeled 5d electrons with superscripts in matrices (5.96) and (5.98) to indicate their orbital properties. These 5d electrons are all buried under 6s electrons. If we use elements to represent their characteristic electrons, then there is a one-to-one correspondence between the characteristic waveforms and the chemical elements.

$$\begin{pmatrix} 5d^{4s} & 5d^{4s} \\ 5d^{4px} & 5d^{4px} \\ 5d^{4py} & 5d^{4py} \\ 5d^{4pz} & 5d^{4pz} \\ 5d^{2s} & 5d^{2s} \end{pmatrix} = \begin{pmatrix} La & Hf \\ Ta & Os \\ W & Ir \\ Re & Pt \\ Au & Hg \end{pmatrix} = \begin{pmatrix} \overline{\Gamma}_0 & \overline{\Gamma}_4 \\ \overline{\Gamma}_1 & \overline{\Gamma}_5 \\ \overline{\Gamma}_2 & \overline{\Gamma}_6 \\ \overline{\Gamma}_3 & \overline{\Gamma}_7 \\ \overline{\Phi}_0 & \overline{\Phi}_4 \end{pmatrix}. \tag{5.99}$$

Inner-transition 4f electrons are characterized by spherical quantities cascading to even a deeper level close to the nucleus.

$$\begin{pmatrix} 4f^{4px} & 4f^{4px} \\ 4f^{4py} & 4f^{4py} \\ 4f^{4pz} & 4f^{4pz} \\ 4f^{4s} & 4f^{4s} \\ 4f^{2px} & 4f^{2px} \\ 4f^{2py} & 4f^{2py} \\ 4f^{2pz} & 4f^{2pz} \end{pmatrix} = \begin{pmatrix} Ce & Pm \\ Pr & Sm \\ Nd & Eu \\ Gd & Tb \\ Dy & Tm \\ Ho & Yb \\ Er & Lu \end{pmatrix}. \tag{5.100}$$

They reflect the possible combinations of gateway functions $\dot{\Gamma}$ and $\dot{\Theta}$. Fourteen 4f electrons may be categorized into three major types: $4f^{4p}$, $4f^{4s}$, and $4f^{2p}$, which correspond to chemical elements in equation (5.100).

Under the framework of function H_0, gateway function $\dot{\Gamma}$ might take eight values as $\overline{\Gamma}_i$; and within each value $\overline{\Gamma}_i$, gateway function $\dot{\Theta}$ (or $\dot{\theta}$) might take eight values as $\overline{\Phi}_j$, too. For brevity, eight possible wave functions $\overline{\Gamma}_i$ are directly placed at the position of gateway function $\dot{\Gamma}$, and eight possible wave functions $\overline{\Phi}_j$ are directly placed at the position of gateway function $\dot{\Theta}$ as follows.

$$H_0 = A_5 A_6 (\cos\Gamma \cos\gamma - \begin{pmatrix} \overline{\Gamma}_0 & \overline{\Gamma}_4 \\ \overline{\Gamma}_1 & \overline{\Gamma}_5 \\ \overline{\Gamma}_2 & \overline{\Gamma}_6 \\ \overline{\Gamma}_3 & \overline{\Gamma}_7 \end{pmatrix} \sin\Gamma \cos\gamma), \tag{5.101}$$

$$\overline{\Gamma}_0 = A_3 A_4 (\cos\Theta \cos\theta - \begin{pmatrix} \overline{\Phi}_0 & \overline{\Phi}_4 \\ \overline{\Phi}_1 & \overline{\Phi}_5 \\ \overline{\Phi}_2 & \overline{\Phi}_6 \\ \overline{\Phi}_3 & \overline{\Phi}_7 \end{pmatrix} \sin\Theta \cos\theta), \tag{5.102}$$

where $\overline{\Gamma}_i$ means $5d^{4s4p}$ electrons, complementary to 4s4p electrons, because they attach to H_0 in the outer spherical layer. Function $\overline{\Gamma}_0$ in equation (5.102) is only an example of $\overline{\Gamma}_i$. Spherical quantities $\overline{\Phi}_j$ attach to $\overline{\Gamma}_i$. However, the combination of both $\overline{\Gamma}_i$ and $\overline{\Phi}_j$ values in matrices (5.101) and (5.102) does not produce 8×8 variants of wave functions. To form valid wave functions, an s-orbital in one variable may combine with an s-orbital in another variable to form an s-nature orbital, but three pairs of p-orbitals in one variable must combine with two s-orbitals in another to form six p-nature orbitals. The combinations of p-orbitals with p-orbitals are not valid within an atom. Mathematically, we shall define a new subscript symbol "$\otimes$" as a gateways combination operator. According to the above combination rule, we write $\overline{\Gamma}_{i\otimes}\overline{\Phi}_j$ to indicate valid combinations of both gateway variables ($\dot{\Gamma}$ and $\dot{\Theta}$) and to indicate that $\overline{\Phi}_j$ attaches to function $\overline{\Gamma}_i$ so that

$$\overline{\Gamma}_{i\otimes}\overline{\Phi}_j = \begin{pmatrix} \overline{\Gamma}_0 & \overline{\Gamma}_4 \\ \overline{\Gamma}_1 & \overline{\Gamma}_5 \\ \overline{\Gamma}_2 & \overline{\Gamma}_6 \\ \overline{\Gamma}_3 & \overline{\Gamma}_7 \end{pmatrix}_{\otimes} \begin{pmatrix} \overline{\Phi}_0 & \overline{\Phi}_4 \\ \overline{\Phi}_1 & \overline{\Phi}_5 \\ \overline{\Phi}_2 & \overline{\Phi}_6 \\ \overline{\Phi}_3 & \overline{\Phi}_7 \end{pmatrix} = \overline{\Gamma}_{0\otimes}\begin{pmatrix} \overline{\Phi}_0 & \overline{\Phi}_4 \end{pmatrix} + \begin{pmatrix} \overline{\Gamma}_1 & \overline{\Gamma}_5 \\ \overline{\Gamma}_2 & \overline{\Gamma}_6 \\ \overline{\Gamma}_3 & \overline{\Gamma}_7 \end{pmatrix}_{\otimes} \begin{pmatrix} \overline{\Phi}_0 & \overline{\Phi}_4 \end{pmatrix}$$
$$+ \overline{\Gamma}_{4\otimes}\begin{pmatrix} \overline{\Phi}_0 & \overline{\Phi}_4 \end{pmatrix} + \begin{pmatrix} \overline{\Gamma}_0 & \overline{\Gamma}_4 \end{pmatrix}_{\otimes} \begin{pmatrix} \overline{\Phi}_1 & \overline{\Phi}_5 \\ \overline{\Phi}_2 & \overline{\Phi}_6 \\ \overline{\Phi}_3 & \overline{\Phi}_7 \end{pmatrix}, \tag{5.103}$$

where the first term on the right-hand side of the equation denotes two $5d^{2s}$ electrons as defined in equation (5.98) previously; the second term contains six $4f^{4p}$ electrons; the third term is two $4f^{4s}$ electrons; and the last term comprises six $4f^{2p}$ electrons. By calculus chain rules, in the second term, the rotations iterating from $\overline{\Gamma}_1$ through $\overline{\Gamma}_7$ electrons are coupled with rotations alternating between both $\overline{\Phi}_0$ and $\overline{\Phi}_4$ platforms; and in the fourth term, the rotations alternating between both $\overline{\Gamma}_0$ and $\overline{\Gamma}_4$ platforms are coupled with the rotations iterating from $\overline{\Phi}_1$ through $\overline{\Phi}_7$ electrons. These platforms are s-nature orbitals of $5d^{2s}$ and $4f^{4s}$ electrons.

Fourteen 4f electrons are confined between two $5d^{4s}$ electrons. The outer electron of La represents the ground state of $5d^{4s}$ energy level, and the outer electron of Hf is the upper bound of $5d^{4s}$ energy level. Because the energy difference between these two states is trivial, fourteen 4f electrons have very close energy levels and therefore exhibit similar physical properties. And for this reason, a 4f electron may easily switch between various neighboring states. For example, the electron configuration of a cerium atom may take the form of $[Xe]6s^2 5d^1 4f^1$, $[Xe]6s^2 5d^2$, or $[Xe]6s^2 4f^2$. The valence of cerium normally takes +3 or +4 in chemical compounds.

Since 5d and 4f electrons are intertwined with 4s4p and 2s2p electrons, their existence brings 6s6p electrons closer to the inner spherical layers. In other words, 5d and 4f orbitals serve as expressways for 6s6p electrons to communicate with the inner 4s4p and 2s2p shells.

They behave like many anchors cast deeply into the inner shells to fasten the floating ship of 6s6p electrons.

5.4.2. The Quaternity Periodic Table of Elements

Spherical quantities for 7s5f6d7p electrons can be constructed through clockwise rotations based on radian angles γ and Γ. Given the initial function of

$$\eta_0 = A_5 A_6 (\cos\gamma\cos\Gamma - \gamma' \sin\gamma\cos\Gamma), \tag{5.104}$$

The 7s7p electron octet except η_0 may be expressed succinctly by

$$\eta_i = q^i \eta_0, i = 1,2,\ldots,7 \tag{5.105}$$

where q is reciprocal to p serving as a clockwise rotation operator in the oganesson shell, hence $q^i\eta_0$ indicates clockwise rotations upon η_0 for *i* times. Spherical quantities for 6d and 5f orbitals can be obtained in a similar way to those for 5d and 4f orbitals. We shall not elaborate here again. Like 4f electrons, 5f electrons pass two gateways and are inductively intertwined with 5p5s3p waveforms. They serve as mortar joints among 3s3p, 5s5p, and 6d orbitals. Their energy levels are confined between those of Ac and Rf elements.

Although d- and f-orbitals span large space and time ranges, they are still sufficiently explained by the four-dimensional spacetime framework. Quaternity space is scalable with an outer spherical layer being more complex than an inner spherical layer. All electrons are divided into four spherical layers. Spherical quantities for electrons in an outer spherical layer are constructed upon those in the inner spherical layer through dimension encapsulation mechanism.

Table 5.1. Periodic table of elements based on quaternity spacetime, which shows the enclosure of p-orbitals between s-orbitals in general except f-orbitals

scalors		polors		metors		vitors		scalors	
H									He
Li	Be	B	C	N	O	F	Ne	Na	Mg
		Al	Si	P	S	Cl	Ar	K	Ca
Sc	Ti	V	Cr	Mn	Fe	Co	Ni		
Cu	Zn	Ga	Ge	As	Sn	Br	Kr	Rb	Sr
Y	Zr	Nb	Mo	Tc	Ru	Rh	Pd		
Ag	Cd	In	Sn	Sb	Te	I	Xe	Cs	Ba
La		*Ce*	*Pr*	*Nd*	*Pm*	*Sm*	*Eu*	*Gd*	
	Tb	*Dy*	*Ho*	*Er*	*Tm*	*Yb*	*Lu*		Hf
		Ta	W	Re	Os	Ir	Pt	Au	
	Hg	Tl	Pb	Bi	Po	At	Rn	Fr	Ra
Ac		*Th*	*Pa*	*U*	*Np*	*Pu*	*Am*	*Cm*	
	Bk	*Cf*	*Es*	*Fm*	*Md*	*No*	*Lr*		Rf
		Db	Sg	Bh	Hs	Mt	Ds	Rg	
	Cn	Nh	Fl	Mc	Lv	Ts	Og		

When considering spherical quantities of an electron in the outer layer, the curled dimensions of the inner layer can be ignored to a certain extent. Each spherical layer is relatively independent, but there are transition messengers passing gateways between spherical layers to deliver messages for better communication. For example, 3d electrons attach to 4s4p electrons and are complementary to 1s2p2s electrons in spacetime. While s- and p-electrons stay within the same shell, d- and f-orbitals are long-range ones inductive with the inner shells of the atom. Because 3d electrons are complementary to 1s2p2s electrons and the other d-electrons have similar properties, we may treat them as composite orbitals of $s^2p^6s^2$ to a certain extent. In the same way, f^{14}-electrons can be regarded as $p^6s^2p^6$-like orbitals, e.g., 4f electrons form a weak rope with 4p4s2p electrons combined. In this manner, we organize all elements into a periodic table (Table 5.1) according to quaternity spacetime.

The periodic table shows that p-orbitals are enclosed by s-orbitals. Such an enclosure of p-orbitals ensures the stability of atoms except lanthanides and actinides that feature anti-enclosure. Of these two inner-transition metal series, only Gd, Tb, Cm, and Bk have the scalor properties, the other twenty-four elements being the polors, metors, and vitors in nature. If we regard s-orbitals as round spheres, then p-orbitals are ovals, d-orbitals are narrow ovals, and f-orbitals are even narrow ovals. Because of their severe deviations from spheres, f-orbitals are so narrow and long-ranged that they are susceptible to decay like the instability of a comet surrounding the sun along a narrow ellipse.

Table 5.1 is not intended to replace the traditional periodic table of elements, but it is an alternative view in quaternity spacetime. The p-block elements are fully-packed in the shaded region whereas some s-block elements have two possible arrangements in the table. For example, Cu and Zn can also be placed in the slots under K and Ca. This agrees with their chemical properties in those columns. Cu and Zn often take +1 and +2 valances respectively in chemical compounds. Their chemical properties are somewhat similar to those of K and Ca, but not exactly the same due to their d-orbital nature, which determines them as transition metals.

Despite the primary difference between d-orbitals and s- and p-orbitals, elements within the same column still show certain similarity in their behavior. For example, manganese is a useful alloying element to achieve high strength stainless steels. It has been found that the addition of manganese increases the solubility of nitrogen in austenitic steels [11], somewhat reflecting the attraction of nitrogen by manganese. This attraction lies in the inductive effect of their complementary waveforms because Mn and N are in the same column in Table 5.1. This partly uncovers the secret of good solubility with similar molecules as the solvent and the solute. Electronic attraction between complementary orbitals is a key factor to good solubility. Can this also explain the mechanism of hydrophobic forces acting between organic macromolecules in water under water repulsion? The hydrophobic effect is the tendency of nonpolar substances to aggregate in aqueous solution and exclude water molecules.

5.4.3. Electronic Interactions and Waves in Atoms

Even though we have formulated electronic orbitals for all kinds of electrons, the forces that bind them together in atoms are not clear yet. They deserve further explanations. When electric charges are all treated as static, the electrostatic Coulomb force holds; but when electrons are moving and transforming in spacetime, Maxwell's electrodynamic equations work instead (cf. Section 10.3). Note that an electron within an atom might not exist as a particle all the time, but manifest as a changing electric field or a changing magnetic field. Geometrically, an electron might exist as a point, an arc, a surface, or a solid in space; and chemically, we talk about an electron cloud. Because in quaternity spacetime, the *curl* operator is equivalent to $\partial/\partial l$ in dynamic calculus, the rotation of a spherical quantity is equivalent to Faraday's law (equation 1.21). The duality equation is actually a combination of two Maxwell's equations (cf. Equations 10.13 and 10.14). Please also refer to section 9.3 for the mathematical expression of forces.

To clarify, a force is not a mysterious quantity or the power of an invisible hand. In its primitive form, a force requires a certain way of message exchanges between acting particles. When two people exchange information or goods, they are likely to make friends if the trade benefits both and is meaningful to their livelihoods. The interdependence of entities is a manifestation of forces. The message exchange is an essential aspect of force expression or mechanism. Without messengers or force carriers to deliver messages, there would be no forces. We previously introduced virtual photons as the vehicle of electron interaction in section 3.5. A virtual photon delivers an electronic wave. Here we shall characterize the message exchange within atoms by waves, the theme that organizes various electrons in inert atoms in a harmonious order. Forces are implemented by the wave behavior of electrons.

Within the atomic sphere, the most significant force is perhaps the interaction between the nucleus and electrons, or more specifically the force between the nucleus and both 1s electrons because these two electrons form the inner core that delivers wave signals toward outer spherical layers. The interaction between 1s electrons and the nucleus might involve the continual annihilation and recreation of electrons at the nucleus (see Figure 1.16). A 1s electron disappears and is immediately created again with a fresh electronic state that carries messages from the nucleus. Imagine throwing a pebble into a still pond and watching the circular ripples moving outward. Both 1s electrons may be regarded as electromagnetic waves confined within the helium shell. This initial pulse must be strong enough to generate waves that eventually propagate over the entire atomic sphere.

In a neon atom, the nucleus and electrons constitute an electromagnetic wave generator. The waves generated as both 1s electrons are soon delivered to 2s2p electrons through dimension encapsulation mechanism. Wave functions of 2s2p are built upon 1s wave functions. Electronic waves in the helium shell drive electronic waves in the neon shell by dimension encapsulation mechanism, and 2s2p wave functions harness wave energy from 1s electrons. Four cycles of 1s electrons deliver four thrusts upon 2s2p electrons, producing four-dimensional waves that propagate from a zero-dimensional 2s sphere to one-dimensional $2p_x$ poles, to two-dimensional $2p_y$ semicircular arcs, to three-dimensional $2p_z$ hemispherical surfaces, and then fall back to a four-dimensional 2s sphere. The wave signals then reflect back from 2s2p toward 1s through the calculus chain rules, reinforcing the original 1s waves. As shown in equation (2.16), dimension factor u of 2s2p wave functions denotes the rotation of 1s wave functions ($u = v\Omega_0$). Thus, 1s electronic waves propagate outward into 2s2p waves that echo back as 1s waves again cyclically forming a resonant medium within a neon atom [12-14].

In a larger atom such as radon, 2s2p and 3s3p electronic waves are driven by 1s waves, and 4s3d4p and 5s4d5p are in turn generated by 2s2p and 3s3p waves respectively, and finally the outskirt 6s4f5d6p electrons are built upon the 4s4p shell. Since electronic orbitals are waves in nature, the entire atom is a dynamic multi-dimensional wave that propagates out from the central nucleus to its outer electrons and reflects backward through d- and f-orbitals. A dimension encapsulation represents waves delivered outward; and calculus chains represent waves reflected inward as d- and f-orbitals. For example, 3d-orbitals are wave signals that are reflected from 4s electrons toward the inner neon and helium shells. $3d^{2s2p}$ waves complement 2s2p waves; and $3d^{1s}$ waves complement 1s waves.

A significant role of 3d-orbitals is the reinforcement of contact between electrons in the outer and inner spherical layers. Wave functions of 4s4p electrons are based on the dimension encapsulation upon 2s2p electrons. Both 4s spherical orbitals communicate with the nucleus through 2s sphere and 1s sphere stepwise. However, the force induced by message exchanges between 4s4p and the inner spherical layer through dimension encapsulation mechanism is not strong enough to bind all 4s4p electrons to the atom. That is why 3d-orbitals are involved. On one end, 3d electrons directly connect with 4s4p sphere; and on the other end, they interact with electrons in both neon and helium shells. A 3d-orbital is like a tendril dropping down from 4s spherical branch and tightly grasping the inner cores of 2s2p1s electrons. This helps stabilize the system.

Compared with d-orbitals that deliver strong messages, f-orbitals implement forces that are relatively weak. The weakness is revealed by the instability of inner-transition metals that are susceptible to decaying. Atoms in the last period of the traditional periodic table have electrons with maximum space volume and hence minimum time stability. It might be said that their outer electrons have higher potential energy with lower kinetic energy than their inner layer electrons. These atoms are all radioactive due to their need for a delicate balance between their various shells that is often broken.

5.4.4. The Atomic Shell Octet

We shall extend the significance of a spherical quadrant to an atomic shell as a whole. Within the oganesson atom with 118 protons, the atomic shells in four spherical layers can be organized into an octet as shown in Table 5.2. This octet has the characteristics of polors, metors, vitors, and scalors in certain manifestations. The eight shells split into two groups with H, Ne, Kr, and Rn in one group having counterclockwise rotation tendencies and Og, Xe, Ar, and He in another group featuring clockwise rotations, both groups forming a rope structure within an Og atom (Figure 5.16). Note that hydrogen is not an inert atomic shell, but it is a representative atomic shell of a 1s electron that is in symmetry with another 1s electron in a helium shell.

A helium shell has s-orbitals only. A dimension encapsulation upon the helium shell leads to wave functions of a neon shell, which contains s- and p-orbitals. A dimension encapsulation upon the neon shell leads to wave functions of a krypton shell, which contains s-, p-, and d-orbitals. And a dimension encapsulation upon the krypton shell brings f-electrons in addition to s-, p- and d-electrons in a radon shell. Thus it is clear that each dimension encapsulation leads to one increment of orbital types in the outer spherical layer. The helium shell is the most stable atomic shell. The stability of the atomic shells is relaxed stepwise with adding orbital types upon dimension encapsulations. Each of these dimension encapsulations is actually a rotatory operation marked on the right-hand side of Figure 5.16. For example, as was explained in section 2.2.4, a dimension encapsulation upon wave function

$-C_1C_2v \sin\alpha \sin\beta$ results in $A_1A_2 \cos\Psi \cos\psi$. The dimension encapsulation is a rotatory operation because a cosine operator may function as a differential or integral operator. Frankly, it took me a long time to thoroughly realize that cosine is a calculus operator.

Table 5.2. The atomic shell octet within the largest atom, $_{118}Og$

Spherical quantities	Spherical layer	Atomic Shells	Electronic orbitals	Rotation directions
Polors	1	H, He	$1s^1$, $1s^2$	↑,↓
Metors	2	Ne, Ar	$2s^22p^6$, $3s^23p^6$	↑,↓
Vitors	3	Kr, Xe	$4s^23d^{10}4p^6$, $5s^24d^{10}5p^6$	↑,↓
Scalors	4	Rn, Og	$6s^25d^14f^{14}5d^96p^6$, $7s^26d^15f^{14}6d^97p^6$	↑,↓

On the other hand, a dimension encapsulation upon a helium shell may also be performed on the left-hand side of Figure 5.16, which leads to an argon shell. A dimension encapsulation upon the argon shell leads to a xenon shell; and a dimension encapsulation upon the xenon shell leads to an oganesson shell. These dimension encapsulations are clockwise rotatory operations ascending from inner shells to outer shells. Because the calculus chain rules in the reverse direction from the Og shell to the Xe, Ar and He shells have the same clockwise rotation direction to dimension encapsulations, there are clockwise rotation relationships between atomic shells on the left-hand side of Figure 5.16 in four spherical layers through calculus chain rules. At both ends, there is a rotation relationship between a 6s electron in a radon shell and a 7s electron in an oganesson shell (cf. Equation 5.34 and Figure 5.6 for the similar relationship between 2s and 3s electrons); and there is a clockwise rotation relationship between both electrons in a helium shell. Thus we have a rotation cycle of the atomic shells within an oganesson atom like that of Figure 3.5.

$$H \mapsto Ne \mapsto Kr \mapsto Rn \mapsto Og \mapsto Xe \mapsto Ar \mapsto He \mapsto H \tag{5.106}$$

where each arrow $\mapsto$ indicates a counterclockwise or clockwise rotatory operation from one atomic shell to another. The first four processes are counterclockwise rotations implemented by dimension encapsulation mechanism whereas the other four processes are clockwise rotations implemented by calculus chain rules. This octet cycle is marked in Figure 5.16 as an almond-shaped circle.

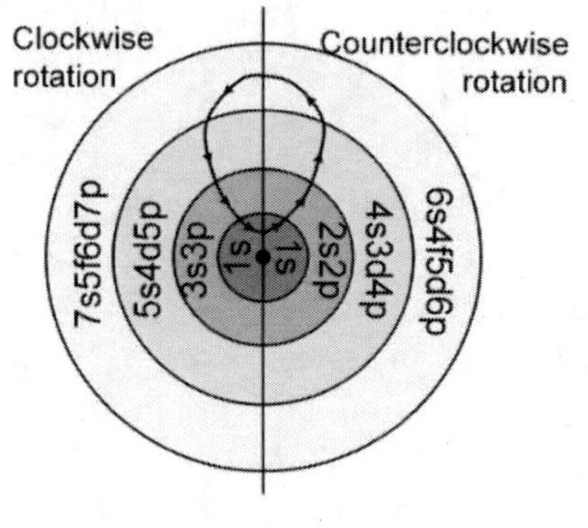

Figure 5.16. The atomic shell octet in four spherical layers of an oganesson atom with an almond-shaped circle indicating a transformation cycle.

The atomic shell octet is a conservative system similar to the configuration of the electron octet within a neon shell. Due to the octet arrangement, various shells can cooperate with each other harmonically in an oganesson atom. This implies that there are only 118 possible chemical elements because the octet arrangement is a full and stable configuration. We believe that oganesson is the last element. Unfortunately, with wrong guidelines, recent attempts by some physicists to synthesize the chemical element of atomic number 119 are doomed to failure.

Conclusion

We have formulated wave functions for all electrons in their primitive inert atoms in quaternity spacetime in a relentless effort (Table 5.3). Each electron is unique in terms of the waveform and role in delivering force. Not any two electrons are exactly the same. This is in good agreement with the Pauli exclusion principle that no more than one electron can occupy a single quantum state. Some Greek letters for spherical quantities are the same, but they reflect different orbital geometries for different kinds of electrons. For examples, spherical quantities in the shaded areas indicate 4f and 5f electronic orbitals. The organization of electrons shows some patterns: p-electrons are confined by s-electrons in a spherical layer; and d- and f-electrons pass gateways to deliver messages between electrons in the outer and inner spherical layers. Note that in a krypton shell, 3d electrons have a principal quantum number (n=3) lower than that of 4s4p electrons (n=4) and are sometimes referred to as belonging to an inner spherical layer for the sake of convenience. All electrons are interwoven, bringing about harmonic oscillating waves over the whole atomic sphere. They interact with each other in such a way that keeps electronic waves harmonic and total energy low.

The theory of quaternity provides a unique observation on the atomic structure. It has given us much more detailed explanations on electronic arrangement and motion than quantum mechanics. Among other things, it is revealing to see that chemical elements corresponding to their outmost characteristic electrons can be tabulated under the fields of the scalors, polors, metors, vitors, and scalors into a quaternity periodic table of elements. This alternative approach is fully compatible with established chemistry and physics and provides further insight into elemental behavior.

The elucidation of the electronic pattern within inert atoms is perhaps one of the greatest achievements of mathematical derivation combining calculus, trigonometry, and geometry together. Dimension encapsulations reflect layered structure of electrons within a large atom, while the principle of rotatory operations on spherical quantities brings order and harmony to electrons within each spherical layer. Orbital wave functions follow a pattern of self-similarity within every spherical layer. No matter how many spherical layers an atom may have, all layers are spacetime continuous via dimension encapsulation mechanism outwards and calculus chain rules inwards. The theory of quaternity is so elegant and wonderful in the mathematical expression and conceptual interpretation of electrons. The striking beauty that it reveals is the best proof of its correctness for only truth has so unparalleled beauty and exactness.

Table 5.3. The periodic table of elements with spherical quantities for their characteristic electrons in their low energy states. The shaded areas indicate f electrons

scalors		polors		metors		vitors		scalors	
H Ω_0									He Ω_2
Li Φ_0	Be Φ_4	B Φ_1	C Φ_2	N Φ_3	O Φ_5	F Φ_6	Ne Φ_7	Na φ_0	Mg φ_4
		Al φ_1	Si φ_2	P φ_3	S φ_5	Cl φ_6	Ar φ_7	K Γ_0	Ca Γ_4
Sc $\overline{\Phi}_0$	Ti $\overline{\Phi}_4$	V $\overline{\Phi}_1$	Cr $\overline{\Phi}_2$	Mn $\overline{\Phi}_3$	Fe $\overline{\Phi}_5$	Co $\overline{\Phi}_6$	Ni $\overline{\Phi}_7$		
Cu $\overline{\Omega}_0$	Zn $\overline{\Omega}_2$	Ga Γ_1	Ge Γ_2	As Γ_3	Sn Γ_5	Br Γ_6	Kr Γ_7	Rb γ_0	Sr γ_4
Y $\overline{\varphi}_0$	Zr $\overline{\varphi}_4$	Nb $\overline{\varphi}_1$	Mo $\overline{\varphi}_2$	Tc $\overline{\varphi}_3$	Ru $\overline{\varphi}_5$	Rh $\overline{\varphi}_6$	Pd $\overline{\varphi}_7$		
Ag $\overline{\Omega}_2$	Cd $\overline{\Omega}_0$	In γ_1	Sn γ_2	Sb γ_3	Te γ_5	I γ_6	Xe γ_7	Cs H_0	Ba H_4
La $\overline{\Gamma}_0$		*Ce* $\overline{\Gamma}_1$	*Pr* $\overline{\Gamma}_2$	*Nd* $\overline{\Gamma}_3$	*Pm* $\overline{\Gamma}_5$	*Sm* $\overline{\Gamma}_6$	*Eu* $\overline{\Gamma}_7$	*Gd* $\overline{\Phi}_0$	
	Tb $\overline{\Phi}_4$	*Dy* $\overline{\Phi}_1$	*Ho* $\overline{\Phi}_2$	*Er* $\overline{\Phi}_3$	*Tm* $\overline{\Phi}_5$	*Yb* $\overline{\Phi}_6$	*Lu* $\overline{\Phi}_7$		Hf $\overline{\Gamma}_4$
		Ta $\overline{\Gamma}_1$	W $\overline{\Gamma}_2$	Re $\overline{\Gamma}_3$	Os $\overline{\Gamma}_5$	Ir $\overline{\Gamma}_6$	Pt $\overline{\Gamma}_7$	Au $\overline{\Phi}_0$	
	Hg $\overline{\Phi}_4$	Tl H_1	Pb H_2	Bi H_3	Po H_5	At H_6	Rn H_7	Fr η_0	Ra η_4
Ac $\overline{\gamma}_0$		*Th* $\overline{\gamma}_1$	*Pa* $\overline{\gamma}_2$	*U* $\overline{\gamma}_3$	*Np* $\overline{\gamma}_5$	*Pu* $\overline{\gamma}_6$	*Am* $\overline{\gamma}_7$	*Cm* $\overline{\varphi}_0$	
	Bk $\overline{\varphi}_4$	*Cf* $\overline{\varphi}_1$	*Es* $\overline{\varphi}_2$	*Fm* $\overline{\varphi}_3$	*Md* $\overline{\varphi}_5$	*No* $\overline{\varphi}_6$	*Lr* $\overline{\varphi}_7$		Rf $\overline{\gamma}_4$
		Db $\overline{\gamma}_1$	Sg $\overline{\gamma}_2$	Bh $\overline{\gamma}_3$	Hs $\overline{\gamma}_5$	Mt $\overline{\gamma}_6$	Ds $\overline{\gamma}_7$	Rg $\overline{\varphi}_0$	
	Cn $\overline{\varphi}_4$	Nh η_1	Fl η_2	Mc η_3	Lv η_5	Ts η_6	Og η_7		

Note from Author

We have described the dimensional structure and law of motion for electrons in various inert atoms at a stretch. The theory of quaternity has been developed by a single soul from a unique perspective over a long period of hectic days. The discovery of natural law is so wonderful and engrossing that sometimes I need to be reminded to slow down or stop the ongoing research to enjoy a light weekend that we deserve. Now that the atomic theory is coherent and complete, it is time to take a break, have a cup of coffee, and look around the bulletin board to amuse yourselves. Wow!

After the class break, it is also the right time for one to ponder over the application and implication of the fresh spacetime theory that best characterizes the general harmonic oscillation of diverse entities. As one continues to pursue the approach of dynamic calculus of spherical quantities to natural science, he or she will soon discover the even greater beauty of nature lying ahead and gradually come to appreciate the extraordinary power and richness of the quaternity theory that effectively breaks new ground in science. After discovering the law of nature, life when in sickness becomes more bearable because I can constantly enjoy the beauty of quaternity space, and life when in solitude is never boring because I always have the company of the natural law in dynamic calculus. The more we learn the beautiful law of nature, the better we appreciate our life's every moment.

Bulletin

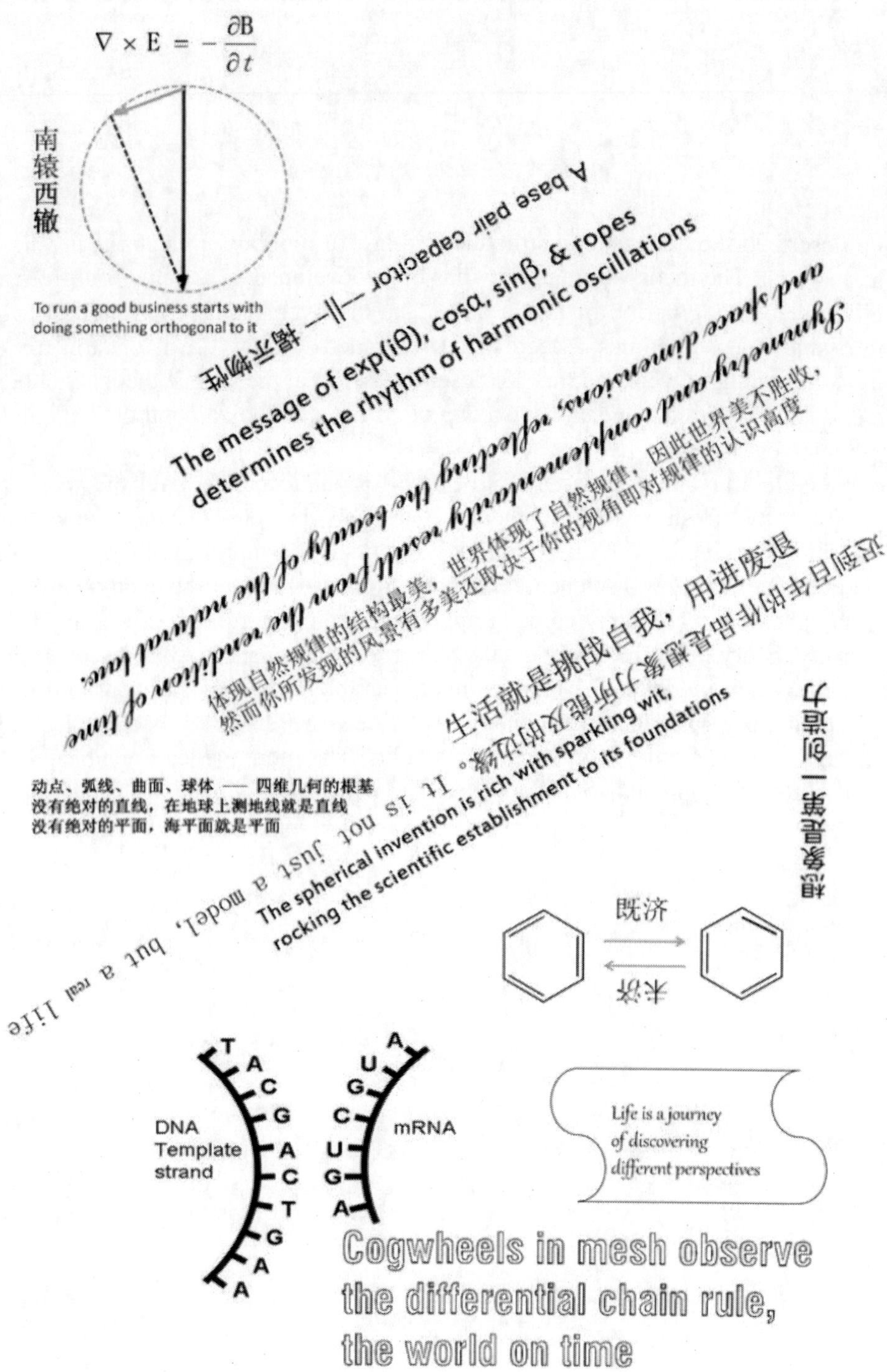
$\nabla \times E = -\frac{\partial B}{\partial t}$
南辕西辙
To run a good business starts with doing something orthogonal to it
A base pair capacitor —||—
The message of exp(iθ), cosα, sinβ, & ropes determines the rhythm of harmonic oscillations
Symmetry and space dimensions, reflecting the beauty of the natural law,
and complementarity result from the rendition of time
体现自然规律的结构最美，世界体现了自然规律，因此世界美不胜收，
然而你所发现的风景有多美还取决于你的视角即对规律的认识高度
生活就是挑战自我，用进废退
想象是第一创造力
动点、弧线、曲面、球体 — 四维几何的根基
没有绝对的直线，在地球上测地线就是直线
没有绝对的平面，海平面就是平面
It is not just a model, but a real life
The spherical invention is rich with sparkling wit, rocking the scientific establishment to its foundations
既济
未济
DNA Template strand
mRNA
Life is a journey of discovering different perspectives
Cogwheels in mesh observe the differential chain rule, the world on time

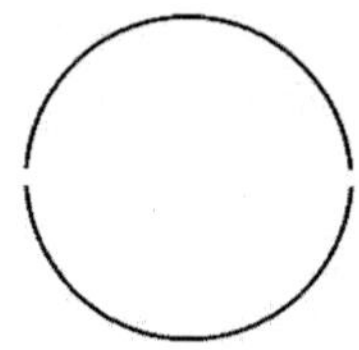

On Structure with Law

On this beautiful square land,
Entities are as diverse as the environment could demand.
Given a neon shell for four space dimensions to stand,
In which successive direction should an electron expand?
That wonders a humble man.

Under this celestial domed span,
Life evolves gradually as the law of nature would command.
Each circular dimension is traversed by harmonic oscillation and
This process follows dynamic calculus in God's hand,
Which the man tries hard to understand.

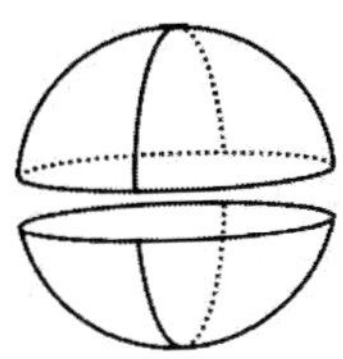

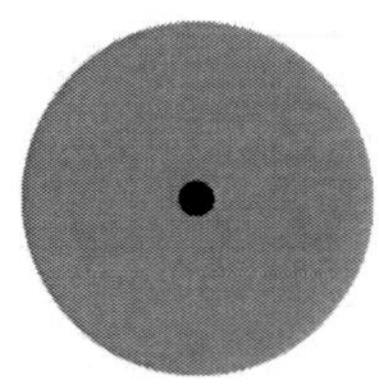

Part II. The Natural Pattern

We shall set out from the theoretical basis of Part I to further explore the unknown spacetime realm in Part II. Part I has defined the motion of electrons in various atoms with spherical quantities in dynamic calculus; and Part II will continue to pursue the spherical view for the characterization of molecules, cells, organisms, and the earth. For example, both 1s electrons within a helium shell constitute a two-dimensional system, which is a common characteristic of life phenomena. A husband and a wife are two dimensions of the family; and plants and animals are two kingdoms of the advanced lives. The interplay and transformation between both dimensions are the eternal theme of culture as well as nature. The spherical view describing the atomic structure may be applied to studying the biological pattern. Dynamic calculus of spherical quantities is the law of nature. We shall demonstrate how powerful this instrument is in bringing a sound perspective into natural sciences.

The principal aim of natural sciences is generally considered to be the search for order in our observations of the world around us. Starting from quaternity spacetime in a neon shell, Chapter 6 explains the configurations and reaction mechanisms of several organic molecules by the anisotropy of 2s2p orbitals within carbon atoms forming the molecules. The explanation is more natural than orbital hybridization. The anisotropy of 2s2p orbitals is the origin of molecular chirality when a carbon atom is bonded to four different substituents. It is deduced that a chiral carbon may transfer electrons along a selective pathway from a bonded substituent to another. Based on the chiral property, Chapter 7 characterizes a DNA molecule as stepwise *LC* oscillatory circuits. Charge transfer from one base pair to another through the phosphate bridges observes harmonic oscillation. If the electrical characteristic of a DNA molecule obeys dynamic calculus, then a cell determined by the DNA molecule must also obey dynamic calculus. Chapter 8 examines the relationship between biochemical cycles and the relationship between cellular organelles. It is suggested that these relationships within a cell are similar to those between electronic orbitals within atoms. They are governed by the same natural law. Chapter 9 further explores the possibility of unifying planetary orbits with electronic orbitals under the framework of uniform circular motion. It also establishes the connection between the central force density and the wave function of electrons within a helium atom.

After defining spherical quantities in dynamic calculus as the law of nature, it is a logical step to arrange physical quantities into a periodic table according to their spacetime dimensions. From the intricate relationships between various physical quantities, Chapter 10 tries to clarify the spacetime dimension of each physical quantity so as to make a periodic table of physical quantities. This rudimentary work provides sharp insight into the laws of physics – the rules and properties of one or more physical quantities, or the relationships that govern two or more physical quantities. Last but not least, Chapter 11 defines the core concepts of the traditional Chinese medical theory in strict geometry and dynamic calculus.

The mathematical definition with certainty and precision is revolutionary as compared with the traditional Chinese medical theory in philosophy that invariably produces doubts and disputes. The spherical view coincides with the traditional Chinese medical theory and is expected to revitalize the tradition effectively in generations to come.

The discovery of the atomic structure provides a model for describing the natural pattern in general. Part II undertakes to recast the gamut of knowledge from molecules to cells to organisms in accordance with the spherical view. Compared with Part I, this extensional research is more pragmatic and captivating, through the analogy among atomic structures, molecular structures, cellular models and organisms, hiking toward the unification of the world order. Hypothetical as it may be occasionally, it is nevertheless valuable because it furnishes fresh methodology and sharp vision for the expedition to the fundamental laws of the universe, broadening our conventional mindset. After all, science is the unreserved communication between scientists. Part I focused on mathematics and physics; and Part II is mainly about chemistry, biology and medicine as well as physics. We define matter and life with dynamic calculus instead of static algebra, focusing on the evolving processes rather than mechanical frames. From various disciplines, we fashion only the primary order of Nature coherently. So far, the readers are advised to peruse Part I before continuing.

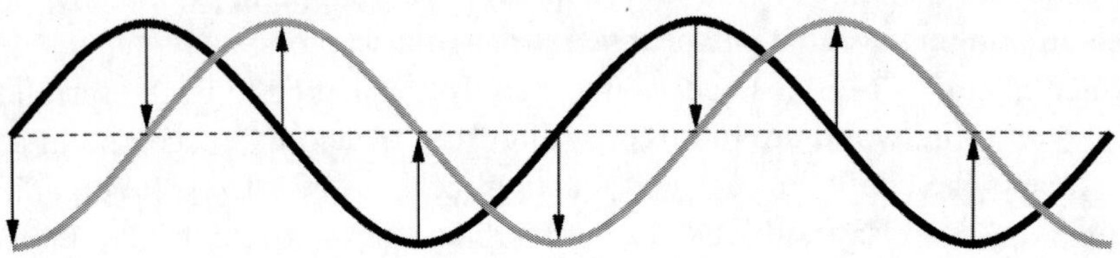

Motif 2. Oscillations, waves, rotations, circulation, rope, induction, propagation, orthogonality, and complementarity.

Chapter 6

Anisotropic 2p Orbitals and a Chiral Carbon

The outer electrons of an atom determine the chemical property of the atom in forming inorganic and organic molecules. This chapter extends the theory of quaternity from the atomic structure to the molecular level. The discovery of $2s2p_x2p_y2p_z$ geometries corresponding to four dimensions in quaternity space has enormous significance in organic chemistry. By analyzing the structure of carbon atoms in organic molecules as basic as methane, ethene, ethyne, and benzene, we provide strong evidence of the anisotropic electron configuration $2s^12p_x{}^12p_y{}^12p_z{}^1$ of the carbon atoms that form the molecules, as an alternative to orbital hybridization. The anisotropic electron configuration of carbon atoms explains well the reaction mechanisms of nucleophilic substitutions of alkyl halides and of electrophilic substitutions of mono-substituted benzenes. On further analyses, we find that the anisotropic electron configuration gives the characteristic of an asymmetrical carbon for chirality. A chiral molecule is a molecule that is nonsuperposable on its mirror image. A pair of chiral enantiomers rotate the plane of polarization of plane-polarized light an equal amount, but in opposite directions. We suggest that the spacetime differences of the 2s, $2p_x$, $2p_y$, and $2p_z$ orbitals are responsible for the optical rotation. Light bends due to a density gradient of electrons within atomic spacetime. By the same principle, chiral carbon centers also play a decisive role in directing the electron flow through itself in biomolecules. We illustrate the last point by the examples of C1', C3', and C4' stereocenters in a deoxyribonucleoside.

6.1. Introduction to Anisotropic 2p Orbitals

Quantum mechanics depicts 2p orbitals as being equally distributed in the X, Y, and Z directions in Euclidean space. Such an intuitive assumption violates the true orthogonal principle and has escaped scientific scrutiny for nearly a century. For quantities to be orthogonal, they must belong to different spacetime dimensions as distinguished by the polor, metor, vitor, and scalor (Table 6.1). Three quantities being equally distributed in the X, Y, and Z orientations do not constitute orthogonality. Assuming that electron clouds distribute in three spatial orientations equally and statically cannot explain the orientation of 2p orbitals in carbon atoms that form methane, ethene, ethyne, and benzene. Hypothetical orbital hybridization has to be supplemented to account for this. Starting from the definition of the geometries of anisotropic 2p orbitals, this section tries to explain quaternity space in carbon atoms that form various molecules. For example, how to explain the chemical structure of basic organic compounds, such as methane, ethene, ethyne, and benzene, without appealing to orbital hybridization? How does quaternity space in a carbon atom facilitate the S_N2 reaction mechanism in nucleophilic substitution of alkyl halides? Does the quaternity theory agree with established organic chemistry in predicting the consequence of electrophilic substitutions of mono-substituted benzenes? By applying quaternity space to carbon atoms, we shall answer these questions in a coherent manner. Successful applications are the best evidence of anisotropic 2s2p orbitals.

Table 6.1. Spherical quantities and geometrical shapes of eight electrons in a neon shell with reference to Figure 3.5. Electrons in the negative direction of quaternity axes are in space and time reversal

Quaternity axis	Spherical quantities	Space dimension	Orbital type	Geometrical shapes radiating from the central nucleus
1	Φ_0, Φ_4	0, 4	2s	Two concentric spheres
u	Φ_1, Φ_5	1	$2p_x$	Two poles
u^2	Φ_2, Φ_6	2	$2p_y$	Two flat sectors
u^3	Φ_3, Φ_7	3	$2p_z$	Two hemispheres

6.1.1. Geometries of 2p Orbitals

In traditional organic chemistry, 2p orbitals are known to be dumbbell-shaped, aligned along the X, Y, and Z axes. So they are assigned $2p_x$, $2p_y$, and $2p_z$ orbitals (Figure 6.1). It is believed that the 2p orbitals have a higher energy than 2s orbitals, and $2p_x$, $2p_y$, and $2p_z$ orbitals have equal energy. Following the aufbau principle, electrons will fill up lower energy atomic orbitals before entering higher energy orbitals; based on the Pauli exclusion principle, each orbital is allowed a maximum of two electrons of opposite spins; and in accordance with Hund's rule, when orbitals of equal energy are available, electrons will occupy separate orbitals before pairing up. These principles determine that the electron configuration of a carbon atom is $1s^2 2s^2 2p_x{}^1 2p_y{}^1$. An atomic orbital refers to an unoccupied spacetime dimension in an atomic sphere. When occupied by an electron, an atomic orbital becomes an electronic orbital. An electronic orbital is an electron in motion.

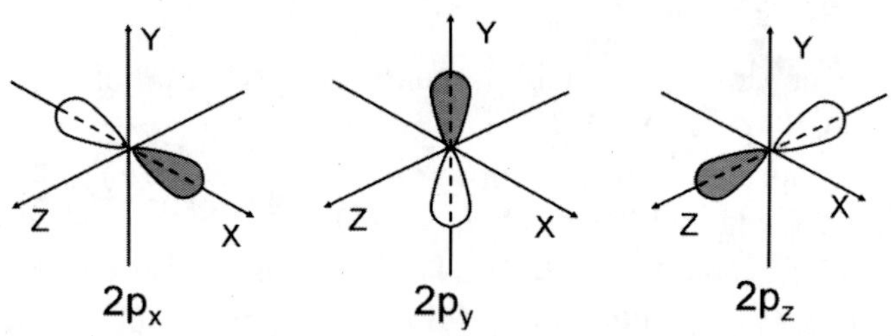

Figure 6.1. Schematic diagram of 2p orbital geometries in traditional opinion under Cartesian coordinates.

In contrast to the traditional view, the quaternity theory describes 2p orbitals as two polors, two metors, and two vitors, which have different dimensions in spacetime. As shown in Figure 6.2, $2p_x$, $2p_y$, and $2p_z$ orbitals possess different space dimensions in Euclidean space. A $2p_x$ is a polor, a line segment pointing from the nucleus to either pole; a $2p_y$ is a metor, a planar sector spreading out from the nucleus to a semicircular arc; and a $2p_z$ is a vitor of hemispherical shape. The shapes plotted here are those geometrical shapes in Figure 2.14 connecting to the central nucleus to reflect the charge attraction between the nucleus and the electrons that form chemical bonds. Electronic orbitals along the X, Y, and Z directions are different in geometry, so they are termed anisotropic 2p orbitals. The $2p_x$ orbital is smaller and denser than the $2p_y$ orbital whereas the $2p_z$ electron cloud is larger and sparser than the $2p_y$ electron cloud. A 2s orbital is spherical in shape and has a maximum volume with minimum density. According to quaternity, electrons within a neon shell are in dynamic motion. Each electron is evolving through various states cyclically in the arrow direction shown in Figure 6.3. Although the 2s orbital is the first orbital of the neon shell, electrons will not occupy the orbital statically all the time. Four outer electrons of a carbon atom typically occupy 2s, $2p_x$, $2p_y$, and $2p_z$ orbitals, respectively. Thus, it is rational to deduce that the excited electron configuration of a carbon atom is $1s^22s^12p_x^12p_y^12p_z^1$.

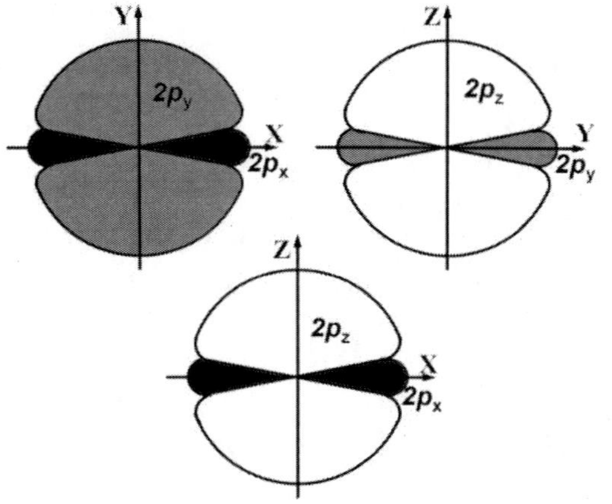

Figure 6.2. Profile diagrams of the shapes of 2p orbitals in three-dimensional Euclidean space where $2p_x$ orbitals are two one-dimensional sticks (black); $2p_y$ orbitals are two flat sectors (gray); and $2p_z$ orbitals are two hemispheres (white) in quaternity.

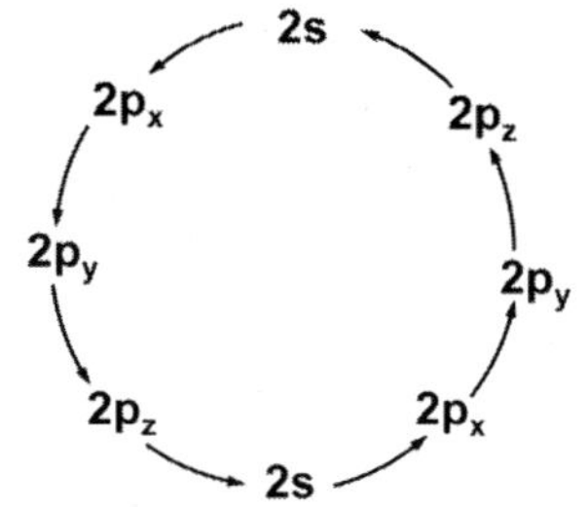

Figure 6.3. Electronic transformation cycle within a neon shell.

The different perceptions of 2p orbitals in a carbon atom lead to different accounts on the nature of covalent bonds in molecules such as methane, ethene, ethyne, and benzene. The electron configuration of carbon $1s^22s^22p_x^12p_y^1$ implies that a carbon atom has two unpaired electrons. However, it is known that a carbon atom forms four bonds with other atoms. The traditional opinion of carbon electron configuration cannot explain the geometric structures of those basic molecules. Orbital hybridization is traditionally hypothesized to describe the organic molecules case by case. Specifically, sp^3 hybridization is hypothesized to explain the tetrahedral shape of methane; sp^2 hybridization is hypothesized to explain the planar shape of an alkene such as ethene ($H_2C{=}CH_2$); and sp hybridization is hypothesized to explain the linear shape of ethyne. Although orbital hybridization has been written in the textbook of elementary organic chemistry, we don't think it is accurate. In the following sections, we shall apply the electron configuration of a carbon atom, $1s^22s^12p_x{}^12p_y{}^12p_z{}^1$, in quaternity space to the explanation of various molecular structures and reaction mechanisms. The purpose is to provide an alternative, simple, and unified account of carbon atoms in organic chemistry.

6.1.2. Covalent Bonds in Methane, Ethene, and Ethyne

We shall describe molecular structures by quaternity space, as an alternative to orbital hybridization. As was mentioned previously, electronic orbitals are dynamic ones whose geometries change continuously as radian angles Ψ and ψ vary. Four electrons in the outer shell of a carbon atom tend to occupy orbitals of $2s^12p_x{}^12p_y{}^12p_z{}^1$. Because four orbitals are different in space dimension, there are various spatial flexibilities associated with them. Specifically, a 2s orbital is in a spherical shape and hence might ramble around the entire sphere; a $2p_x$ orbital has a rectilinear shape connecting the nucleus to either pole, and is restricted to the unidimensional radius of the sphere; a $2p_y$ orbital is a flat sector and therefore might scissor along the semicircle; and a $2p_z$ orbital is a hemisphere and has the highest degree of freedom in space among the three types of 2p electrons. In addition, a free electron might evolve dynamically from 2s to $2p_x$, to $2p_y$, and to $2p_z$ in sequence (Figure 6.3). These geometrical restrictions and cycling transformation determine their possible orbital orientations in forming covalent bonds between atoms.

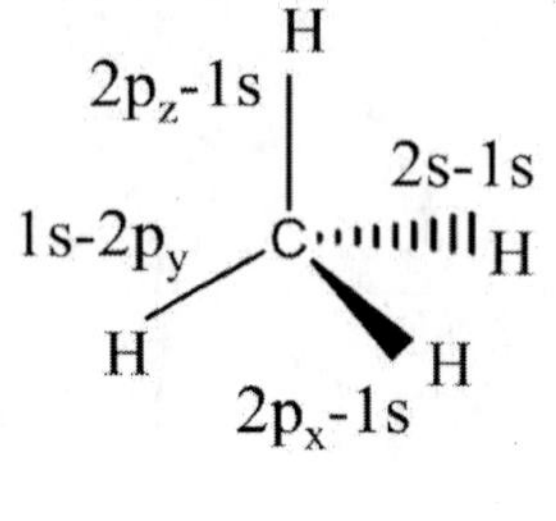

Figure 6.4. A methane molecule with four types of C–H covalent bonds at any moment.

The anisotropic shapes of 2p orbitals were derived from dimensional analyses of a spherical layer. They have broad support from organic chemistry. The shapes explain the σ-bond and π-bonds in methane, ethene, and ethyne reasonably well. In a methane molecule, a carbon atom uses a 2s electron, a $2p_x$ electron, a $2p_y$ electron, and a $2p_z$ electron to form covalent bonds with four hydrogen atoms (Figure 6.4). Because of the flexibility of $2p_y$, $2p_z$, and 2s electronic orbitals in space, each bond is restricted to a corner of a tetrahedron under electrical repulsion; but the four covalent bonds are all different due to different nature of the four electrons in the carbon atom. However, because the electrons might evolve dynamically from 2s to $2p_x$, to $2p_y$, and to $2p_z$ in sequence, each covalent bond switches among four types of bonds cyclically. The restless bond type switch explains why four covalent bonds have an equal length upon rough measurement. Infrared spectroscopy reveals four fundamental frequencies of absorption (3018.9 cm^{-1}, 2913.0 cm^{-1}, 1533.3 cm^{-1}, and 1305.9 cm^{-1}) of various

C−H vibrations in methane including bond stretching, scissoring, rocking, and wagging motion characteristic of $2p_x$, $2p_y$, $2p_z$, and 2s orbital flexibilities respectively. Due to its active infrared absorption, methane is a greenhouse gas.

In an ethene molecule, two carbon atoms form a stable $2p_x$-$2p_x$ σ-bond and a $2p_y$-$2p_y$ π-bond. Geometrically, the π-bond is the partial sideways overlap of two $2p_y$ flat sectors in one atom with two $2p_y$ flat sectors in another atom (Figure 6.5). This is slightly different from the traditional concept of a π-bond. The π-bond has less overlap than the σ-bond and is an unsaturated covalent bond because only one electron occupies each $2p_y$-$2p_y$ overlapping region. In addition, each carbon atom uses its 2s and $2p_z$ orbitals to form two covalent bonds with two hydrogen atoms respectively. The geometrical orientation of $2p_y$ electrons to accommodate the π-bond and the geometrical flexibility of 2s and $2p_z$ under electrical repulsion of the σ-bond and π-bond explain its planar molecular conformation. To illustrate, if the σ-bond and π-bond are on a vertical plane, then the 2s-1s and the $2p_z$-1s C−H bonds are on a horizontal plane.

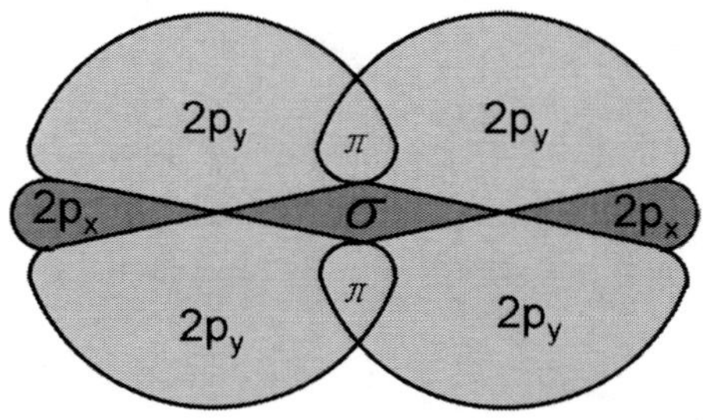

Figure 6.5. In an ethene molecule, a σ-bond is in the middle and a π-bond involves two electrons colonizing a pair of overlapping sector regions above and below the σ-bond.

In an ethyne molecule, two carbon atoms form a $2p_x$-$2p_x$ σ-bond, a $2p_y$-$2p_y$ π-bond, and a $2p_z$-$2p_z$ π-bond. These two π-bonds are directionally specific and have different energy, one being two flat sectors partially overlapping two flat sectors sideways, the other being two hemispheres partially overlapping two hemispheres sideways. As illustrated in section 3.3, only two similar type orbitals with contrary clockwise rotation directions form a stable electron pair, and if the pair of electrons are from two carbon atoms, they constitute a valid covalent bond. Both π-bonds in an ethyne molecule surround the σ-bond on four sides. The C−H bonds in ethyne are oriented toward the opposite sides of the three bonds due to electrical repulsion, giving it linear molecular structure. Overall, any organic molecular structure can be adequately illustrated by the combined effects of the original 2p orbital geometries, their flexible mobility, and electrical repulsion. Orbital hybridization is not necessary in the quaternity theory.

Because quantum mechanics cannot explain geometrical arrangements of methane, ethene, and ethyne molecules, orbital hybridization is fabricated to account for them. Although orbital hybridization is well-accepted, there are at least three defects associated with it. First, orbital hybridization is a pure speculation based on experimental data. One can describe the result of sp^2 hybridization according to the molecular structure of ethene, but cannot explain why orbitals should hybridize in that way. How electrons adapt to the process of orbital hybridization remains a mystery. Second, it is easy to see that two traditional dumbbell-shaped $2p_y$ orbitals are too far apart to overlap sideways for forming a C−C π-bond in an ethene molecule. Third, orbital hybridization cannot explain the bond energy difference between two π-bonds in an ethyne molecule. The difficulties and limitations of quantum mechanics underline the success of quaternity. Four spherical quadrants match exactly with the 2s, $2p_x$, $2p_y$, and $2p_z$ orbitals. The suitability of quaternity space for explaining the geometries of organic molecules is the basic evidence of quaternity theory.

6.1.3. Nucleophilic Substitutions of Alkyl Halides

A nucleophilic substitution of an alkyl halide involves the substitution of the halogen atom with a different nucleophile. The reaction might follow two types of mechanisms, namely S_N1 and S_N2. We shall not present the subject systematically here, but only present the context of the S_N2 mechanism for illustrating how the electron configuration $2s^1 2p_x^1 2p_y^1 2p_z^1$ of the carbon center facilitates the reaction. The S_N2 mechanism is a concerted reaction that proceeds directly from reactant to product without forming any detectable intermediates.

The incoming nucleophile forms a bond to the carbon center at the same time as the C–X bond is broken. The incoming nucleophile attacks the carbon center from one side of the molecule and the outgoing halide departs from the other side. This is called back-side displacement. As a result, the reaction center is inverted during the process. For example, a hydroxide ion displaces an iodide ion in the following reaction (Figure 6.6).

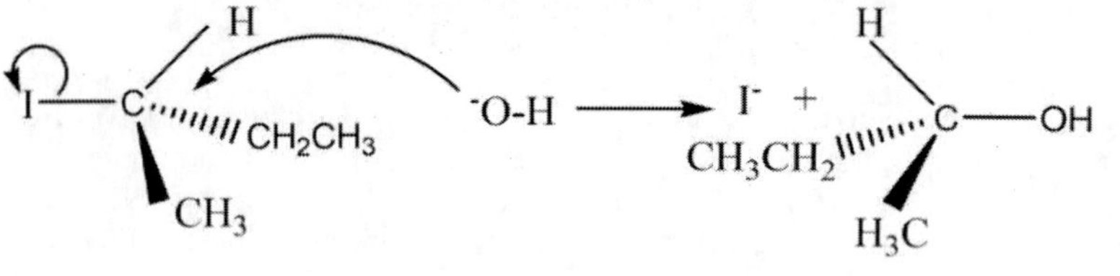

Figure 6.6. Walden inversion of an asymmetrical carbon center in the S_N2 reaction.

If the electron configuration of the carbon center in (S)-2-iodobutane is sp^3 orbital hybridization, then there is no empty atomic orbital on the other side of the iodine for the hydroxide ion to approach (except an antibonding molecular orbital) and so we must expect the hydroxide ion to attack the carbon center from any direction. Thus, sp^3 orbital hybridization in a carbon atom cannot explain the back-side displacement satisfactorily.

We now turn to the quaternity viewpoint of $2s^1 2p_x^1 2p_y^1 2p_z^1$ electron configuration of the carbon center. Because the iodine group is more electronegative than the –H, $-CH_3$, and $-CH_2CH_3$ groups in the molecule, it bonds to the densest orbital ($2p_x$ orbital) in carbon. The other three groups bond to 2s, $2p_z$, and $2p_y$ orbitals in the carbon center, respectively. Because of electrical repulsion, four substituent groups are arranged toward tetrahedral apices. Why does the hydroxide ion attack the carbon center from the other side of the iodine group? Because in the molecule, there are another three empty atomic orbitals $2p_x 2p_y 2p_z$ in the carbon center on the other side of the occupied $2p_x^1 2p_y^1 2p_z^1$ orbitals respectively and an empty spherical 2s orbital around the center. The hydroxide ion preferably attacks the empty $2p_x$ atomic orbital in the carbon center because a negative charge approaching the atomic sphere would best match the polor with a maximum density and minimum volume. The reaction results in Walden inversion of the chiral carbon center from S to R configuration.

Thus the quaternity viewpoint of 2p orbitals explains the back-side displacement of the S_N2 mechanism reasonably well. The explanation is also simple and natural.

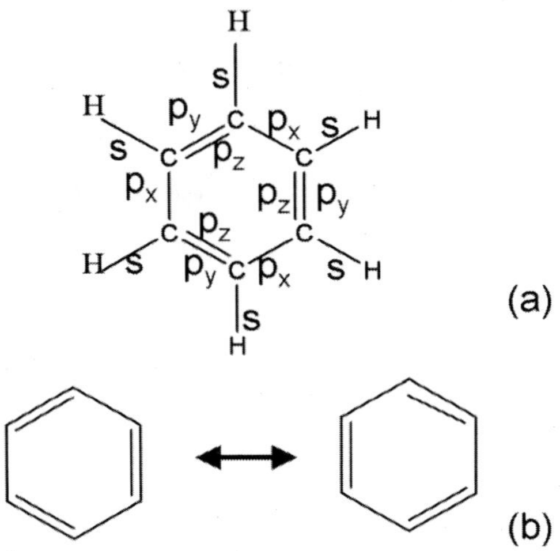

Figure 6.7. The nature of covalent bonds in benzene (a) and its resonance structures (b).

6.1.4. Covalent Bonds in Aromatic Rings

It is traditionally believed that all the carbon atoms in benzene undergo sp^2 hybridization. Each carbon forms three σ-bonds (C–H, C–C, and C–C) with the other atoms. There is also a 2p orbital left over from the sp^2 hybridization in each carbon atom. The unhybridized 2p orbitals of six carbons overlap together, each contributing one electron for a total of six π electrons. These six π electrons are said to be delocalized around the aromatic ring, increasing its stability.

Quaternity provides an alternative explanation for the structure of benzene. With an electron configuration of carbon $2s^1 2p_x^1 2p_y^1 2p_z^1$, each carbon in benzene forms a 2s-1s covalent bond with a hydrogen atom, a $2p_x$-$2p_x$ covalent bond with carbon on one side, and two ($2p_y$-$2p_y$ and $2p_z$-$2p_z$) π bonds with carbon on the other side (Figure 6.7a). The $2p_x$-$2p_x$ covalent bonds alternate with the $2p_y$-$2p_y$ and $2p_z$-$2p_z$ double bonds around the ring, giving it aromatic stability. However, because each carbon is bonded by similar atoms, the double π bonds and $2p_x$-$2p_x$ σ bond might switch their positions rapidly, giving two resonance structures as electrons transform dynamically and cyclically (Figure 6.7b). The resonance structures render equivalent the lengths of six carbon-carbon bonds around the hexagonal ring.

6.1.5. Electrophilic Substitutions of Mono-Substituted Benzenes

Substituents on an aromatic ring can activate or deactivate the ring toward further electrophilic substitution. They can also direct substitution either to the *meta* position or to the *ortho* and *para* positions. In this section, we shall explain these phenomena by quaternity space in carbon atoms. Aniline and nitrobenzene serve as two typical examples of mono-substituted benzenes. In an aniline molecule, because nitrogen is more electronegative than carbon, the N-C bond is a $2p_x$-$2p_x$ covalent bond. This disturbs the original structure of the aromatic ring, around which a $2p_x$-$2p_x$ bond alternates with the double π bonds of $2p_y$-$2p_y$ and $2p_z$-$2p_z$ (Figure 6.7). As shown in Figure 6.8(a), after the $2p_x$ orbital of C1 has connected to N, there is only a 2s orbital available from C1 for C6, so the covalent bond (2s-$2p_x$) between C1 and C6 is mismatched. Only orbitals of similar type form a stable covalent bond. Thus the amino group destabilizes the ring especially at the *ortho* position. We explain the inductive effect of amino group on benzene by the availability of remaining orbitals from C1 for proper covalent bonds with C2 and C6. Moreover, the nitrogen atom of aniline is also able to donate lone-pair electron density to the intermediate cations arising from electrophilic aromatic substitution at the *ortho* and *para* positions. Thus the amino group activates the ring toward further electrophilic substitution at the *ortho* and *para* positions.

A nitro group attaches to the ring to form nitrobenzene through a nitrogen atom with a formal positive charge. As shown in Figure 6.8(b), because the $2p_x$, $2p_y$, and $2p_z$ orbitals of nitrogen are bonded to oxygen, which is more electronegative than carbon, the N-C bond is a 2s-2s covalent bond. Comparing Figure 6.7(a) with Figure 6.8(b), we find that the nitro group does not disturb the original bonding structure of the aromatic ring. Instead, the positive charge absorbs electron density around the ring and hence deactivates further electrophilic substitution.

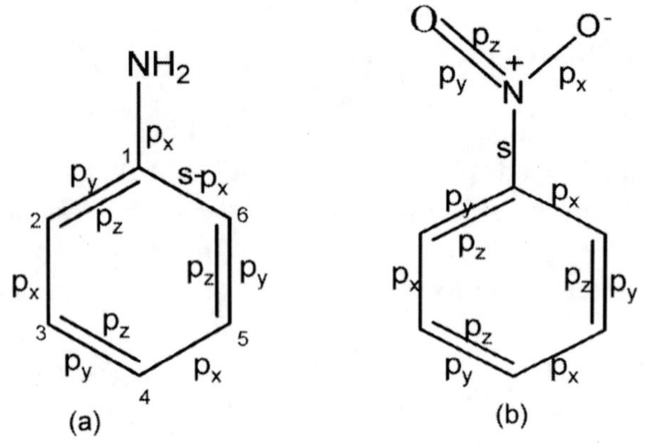

Figure 6.8. The nature of covalent bonds in aniline and nitrobenzene.

The directing effect of the nitro group depends on the stability of intermediates formed during electrophilic substitution. From Figure 6.9, it is easy to see that destabilization is greatest for the intermediate arenium ions that lead to the *ortho* and *para* products because one of the three resonance structures for the arenium ion has a positive charge on the carbon atom bearing the positively charged nitrogen atom of the nitro group. The intermediate leading to the *meta* substitution is less destabilized because none of its three resonance structures has this proximity of positive charges. Thus the nitro group deactivates the ring and directs electrophilic substitution at the *meta* position. It is clear that sp^2 hybridization of carbon cannot explain the resonance structures of benzene, which are crucial in predicting the reactions. The resonance structures are based on alternate single and double bonds around the aromatic ring. With these, quaternity explanation of the chemical property of aniline and nitrobenzene agrees with established organic chemistry and provides fresh insight into chemical reactions.

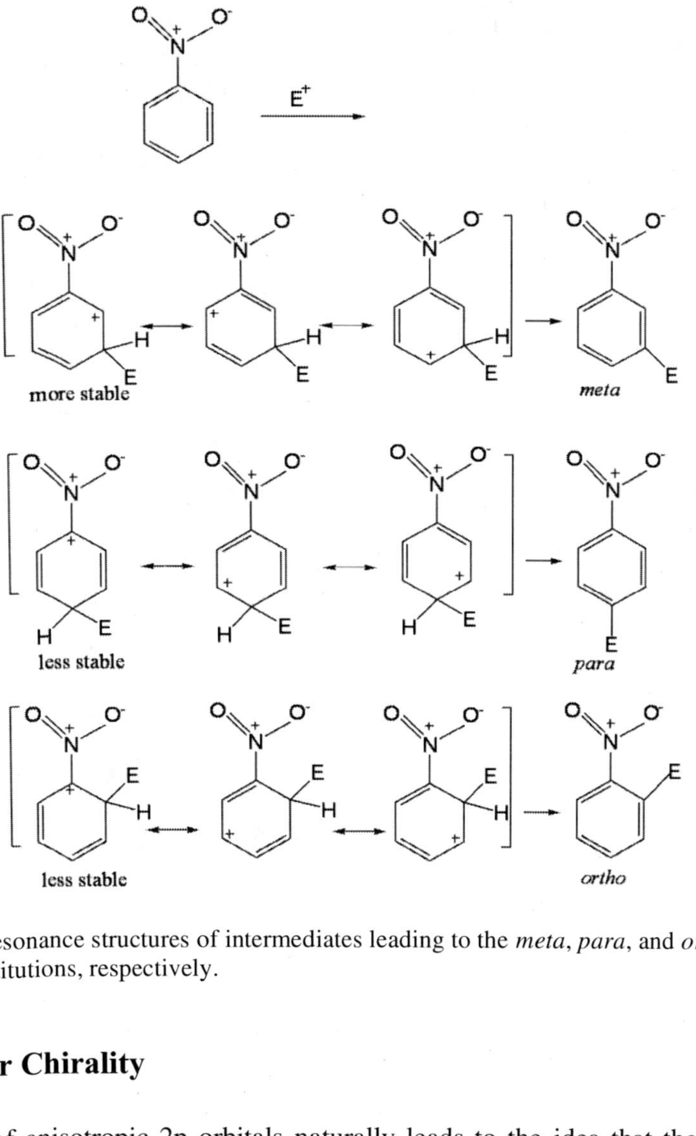

Figure 6.9. The resonance structures of intermediates leading to the *meta*, *para*, and *ortho* aromatic electrophilic substitutions, respectively.

6.2. Molecular Chirality

The discovery of anisotropic 2p orbitals naturally leads to the idea that they might form a handed structure in a tetravalent carbon atom. Here we suggest that molecular chirality results from the electronic orbital arrangement within the atomic sphere of carbon. Although it is generally correct to deduce the existence of molecular chirality based on whether a carbon atom is bonded to four different substituents, the four substituents of different electrophilic powers just fix and amplify the internal chirality of the electron configuration of the central carbon, giving it stable optical activity. Inspired by the dynamic transformation of electrons within the atom, we further suggest that selective electron transport through a stereocenter follows the electronic circulation pattern of $2s \mapsto 2p_x \mapsto 2p_y \mapsto 2p_z \mapsto 2s$. Given the handedness of a chiral molecule, both the sign of its optical rotatory effect and the pathway of selective electron transfer through the carbon center can be predicted to a large extent theoretically. For the first time, we discovered the biological role of chiral carbons in directing the electron flow in DNA.

6.2.1. The Density Gradient of 2s2p Orbitals

It has been established that $2s2p_x2p_y2p_z$ electrons within a neon shell take different shapes as well as orthogonal orientations. As was shown in Figure 2.14, a $2p_x$ orbital is a point displaced from the nucleus; a $2p_y$ orbital is a two-dimensional semicircular arc; a $2p_z$ orbital is a three-dimensional hemispherical surface; and a 2s orbital is a four-dimensional solid sphere. Because all electrons are in dynamic motion, these geometries are only typical illustrations of electrons at a certain moment. The geometries and dynamic motion of the electronic orbitals determine that a carbon atom with an outer electron configuration of $2s^1 2p_x^1 2p_y^1 2p_z^1$ is a potential chiral center.

Figure 2.14 is only a snapshot of 2s2p orbital geometries in their exact dimensions. Figure 2.12 shows these orbitals in dynamic motion, which we shall illustrate in more detail here. First, orbital transformation from 2s to $2p_x$ is the gradual displacement of the whole sphere away from the nucleus and the condensation of the electron in the meantime (Figure 6.10). As a point E traces along a hypothetical dashed semicircle OEB, the center of the sphere moves to point A and the radius of it decreases from OB to AB where point A is the vertical projection of point E onto line OB. At any moment, the electronic orbital can be regarded as a volatile homogeneous medium in its occupied space, which we may call an electron cloud. But its time series images (Figure 6.10b) demonstrate a clear density gradient of the dynamic electron. The tiny dislocated point B represents a full $2p_x$ orbital at the rotation terminal.

Second, let's examine various phases of electronic transformation from both $2p_x$ to both $2p_y$ orbitals. As shown in Figure 6.11, electronic transformation from a pair of dense points (b) toward a pair of semicircular arcs (f) experiences middle phases (c)(d)(e) during the continuous extension process. If we capture this transformation by overlaying several of its middle stage images together into (a), then it is easy to see that the dynamic process renders each $2p_y$ orbital tapered in density. The density gradient is schematically represented by the relative thickness of the arcs and their corresponding lengths. The longer the arc is, the thinner the thread is. The density gradient is directional along the orbital. Directionality refers to dynamic geometries whose head and tail are distinguishable in time series.

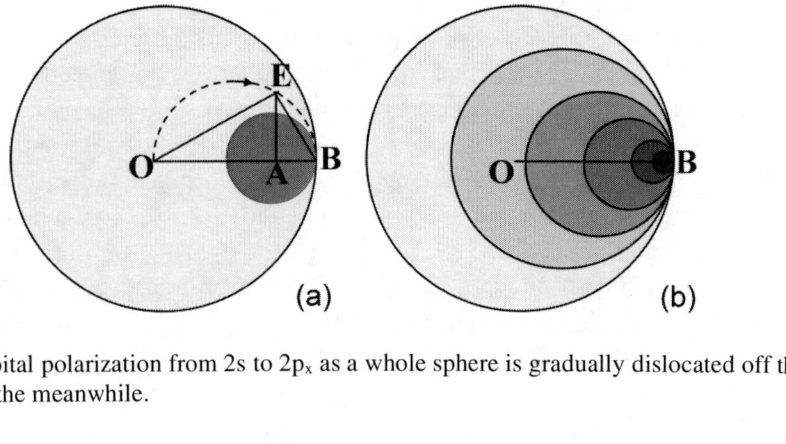

Figure 6.10. Orbital polarization from 2s to $2p_x$ as a whole sphere is gradually dislocated off the center and contracts in the meanwhile.

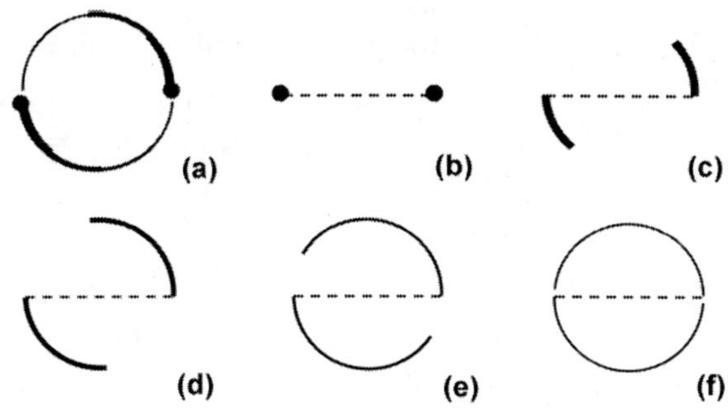

Figure 6.11. Electronic transformation from a pair of $2p_x$ orbitals (b) to a pair of $2p_y$ orbitals (f) through middle phases (c)(d)(e) in time series and their overlaying images (a).

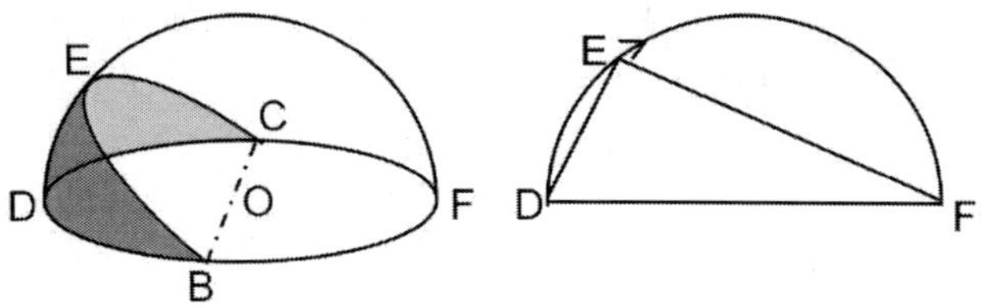

Figure 6.12. Electronic transformation from $2p_y$ to $2p_z$ orbitals as semicircle BDC sweeps through the hemispherical surface around BC axis.

Third, electronic transformation from $2p_y$ to $2p_z$ also reveals a density gradient of the orbital. As shown in Figure 6.12, a $2p_z$ orbital experiences geometric transformation of the shaded spherical lune BDCE widening from arc BDC to the entire hemispherical surface BDCEF as point E traces from point D toward F along the semicircle DEF. During the spreading process, the density of electron decreases continuously as its covering area increases. Like a $2p_y$ orbital, the expanding process in space exhibits orbital directionality in time series for the $2p_z$ electron.

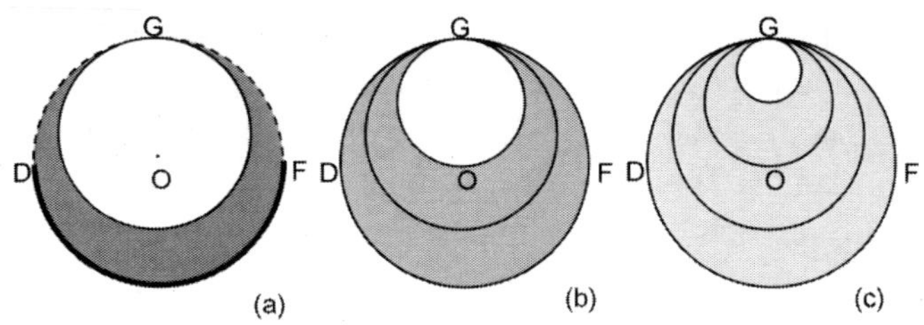

Figure 6.13. The density gradient on gray scale of three snapshots over time during directional orbital implosion from $2p_z$ to 2s.

Finally, orbital transformation from $2p_z$ to 2s can be characterized by the waxing of a crescent solid. As shown in Figure 6.13, a full $2p_z$ orbital is a hemispherical surface represented by DF; and a full 2s orbital is the whole sphere O. The transformation from a $2p_z$ orbital to a full 2s orbital involves the gradual inward filling of the electron cloud in the sphere. The electron cloud is diffusing inward over the atomic sphere during the process, which naturally leads to an orbital of ever sparser density with time. Thus the dynamic 2s orbital also displays a density gradient in time series. For a quantitative characterization of the density distribution of a time dimension in space, please refer to Figure 1.17.

The foregoing description indicates that four harmonic oscillation processes of 2s $\mapsto$ $2p_x$ $\mapsto$ $2p_y$ $\mapsto$ $2p_z$ $\mapsto$ 2s constitute a continuous geometric evolution loop. Each process features a gradient orbital in time series. The geometrical shapes of four orbital types are different. In their dynamic motion, a 2s electron is transforming into a $2p_x$ electron, and a $2p_x$ electron is transforming into a $2p_y$ electron while a $2p_y$ electron is transforming into a $2p_z$ electron, which in turn is evolving into a 2s electron. The geometries and dynamic motion of the electrons determine that a carbon atom with four outer electrons of $2p_x^1 2p_y^1 2p_z^1 2s^1$ possesses non-uniform electronic density within the atomic sphere, i.e., the density of a $2p_x$ electron is higher than that of a $2p_y$ electron, and the density of a $2p_y$ electron is higher than that of a $2p_z$, electron, which in turn is higher than that of a 2s electron. The continuous gradient of electronic density in the atomic sphere formed by various orbitals is the physical reason for carbon chirality. Incident light bends due to the existence of the continuous density gradient of electron clouds within the atomic sphere.

6.2.2. The Origin of Molecular Chirality

A molecule that is non-superimposable on its mirror image is chiral. A chiral molecule rotates the plane of incident polarized light and is said to be optically active. A carbon atom bonded by four different substituent groups constitutes a chiral center or stereocenter. In spite of the fact that thousands of compounds with one chiral center have been found to be optically active with no exceptions, the notion of chirality according to the existence of a dissymmetric atom in a molecule is empirical. There must be a fundamental reason underlying the optical phenomenon. Here we investigate the cause of chirality at an atomic structure level, deeper than the level of molecular configurations. We propose that chirality of a molecule originates from the electron configuration ($2p_x^1 2p_y^1 2p_z^1 2s^1$) of the outer shell of a carbon atom. The optical activity of a chiral center is a phenomenon of light refraction.

As was described in section 6.1.2, in a methane molecule, the carbon atom uses a $2p_x$ electron, a $2p_y$ electron, a $2p_z$ electron, and a 2s electron to form four covalent bonds with four hydrogen atoms respectively. Thanks to the spatial flexibility of the last three orbital types, each hydrogen atom is restricted to a corner of a tetrahedron under electrical repulsion between the covalent bonds. However, due to dynamic transformation of the electrons (see Figure 6.3), the nature of each covalent bond normally switches among four types of C–H bonds periodically, e.g., from a $2p_x$-1s bond to a $2p_y$-1s bond, to a $2p_z$-1s bond, to a 2s-1s bond, and back to the original $2p_x$-1s covalent bond. The rapid dynamic cycling renders a methane molecule achiral or ambi-chiral because the molecule is switching instantly between two opposite kinds of chiral configurations, which are characterized as follows.

As shown in Figure 6.14, because three types of 2p orbitals are mutually perpendicular in the absence of electrical repulsion, if we set up a Cartesian coordinate system in such a way that its origin sits at the nucleus, the X-axis points to the orientation of $2p_x$ and the Y-axis points to the orientation of $2p_y$, then $2p_z$ would be along the Z-axis protruding either toward the negative direction (Figure 6.14a) or toward the positive direction (Figure 6.14b). Therefore, three orbitals of $2p_x$, $2p_y$, and $2p_z$ have two kinds of handedness corresponding to the left-hand rule and right-hand rule of the X, Y, and Z-axes in the Cartesian coordinate system.

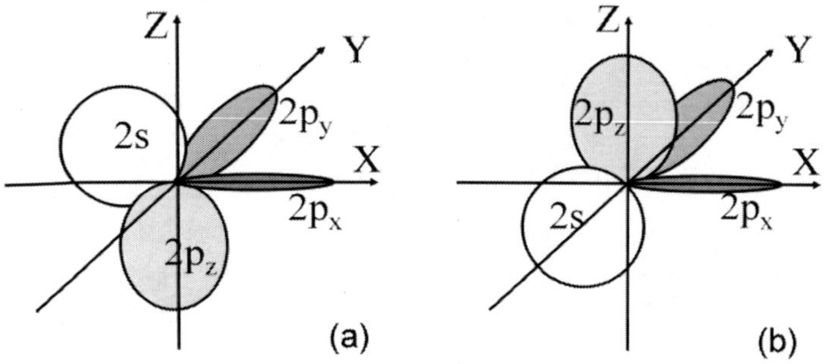

Figure 6.14. Two opposite kinds of chiral orientations due to anisotropic 2s2p orbitals in Cartesian coordinates.

Under electrical repulsion, the angles between the three 2p orbitals would be larger than 90° without changing their relative handedness, and 2s orbital would be oriented to a position farthest away from the three due to its high spatial flexibility. Hence anisotropy of 2p orbitals leads to anisotropy of 2s2p orbitals. We use both terms interchangeably. A carbon atom with outer electrons of $2p_x^1 2p_y^1 2p_z^1 2s^1$ has two possible chiral configurations. Because of the dynamic transformation of orbitals in the directions of $2s \mapsto 2p_x \mapsto 2p_y \mapsto 2p_z \mapsto 2s$, a methane molecule exhibits both chiral properties with time. Many CH_4 molecules amount to a racemic mixture at any given moment and appear achiral.

A methane molecule does not show a stable chiral property because four hydrogen atoms that are bonded to $2p_x^1 2p_y^1 2p_z^1 2s^1$ have similar actions on them. Only when the four types of electrons within a carbon atom are bonded to four different substituent groups does the compound show a stable chiral property because the four substituents, having different electronegativities and electrical polarizabilities, not only inhibit the free switch of $2p_x^1 2p_y^1 2p_z^1 2s^1$ orbitals among themselves within the carbon, but also fix and reinforce the handedness of the four orbital types. Thus, even though the judgement of a chiral compound based on the handedness of the entire molecule is apparently correct, the cause of chirality indeed rests with the anisotropy of $2p_x^1 2p_y^1 2p_z^1 2s^1$ electronic orbitals. Corresponding to Figure 6.14, we depict both kinds of orbital handedness in Figure 6.15 for clarity, but bear in mind that the atomic sphere is a continuous spacetime medium of four dynamic orbitals. It is that chirality at the orbital level extending to the molecular level that accounts for its optical activity. Had it not been caused by an orbital gradient, chirality merely on the molecular scale might be too macro to produce any effects on light refraction.

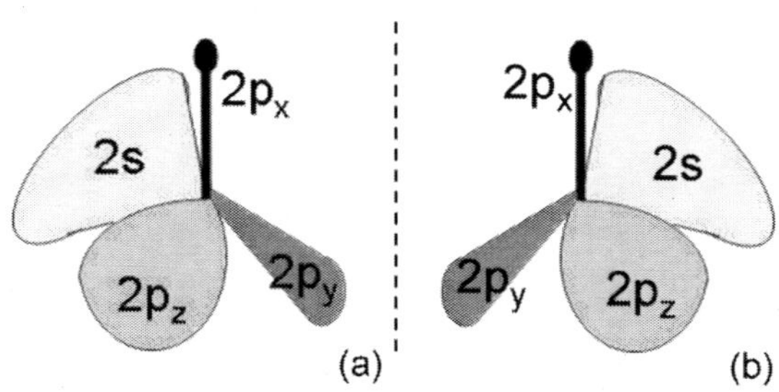

Figure 6.15. Schematic diagram of enantiomers owing to chiral carbon centers with $2p_x^1 2p_y^1 2p_z^1 2s^1$ orbitals of different volumes (size) and densities (shade). Note that (a) is levorotatory with a negative $[\alpha]_D$ value whereas (b) is dextrorotatory with a positive $[\alpha]_D$ value.

6.2.3. The Physical Principle of Optical Rotation

Figure 6.15 delineates two enantiomers with four types of electronic orbitals in discrete shapes for clarity, but in fact the four types of orbitals form a continuous medium for chirality. For an electronic orbital, the smaller the volume is, the denser the electron cloud is. Light bends as it travels through a medium of gradient density by the principle of refraction. For example, as shown in Figure 6.16, when a beam of light enters a transforming $2p_x$ orbital (Figure 6.10b), it will be refracted downwards at the density inclines due to the refractive index increase. Since the density gradient of the medium is a continuous variable, light travels along a smooth curve within the atom as a result, even though it is shown as a broken line in Figure 6.16.

How does a chiral center rotate plane-polarized light? Plane-polarized light may be regarded as being made up of two kinds of circularly polarized light, each having the appearance of a helix propagating around the axis of light motion. One kind is left-handed while the other is a right-handed helix. As plane light enters an uneven medium, it tends to bend toward a denser region because of its relatively higher refractive index. The higher the index is, the slower the speed of light travels through the medium. This would seem to mean that the left- and right-handed circularly polarized components travel at different velocities because each is slowed down to a different extent. However, since both components are inseparable, the consequence is the curving of light's pathway and the rotation of the plane light. This phenomenon is analogous to a car running on the road. When the right wheels are faster than the left wheels, the car turns left and the driver's body rotates a certain degree counterclockwise accordingly. As plane-polarized light undergoes continuous bending in the four-dimensional chiral carbon center, it rotates an angle of its polarized plane. This is a manifestation of light refraction phenomenon.

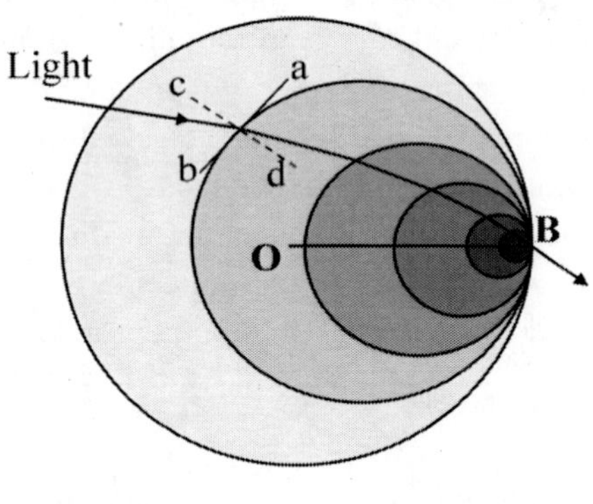

Figure 6.16. Light refraction at density gradients where ab denotes a plane tangential to a refractive index contour and cd indicates the line normal to the plane.

To explain chiral optical activity visually, supposing that the inner region of a chiral structure has a higher refractive index than the outer region, light entering the chiral structure will be refracted toward the right direction all the way through the medium (Figure 6.17). Since a chiral carbon is composed of four orbitals in the transformation pattern of $2s \mapsto 2p_x \mapsto 2p_y \mapsto 2p_z \mapsto 2s$, we assume that electronic orbitals that form a chiral center are arranged in a manner somehow like a screw pattern or a snail shell in spacetime. Light approaching the center will be routed forward in such a way as a screw rod is guided forward by the screw thread, complying with refraction.

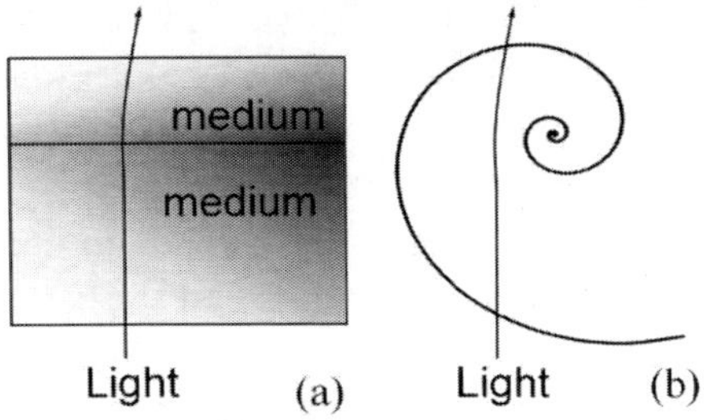

Figure 6.17. Planar illustration of light following (b) the spiral curve of a chiral center to a certain extent due to (a) the existence of a density gradient.

Moreover, the rotation of plane light is in a coherent thread direction of clockwise or counterclockwise even though light might enter the medium at any directions. For example, a coiled spring is right-handed forward when perceived from front and rear; and the spiral line of a field snail remains right-handed forward when perceived from both head and tail (Figure Q10). One remarkable feature of a chiral carbon must be that its screw pattern remains the same when perceived from any directions. There is only a fixed spacetime direction of circular orbital transformation (2s $\mapsto$ $2p_x$ $\mapsto$ $2p_y$ $\mapsto$ $2p_z$ $\mapsto$ 2s) for a specific chirality, like a traffic circle at a junction of thoroughfares that routes the cars in a one-way direction. We shall demonstrate five electron generation cycle within a carbon atom in Figure 11.6. No matter which direction plane light approaches a chiral center, it will follow the screw pattern to rotate a certain degree at that fixed direction, thereby producing a measurable optical rotation magnitude proportional to the chiral compound concentration in the solution.

6.2.4. Chiral Disposition Induced by Electrophilic Groups

The relationship between a chiral carbon center and its specific optical rotatory effect is complicated and defies theoretical generalization. By tradition, one can only know the direction and magnitude of rotation, $[\alpha]_D$, of an asymmetric carbon by experimentation. After discovering the essence of chirality at the orbital level, we wish to shed light on this important issue. First, since the property of a chiral carbon comes of orbital anisotropy, we may derive the relationship between the handedness of 2s2p orbitals and the rotatory direction of plane-polarized light. As was shown in Figure 6.3, electrons transform from $2p_x$ to $2p_y$, from $2p_y$ to $2p_z$, and from $2p_z$ to 2s following the rule of dynamic calculus. It has been demonstrated that electronic transformations are realized by transferring virtual photons (Figure 3.13). Virtual photons travel in the opposite direction to electrons. Therefore plane-polarized light entering a chiral carbon will be routed in a direction contrary to the electronic transformation direction. Viewing from the opposite side of the 2s orbital, we can tell the direction of plane light rotation based on the handedness of three 2p orbitals. Plane-polarized light rotates along the $2p_z \rightarrow 2p_y \rightarrow 2p_x$ direction, bending toward an ever denser region in the continuous electron clouds as was illustrated in Figure 6.17. For example, Figure 6.15(a) is levorotatory having a negative [α]$_D$ value whereas Figure 6.15(b) is dextrorotatory having a positive [α]$_D$ value.

Second, for four different substituents bonded to a carbon atom, we need to define their priority in forming covalent bonds to various orbitals. Because a $2p_x$ orbital has the highest electron density in the 2s2p shell, the most electrophilic substituent would form a σ-bond with it, the second-most electrophilic substituent would bind to a $2p_y$ orbital, the third would attach to a $2p_z$ orbital, and the least electrophilic substituent (which is −H more often than not) attaches to a 2s orbital. The hydrogen atom usually bonds to the 2s orbital of an asymmetric carbon also because 1s and 2s orbitals are similar orbital types that match perfectly in forming a covalent bond.

In terms of attracting an electron cloud, the electrophilic property must include two aspects: high electronegativity and low polarizability. For example, −F ligand bonded to a carbon atom would surely capture its $2p_x$ electron to form a σ-bond. Thus the handedness of 2s2p orbitals is contingent on the relative electrophilic powers of four substituent groups attached to them. This establishes the connection between chirality owing to intrinsic $2p_x^1 2p_y^1 2p_z^1 2s^1$ orbitals and chirality induced by extrinsic substituents bonded to them. It is the different substituents that fix and amplify the chirality of orbitals within the carbon atom, giving it a stable chiral property. This fresh interpretation is in line with established facts.

The tetravalent nature of a carbon atom naturally leads to molecular handedness in any case where four different substituent groups attach to it. The early work by Brewster suggested an empirical method for the prediction of the sign and amount of optical rotation based on bond refractions and polarizabilities of groups in a molecule [15, 16]. The method predicted the direction of optical rotation using an asymmetric screw pattern of electron polarizability, the magnitude of rotation being related to the refractions of the atoms making up the pattern. Under Fischer projection of Figure 6.18(a), the tetrahedral system of xABCD could be described as dextrorotatory when A>B>C>D in terms of polarizability. D is assumed to be a hydrogen atom in most cases. This result agrees with our suggestion that substituent groups from low to high polarizabilities would attach to electrons in the priority order of $2p_x$, $2p_y$, $2p_z$, and 2s within the chiral carbon except the case of –H substituent that always attaches to the 2s orbital in our suggestion (Figure 6.18b).

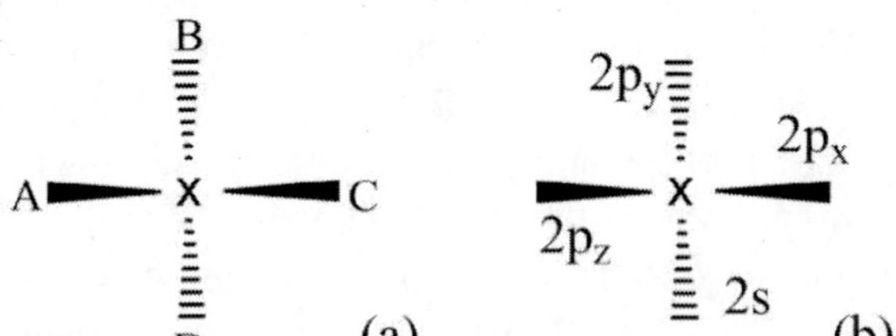

Figure 6.18. Fischer projection of the absolute configuration of xABCD molecule and the bonding orbital types with the stereocenter.

Recently, Zhu et al. devised a matrix model to predict specific magnitudes of optical rotation of acyclic chiral molecules [17]. The matrix elements included the substituents' comprehensive masses and radii, asymmetric factors, and electronegativities of the atoms or groups that are bound to the stereogenic center. Pauling electronegativities were directly used in the matrix and their contribution to the specific rotation was large. The effect of electronegative value on the rotatory direction of plane light in the matrix agrees with our statement that substituent groups from high to low electronegativities would attach to electrons in the priority order of $2p_x$, $2p_y$, $2p_z$, and 2s within the chiral carbon.

The electrophilic power of substituent groups in terms of high electronegativity and low electric polarizability is a chemical property indicative of both extent of electron transfer and stability. If we can arrange various kinds of substituents in the order of electrophilic power, we would be able to predict whether a specific chiral molecule is dextrorotatory or levorotatory, which will be shown to be exceedingly useful in the next section. Qualitatively, the electrophilic power is related to the relative rate constants for reactions of different electrophiles toward a common substrate, usually involving attack at a carbon atom in organic chemistry. It is related to Lewis acidity. Quantitatively, Parr et al. proposed that the electrophilicity index of a specific chemical species should be defined as the square of its electronegativity divided by its chemical hardness [18]. This concept measured the propensity of a substituent group to attract electrons in a covalent bond, which seems to agree with our considerations of both electronegativity and polarizability, except that it tends to assign the index value for hydrogen –H on the high side. How well the electrophilicity index fits into the framework of preferential selection of tetrahedral $2p_x^1 2p_y^1 2p_z^1 2s^1$ orbitals in a chiral carbon remains to be investigated. Obviously, the suggested order of substituents based on the electrophilic power is different from that based on the Cahn-Ingold-Prelog rules. The latter is useful in the nomenclature of R or S chiral molecular configuration but unhelpful in predicting optical rotatory effects, at which the former aims.

In spite of its profound theoretical basis, the determination of specific molecular chirality based on the relative electrophilic powers of four substituent groups attached to a central carbon must prepare to encounter some exceptions for several reasons. First, the chiral configuration of an asymmetric molecule has something to do with its chemical synthetic route. Once a certain handedness has been preserved by four substituents during synthesis, it takes a certain amount of activation energy to overturn its shape later on. Hence the effect of relative electrophilic powers does not always prevail and manifest in chiral compounds. Second, disorder always competes with order in molecular arrangement. Many organic compounds synthesized by chemical methods naturally result in a racemic mixture due to two accessible sites on the central carbon during the reaction, which shows no apparent optical activity. Neither do they comply with the rule of preferential orbital selection by a strong electrophile. Even some biologically produced chiral compounds racemize in time under the driving force of entropy increase. Such racemizations have been exploited in an amino acid dating method [19]. Third, the relative electrophilic powers of four different substituents must be considered in a comprehensive manner. Some substituents are more likely to have $2p_x$ orbitals available for forming a $2p_x$-$2p_x$ σ-bond with the carbon atom, the others tend to expose their $2p_y$ orbitals to form a $2p_y$-$2p_y$ π-bond with the carbon. $2p_y$ orbitals are more polarizable electrically than $2p_x$ orbitals. For example, as was mentioned in section 6.1.5, the nitrogen atom of the amino group has a $2p_x$ orbital available for forming a covalent bond whereas that of the nitro group only has a 2s orbital available. The electrophilic order of various substituents is not invariable. It depends on the specific combination of four different substituents in forming a relatively low-energy chiral configuration. While the general rule is set, case by case analysis is still deemed necessary.

6.2.5. Selective Electron Transfer through Chiral Centers

As was shown in Figure 6.3, electrons transform in the direction of 2s → $2p_x$ → $2p_y$ → $2p_z$ → 2s within a neon shell. When four different substituents fix the orbital types of a carbon center, light hitting the chiral center would follow the reverse direction of electronic transformation and rotate a certain degree due to the gradient of electron density within the atomic sphere. This results in a measurable optical rotation of plane-polarized light passing through the chiral compound medium, $[\alpha]_D$. In this section, we shall first apply this principle to a deoxyribonucleoside molecule and then examine electron transport behavior through various chiral centers within it.

Consider deoxycytidine with three chiral carbon centers in its deoxyribose, C1', C3' and C4'. Four substituents bonded to each chiral carbon are labeled x, y, z, and s in Figure 6.19. Assuming the electrophilic powers of the substituents are in the order x > y > z > s, we may predict the tendency of optical rotatory effect of each chiral center according to the above chiral disposition principle by electrophilic groups. For example, in the C1' atom, the oxygen atom of the ether bond would bind to the $2p_x$ orbital of the chiral carbon; the cytosine would bind to the $2p_y$ orbital of the chiral carbon; the aliphatic chain $-CH_2-$ of C2' would bind to the $2p_z$ orbital of the chiral carbon; and the hydrogen atom would bind to the 2s orbital of it. A substituent of relatively higher electrophilic power binds to an orbital type of relatively higher electron density. Thus the C1' chiral configuration would produce a levorotatory effect on plane-polarized light according to Figure 6.15(a).

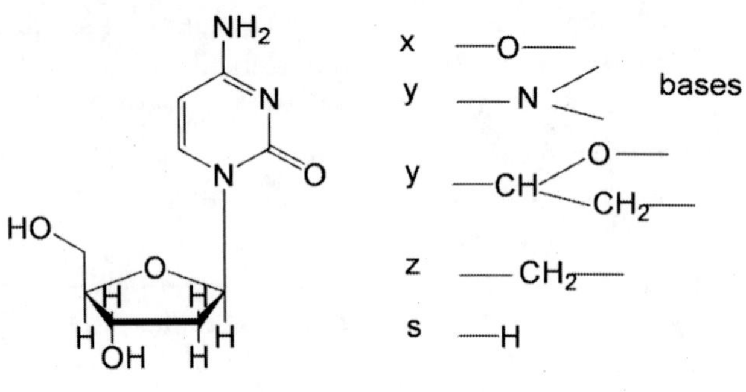

Figure 6.19. Three chiral carbon centers C1', C3' and C4' in a deoxyribose with their substituent groups in the order of relative electrophilic powers x>y>z>s.

Likewise, in the chiral center C3', the hydroxyl group would bind to its $2p_x$ orbital; C4' would bind to its $2p_y$ orbital; C2' would bind to its $2p_z$ orbital; and the hydrogen atom binds to its 2s orbital. Hence C3' chiral center is dextrorotatory. The four substituents are bonded to C4' in a very similar way to those to C3', but they are arranged in a converse pattern for chiral C4' to be levorotatory. Although the optical activity of the entire nucleoside is difficult to predict, we can derive the sign of optical effect of each chiral carbon based on the relative electrophilic strength of the substituents.

If instead of light refraction, there are electrons flowing from C5' to C4', then in which direction should the electrons move at the C4' junction? We suggest that electrons flowing through a chiral center would stick to the transformation pathway 2s → $2p_x$ → $2p_y$ → $2p_z$ → 2s around the central carbon. Because C5' is bonded to the $2p_z$ orbital type of the C4' chiral center, electrons entering C4' are directed toward the 2s direction and then toward the $2p_x$

orbital bonded to the oxygen atom. Hence the negative charges would be routed from C5' to the ether bridge to C1'; and positive holes would travel in the reverse direction (Figure 6.20). Electric current from C4' to C3' is choked by the C4' chiral center. Thus the chiral carbon center transfers electrons through itself along a selective pathway.

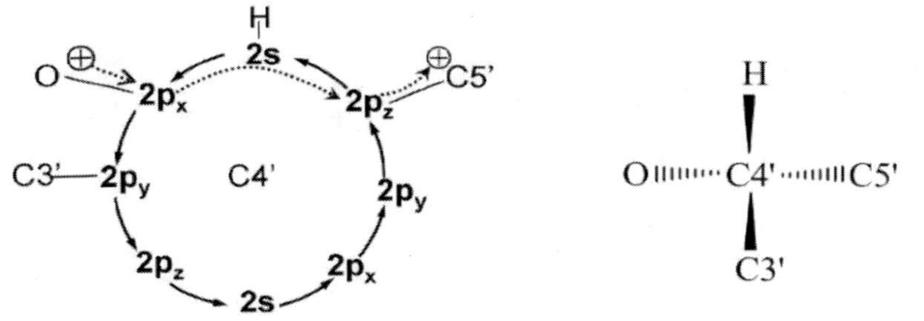

Figure 6.20. Electronic circulation direction (solid arrows) within the C4' chiral atom and selective positive hole migration (dotted arrows) through the center shown with its Fischer projection (right).

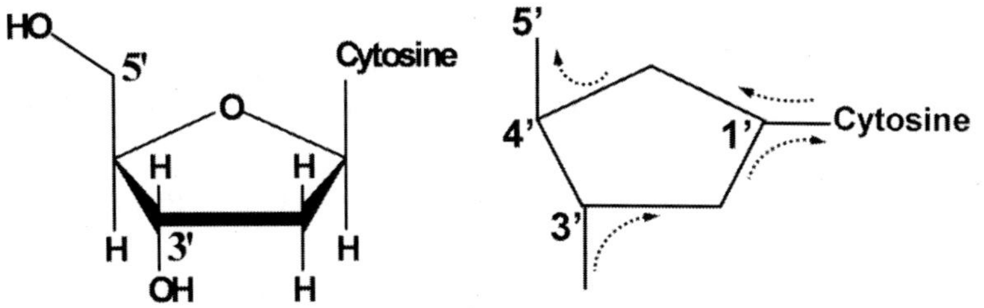

Figure 6.21. Electric currents directed by chiral carbons through a nucleoside. The dashed arrows represent electric current directions.

As electrons arrive at C1' from the ether bridge bonded to the $2p_x$ orbital of chiral C1', they are directed to the N−glycosyl bond that is connected to the $2p_y$ orbital of the chiral center, hence charging the nitrogenous base. Furthermore, when electrons are overflowing from the nitrogenous base toward C1', they would be routed in the C2' direction by the chirality of C1', i.e., from $2p_y$ to $2p_z$ direction. By the same rule, electrons entering C3' from C2' would be routed toward the hydroxyl group rather than C4', i.e., in the $2p_z \rightarrow 2s \rightarrow 2p_x$ direction. Electric current from C3' to C4' is choked by the chirality of C3'.

On the contrary, positive charges travel, representing electric current, in the opposite direction to electrons. We expect that electric current would be routed by the chiral centers in the direction from 3'-hydroxyl group → C3' → C2' → C1' → the nitrogenous base, and then from the nitrogenous base → C1' → ether bridge → C4' → C5' (Figure 6.21). Direct electric current between C3' and C4' is prohibited by the chiral carbons, detour being mandatory.

Besides traveling direction reversal, another significant difference between light refraction and the selective electron flow at a chiral center is that light hitting the atomic sphere immediately swerves out of the atom whereas electrons flowing into the chiral center from any connecting branch must follow the next covalent bond to another connecting branch to exit the center. By the nucleoside example, we have explained the stereo-selectivity of chiral centers in directing electric current.

By the way, for electron transfer direction in a polypeptide, one can easily tell from the relative electrophilic power of substituent groups that electrons will transport through the chiral α-carbon in the direction from the amine group to carbonyl group. This agrees with the fact that the amine group is usually positively charged while the carbonyl group is negatively charged upon the transfer of a proton in the solution.

Living organisms consist almost exclusively of chiral molecules. Specifically, the amino acids, the building blocks of proteins, are found in L-configuration with few exceptions whereas sugars are only found in D-configuration [20]. In explaining why living organisms discriminate one type of enantiomer from the other, it is traditionally stated that biochemical processes in living cells are normally catalyzed by enzymes, which are proteins that fold up into complex structures with an active site for reactants. Only chiral compounds of a specific handedness can be locked into the site and participate in enzyme catalyzed reactions. Although such a stereo-specific explanation sounds logical, the more critical reason probably resides in the mechanism of chiral compounds in selective electron transfer. As was described, a chiral center would direct the electron flow from a substituent branch to another in accordance with 2s2p electronic circulation direction within the chiral atom. A pair of enantiomers, for example, would channel the electron flow in contrary directions and produce different biological consequences. Hence one type of enantiomer cannot substitute another, their electrical properties being different. Living organisms must maintain the homochirality of biochemical molecules to ensure their proper functions in charge transfer, which might contain physiological signals. Racemization of homochiral biological substances leads to stray signals and death.

The discovery of chiral carbons for selective electron transfer in conformity with optical activity undoubtedly breaks new ground for biochemical research. Early in the 1970s, Garay made a hypothesis that biological chiral molecules, such as chlorophylls, during excitation might possess magnetic moment that caused a partial polarization in electron transport [21], but it has received not any attention later on. Not any research work has been carried out in this direction. Modern scientists probing the electron transfer pathways within proteins and nucleic acids are totally unaware of the electron directing role of chiral carbon centers. In a consistent manner, we have set forth the anisotropic orbital theory for delving into the underlying secret of molecular chirality and its selective electron transfer mechanism [22-24].

Conclusion

The quaternity theory indicates that 2s2p orbitals have four varying spacetime dimensions as well as orthogonal orientations and these electrons are circulating by the rule of dynamic calculus. After a brief review of quaternity space, we proposed that a carbon atom has outer electron configuration of $2p_x^1 2p_y^1 2p_z^1 2s^1$. A $2p_x$ orbital is a line segment; a $2p_y$ orbital is a semicircle; a $2p_z$ orbital is a hemisphere; and a 2s orbital occupies the entire sphere. The anisotropic orbitals of a carbon atom possess great geometrical flexibility to accommodate the molecular structure of methane, ethene, ethyne, and benzene. They are also responsible for the reactivity of alkyl halides in the S_N2 mechanism for nucleophilic substitution and the substituent effect of mono-substituted benzenes toward further electrophilic substitution. We have predicted or retrodicted the chemical properties of carbon atoms in organic chemistry based on anisotropic orbitals of $2p_x^1 2p_y^1 2p_z^1 2s^1$. All tend to corroborate the theory of quaternity.

The chirality of carbon centers is the central theme of organic chemistry. This chapter undoubtedly holds important progress in this field. It takes a fresh look at features of dynamic calculus on four dimensions in a carbon atom and asserts that the anisotropy of 2s2p orbitals is the origin of molecular chirality. When a carbon atom is bonded to four different substituents, it becomes a chiral center because the four bonded substituents would differentiate four outer electrons of the carbon atom, thereby preserving its orbital chirality. The handedness of a chiral carbon could be predicted to a certain extent according to the relative electrophilic powers of the four substituents, i.e., substituents from the strongest electrophile to the weakest bind to outer electrons of carbon from the highest electronic density to the lowest in the order of $2p_x 2p_y 2p_z 2s$.

A thorough analysis of 2s2p electronic orbitals reveals that the optical activity of a chiral carbon actually results from light refraction in the uneven spacetime medium, which is otherwise traditionally regarded as a phenomenon independent of refraction. Based on the handedness of a chiral center, one could tell whether a chiral compound is levorotary or dextrorotatory because plane light rotates in the reverse direction to the electronic circulation direction of $2s \rightarrow 2p_x \rightarrow 2p_y \rightarrow 2p_z \rightarrow 2s$ (see section 3.5).

In contrast, free electrons entering a chiral center from one covalent bond will be directed to another covalent branch along the electronic circulation direction. A chiral center, like a special automatic switch valve, is capable of directing the electron flow along macromolecules. We have characterized the crucial role of chiral carbons in selective electron transfer in a deoxyribonucleoside. This original discovery is made by an independent mind through a logical extrapolation of quaternity theory.

Chapter 7

DNA Circuits and Waves

We shall shift our focus from simple organic molecules to complex ones. One of the most conspicuous macromolecules is DNA that is composed of two chains of nucleotides wound around each other. Following the discovery of selective electron transfer through chiral carbons in a nucleoside in section 6.2.5, this chapter further explores the electrical property and wave behavior of a DNA molecule. There have been many experimental reports on the electric conductivity of DNA fibers, providing a mixture of results. Here we sift through the results and present a DNA molecule as stepwise *LC* oscillatory circuits. Based on the sinusoidal signals of *LC* oscillatory circuits, we formulate the base pairs by circular functions and the phosphate bridges of a DNA molecule by spherical quantities in dynamic calculus, giving a DNA molecule a mathematical expression. Oscillatory circuits precisely characterize a DNA molecule on a fundamental level and provide sharp insight into its biophysical significance.

7.1. Introduction

Long after the chemical structure of DNA had been elucidated, the electrical property of the double helix remained a puzzle. Previous studies by direct measurements of electric current in DNA fibers produced various conclusions from no conductivity [25, 26], semi-conductivity [27], good conductivity [28], to superconductivity of a DNA molecule [29]. By appending a photo-oxidant to one end of a DNA duplex and measuring the oxidative damage of guanine doublet site at distances, recent biochemical studies of DNA charge transfer have provided more convincing results. It has been established that very fast charge transport can take place over a short distance (~37 Å) and a slower transport may propagate over a long distance (~200 Å) [30-38]. The results have led to two hypotheses of charge transport mechanism in DNA, one indicating super-exchange, or tunnel, through the sugar phosphate bridge between the bound charge donor and acceptor for the short distance, and the other suggesting charge hopping between discrete bases through the DNA π-stack over the long distance. Although both mechanisms are widely regarded, they are incompatible and rather ill-defined, giving the expression of loose or uncertain charge transfer along a DNA molecule. As genetic substance, DNA must possess well-defined electrical behavior. Here we show a DNA double helix as a stepwise *LC* oscillatory circuitry, in line with the experimental evidence. We demonstrate that a DNA molecule maintains an electron transport mechanism through both strands that matches the fidelity and reliability of its chemical structure. For the first time, we discover the physical meaning of the chemical structure of DNA. The original publication of this section is available at www.springerlink.com (Xu, K. Stepwise oscillatory circuits of a DNA molecule, Journal of Biological Physics, 35, 223-230, 2009, DOI 10.1007 /s10867-009-9149-9).

7.2. DNA Electric Conductivity

DNA is at the heart of various life processes and functions in living organisms, and understanding the intricate electrical behavior of DNA is essential for elucidating various microscopic mechanisms in different fields such as biology, chemistry, and material science. This section characterizes a DNA molecule by a stepwise LC circuitry where each base pair is an electric capacitor, each phosphate bridge is an electric inductor, and each deoxyribose serves as an electric switch. The capacitor is capable of storing electric charges and has a function of conducting alternate current but inhibiting direct current; the inductor is capable of storing magnetic fluxes and has a function of passing direct current but impeding alternate current; and the pentose as an electric switch is capable of routing electric charges along a selective path within the DNA molecule. DNA as stepwise LC circuits is a miracle of nature that allows harmonic oscillations of electrons to proceed along the double helix step by step. The discovery of DNA circuits provides new insights into the underlying mechanisms of genetic mutation and evolution because the environmental factors might cause the electron transports out of phase along both polynucleotide strands by affecting the parameters (i.e., capacitance and inductance values) of the DNA circuits.

7.2.1. DNA Electronic Components and Circuitry

The circuit model regards each base pair as an electric capacitor because the base pair is composed of two heterocyclic amines placed at a semi-conductive distance, capable of storing opposite charges. The hydrogen bonds between the base pair deliver electric pulses but prohibit direct current between them, which is a typical characteristic of electric capacitors. In the double helix, the phosphate bridges are twisted physically like both strands of a rope under torsion. The wound phosphate bridge can be treated as an electric inductor, capable of storing energy as well. The rigid structures of the Watson-Crick parities and of the pentose sugars help in forcing the torsion onto the phosphate bridges mechanically. Charge transport through the strands, like electric current through a coiled wire, produces electric inductance. Hence DNA structure is a circuitry composed of multiple oscillatory *LC* circuits (Figure 7.1).

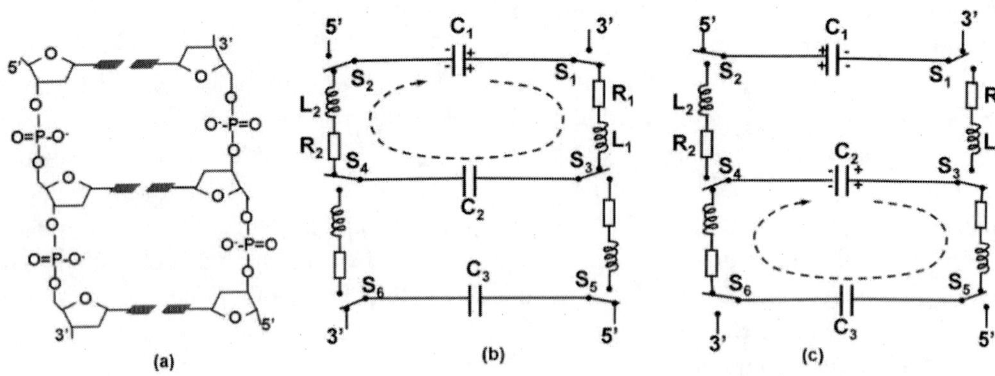

Figure 7.1. The DNA double helix (a) along with stepwise oscillatory circuits (b) and (c), where the dashed arrows indicate electric current direction and the deoxyriboses are represented by electric switches.

The deoxyribose is a key electronic component in the circuit, which serves as a switch at the juncture (Figure 7.2). It has been suggested that a chiral carbon center might transfer electrons toward a selective pathway due to its asymmetric polarizations [39]. Three chiral carbons in a deoxyribose direct electric charges in a selective course (see section 6.2.5). Specifically, as electrons enter C4' from C5', they will be routed through the ether bond by the chirality of C4' because an oxygen atom is more electronegative than carbon in drawing the electrons; and thereafter the electrons must build up to a threshold potential to escape oxygen attraction. In this way, the ether bond delivers electric pulses through itself. When the electrons arrive at C1', they will be directed to the N-glycosyl linkage, which is more electronegative than carbon, hence charging the nitrogenous base; and after electrons overflow out of the base toward C1', they will be transferred via C2' to C3', which in turn directs the electrons toward the oxygen atom connecting the phosphate, and then through the phosphate bridge. Electric current between C3' and C4' is choked by both chiral carbons.

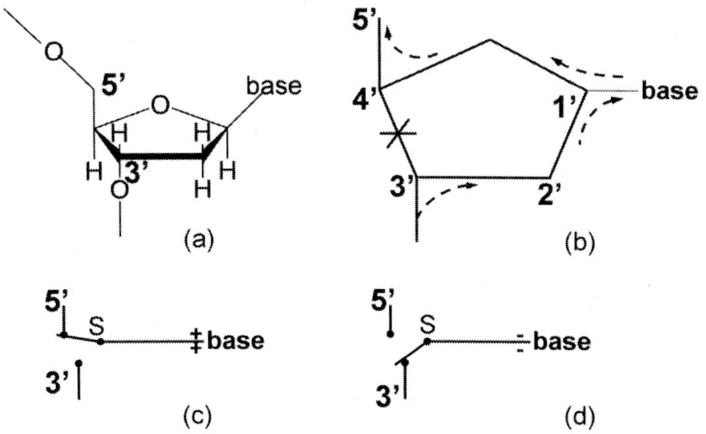

Figure 7.2. Chemical structure of a deoxyribose (a) and its electric switch property (b) where dashed arrows indicate electric current direction. Positive holes at the nucleotide base are routed to C5' (c) and electrons at the nucleotide base to C3' (d).

Positive holes transport in the reverse direction as indicated by the dashed arrows. Thus the role of a deoxyribose is an electric switch. Depending on whether the nucleotide base is positively charged or negatively charged, the switch connects C1' to C5' or otherwise connects C1' to C3' respectively. The anti-parallel alignment of the pentose sugars on both strands determines that electric current forms a stepwise closed loop between every two adjacent base pairs.

In the closed circuit (Figure 7.1b), charges stored in capacitor C_1 would transport through the phosphate inductors L_1 and L_2 to reach capacitor C_2 under the routing of the deoxyriboses. The anti-parallel alignment of the pentose sugars on both strands determines that positive charges transfer along the strand in the direction of 3'→ 5' while negative charges transfer along the strand in the direction of 5'→ 3'. The electric charges flow through a closed circuit so that by Kirchhoff's loop rule, we have a voltage drop relationship:

$$\frac{Q_1}{C_1} = IR_1 + L_1\frac{dI}{dt} + \frac{Q_2}{C_2} + IR_2 + L_2\frac{dI}{dt}, \tag{7.1}$$

where I is the electric current around the circuit unit, Q_1 and Q_2 are charges stored in base pair capacitances C_1 and C_2 respectively, and R_1 and R_2 are representative resistors along the path. Let

$$R_1 + R_2 = R\,, \tag{7.2}$$

$$L_1 + L_2 = L\,, \tag{7.3}$$

$$\frac{1}{C_1} + \frac{1}{C_2} = \frac{1}{C}\,, \tag{7.4}$$

$$Q_1 + Q_2 = Q_0\,, \tag{7.5}$$

then upon differentiation on both sides, equation (7.1) becomes

$$L\frac{d^2 I}{dt^2} + R\frac{dI}{dt} + \frac{I}{C} = 0\,. \tag{7.6}$$

This second-order differential equation is a damped harmonic oscillator. Depending on the relative values of L, R and C, the system may be over-damped, critically damped, under-damped, or simple harmonic. When $R^2 < 4L/C$, the circuit property falls into the last two categories with a current function of

$$I = I_0 e^{-\frac{R}{2L}t} \cos(\varpi t + \phi)\,, \tag{7.7}$$

where I_0 is the maximum current and ϕ is a phase constant that depend on the initial conditions, and ϖ is the angular velocity of the oscillation with a value of

$$\varpi = \sqrt{\frac{1}{LC} - \frac{R^2}{4L^2}}\,. \tag{7.8}$$

The circuit system transfers electric charges from a base pair to another in the stepwise oscillatory processes, i.e., from capacitor C_1 to C_2 (Figure 7.1b) and then from capacitor C_2 to C_3 (Figure 7.1c) along the double helix. Simple as it is, this model is in line with the experimental evidence [30-37] so far established concerning DNA conductivity and reconciles both super-exchange and multi-step hopping mechanisms.

7.2.2. Circuit Model Discussion and Prediction

It has been reported that the rate of electron transfer within a short distance decreases exponentially with increasing distance [31-34]. Such a phenomenon corresponds to the under-damped condition of the circuit system due to the relatively low ($R^2 < 4L/C$) resistance response of DNA backbones to the artificial introduction of a voltage drop between base pairs in the experiments. However, when R value is considerable, the exponential signal of electric current prevails over the sinusoidal cycle in equation (7.7) and vanishes within a few oscillatory cycles. Since the distance between the base pairs of B-DNA is a fixed value of 3.4Å and each stepwise oscillatory cycle takes a certain interval of time for that distance, we may replace the time variable in equation (7.7) with a distance variable Δr. In considering that electric current is a measure of electron transfer rate k, equation (7.7) is equivalent to the Marcus correlation [40, 41]:

$$k \propto e^{-\beta\Delta r} \tag{7.9}$$

where β values between 0.1 and 1.4 $Å^{-1}$ have been estimated for the double helix [31-34]. The dramatic difference can be ascribed to the uncertainty of resistance R that is sensitive to various experimental conditions. The reason for the presence of considerable R value along the circuit is because the phosphate bridges are under a persistent high voltage drop induced artificially in experiments so that inductance partially manifests as resistance in the circuit.

Based on frequency value of 10^{10} s^{-1} measured by Fukui et al. [31], we simulate the under-damped oscillation with parameters of C=0.02pF, L=0.01μH, and R=100Ω in equation (7.8). Assuming three pairs of free electrons and positive holes are initially generated to trigger charge migration from a base pair to another through the stepwise oscillatory circuits, the current function of equation (7.7) is calculated to be

$$I = 34e^{-0.005t}\cos(0.071t), \tag{7.10}$$

where the electric current is in unit of nA and time in picosecond (ps). Supposing at t = 0, capacitor C_1 carries charges in the polarity as shown in Figure 7.1(b), the deoxyribose switches (S_1 to S_4) will route the charges in the dashed arrow direction. After one oscillatory cycle at t = 90ps, the charges reach capacitor C_2 so that switches S_3 and S_4 change their connections. The charges at capacitor C_2 will then be transferred to capacitor C_3 in the next step (Figure 7.1c). The amplitude of the electric current along the double helix decreases exponentially with each oscillatory cycle in the under-damped situation (Figure 7.3). Because each oscillatory cycle takes 90ps for charges to migrate 3.4Å along DNA strands, the β value in equation (7.9) is 0.13$Å^{-1}$ in this case.

In vivo, we suggest that natural electron transport from a base pair to another should incur trivial electric resistance. Even if there exists certain electric resistance along the strands, the thermal energy produced by resistors would immediately be absorbed by both energy storage components of the base pair capacitors and the phosphate inductors. Hence electric resistance can be neglected. Let C=0.02pF, L=0.05μH, and R=5Ω, the circuits decline into a series of LC oscillators that transfer charges step by step along the strands harmonically at a slower pace. The frequency of the harmonic oscillation is slower than that of under-damped oscillation as can be predicted from ϖ formula under the relatively high value of inductance and trivial resistance. It takes about 200ps for electrons to transfer from a base pair to the next, and the amplitude of the electric current remains almost the same in each oscillatory cycle. The electric current through a DNA molecule in vivo can be expressed as:

$$I = 15e^{-0.00005t} \cos(0.032t). \tag{7.11}$$

A comparison of under-damped oscillation and simple harmonic oscillation can be found in Figure 7.3. Thus, harmonic oscillation, the theme of nature, manifests in the stepwise oscillatory circuits of a DNA molecule.

In the stepwise *LC* circuits, an electric current is defined as positive when a base pair capacitor is being charged and negative when discharged in the next cycle down the chain. The stepwise oscillations agree with the evidence that has supported hopping mechanism through the base π-stack [6-14], but the electric current through the sugar phosphate bridges is more reliable than the haphazard charge migration by hopping across the base rungs. Charges stored in capacitor C_1 transport through both strands to capacitor C_2, and will continue to move toward C_3 in a similar sinusoidal manner but at a lagging phase of 2π in the cycle. During the processes, positive holes move in the direction of 3'→5' on one strand while electrons flow in the direction of 5'→3' on the other strand in good synchronization. The mechanisms for charge transport through long and short distances are the same. It takes more cycles of oscillations for charges to transport through longer sequences of base pairs. Without considering the effect of sequence dependence, traveling time is proportional to distance.

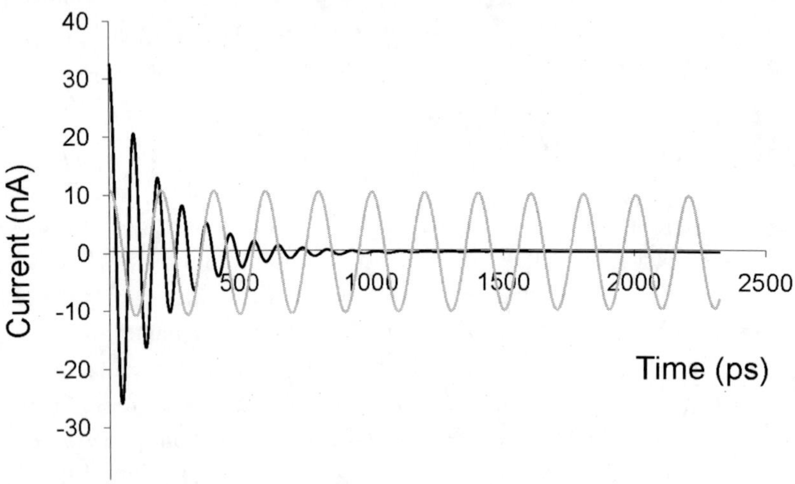

Figure 7.3. Charge transfer of DNA circuits by fast stepwise under-damped oscillations in vitro (black curve) in contrast to slower stepwise simple harmonic oscillations in vivo (gray curve) where the electric current flows 3.4Å in distance during each cycle. Charges undergoing under-damped oscillations have a hasty speed but cannot reach a long distance.

The stepwise oscillatory *LC* circuits determine that a DNA fiber is a special AC wire. The oscillatory frequencies in the range 10^7 to 10^{11} s^{-1} have been reported [42-44]. Single strand turned out to be less conductive than double strands [26], which agrees with the circuitry model prediction because only double strands constitute closed circuits in each grid for proper conduction.

We shall carry a further step to discuss the biological implication of the circuitry. From equation (7.8), we know that ϖ is codetermined by inductance L and capacitance C when R is negligible. Assuming constant inductance for all nucleotides, the harmonic oscillation frequency would be determined by two neighboring capacitance values in the closed circuit, such as C_1 and C_2, which are specific to the base pairs. And this is perhaps the most subtle part of the story for it indicates that charge transfer rate is sequence dependent [32-34, 46]. Because nucleotide bases have ionization potentials in the order of G<A<C<T and electron affinities in the order of C<T<G<A (disregarding the negative sign) [47], it takes the least amount of energy to charge a G:C base pair in the polarity of +G:C- and requires the highest amount of energy to charge a +T:A- capacitor. This means that the capacitances of the base pairs are in the order of +G:C- > +A:T- > +C:G- > +T:A- polarities. Thus there are four distinct capacitance values depending on the base pair and polarity. Since ϖ is determined by two neighboring capacitances in series, it may take 10 possible values ($4^2 - C_4^2$).

It is predicted that charge transport along the strands will produce various frequencies reflecting the identity of the bases. The pattern of charge transport is precisely controlled by the gene sequence. In other words, the gene sequence determines the theme of the electronic music. The +G:C- capacitor has a higher capacitance than the other base pairs and carries a higher amount of charges in experiments whereas +T:A- has a lower capacitance than the other base pairs and is the limiting step or bottleneck in charge transfer along the DNA double strands [35, 41]. However, at long $(+T:A-)_n$ sequences the size of the bottleneck remains the same so that the sequence distance dependence vanishes [41]. When the capacitance of a base pair is low, such as in the case of +T:A-, the angular velocity ϖ tends to be large, so that the current attenuation is large and the base pair becomes a bottleneck of electric current transport. Because +G:C- has a relatively high capacitance, it is easy to trap charges, so it is likely to become the end point of charge transport [35, 41]. In this regard, the circuit model prediction agrees with experimental results completely.

The oscillatory electric current through the double helix is likely to have physiological significance. For example, if a base pair is mismatched at a certain position, then the oscillatory rhythms would be broken. Proofreading enzyme that scans the DNA sequence continually might easily locate the trouble point by the abnormal electric signal [48]. Furthermore, biomolecules are inherently unstable. Only a constant flow of energy prevents them from being disorganized. It stands to reason that the incessant charge vibrations in the genetic substance are vital for living organisms to maintain the integrity of the gene sequence.

From organic chemistry perspective, the base pair is capable of storing a considerable amount of charges in either polarity by at least two conceivable mechanisms. First, both pyrimidine and purine are heterocyclic rings composed of carbon and nitrogen atoms. On the one hand, nitrogen is more electronegative than carbon for attracting higher electron density in covalent bonds with carbon. On the other hand, the nitrogen atom has lone pair electrons to share with carbon under electron deficiency. The combination imparts great flexibility to the nucleotide bases for either holding or releasing electrons. Second, the heterocyclic rings are aromatic chemical substances that tend to generate a diamagnetic ring current. In DNA circuits, the aromatic ring current might reduce the charge saturation of the nitrogenous base through electromagnetic induction, and as a result increase the electric capacitance of the base pair. Both properties enable a base pair to be a good bipolar capacitor. The recognition of a base pair as a valid capacitor provides sharp insight into the electrical property of DNA.

In the stepwise oscillatory circuit, a capacitor must be charged up to a threshold potential for positive charges to overflow through the ether bond of the deoxyribose. And this is mediated by the ether bond shown in Figure 7.4 as an electric switch. The oxygen atom would preferably use a $2p_x$ and a $2p_y$ orbitals to form covalent bonds with carbon atoms in the ether bridge. A full electron octet is a stable configuration of the oxygen atom because eight outer electrons constitute a complete circulating cycle. The electric current through the oxygen atom must comply with the circulation direction of electrons (Figure 7.4). Like a gas-filled tube, once a path is channeled by a positive threshold potential, the ether switch will remain on until the base pair is over-discharged to apply a negative potential of the same magnitude to the ether bond, thereby flipping the oxygen state from Figure 7.4(ON) to 7.4(OFF). To open the channel again requires the base pair to be recharged up to the initial threshold potential to flip the state over again. In the ON state, C1'—O covalent bond is $2p_y$-$2p_y$ overlap while C4'—O covalent bond is $2p_x$-$2p_x$ overlap; in the OFF state, the nature of the covalent bonds is reversed (Figure 7.4). The chirality of C3', C1' and C4' determines that the electric current is unidirectional through the ether bond and the phosphodiester bonds while the base pair is being charged and discharged cyclically. Such an ether bond gateway mechanism ensures that the oscillatory current is in a strict stepwise process.

It has been found that charge migration in DNA is an ion-gated transport depending on the hydrated counter-ions and configurations [49]. This phenomenon indirectly supports our model because any factors that influence the properties of the circuit would modify the efficiency of electron transport in DNA strands, which are especially true in vitro. However, we suggest that in vivo charge migration should be gated by the dependable ether bond instead of by random ions coming from the environment. When the capacitance of a base pair is large, such as in the case of +G:C-, the electric potential, $U=Q/C$, tends to be low, which might be gated by the required threshold potential of the ether bond so that the LC oscillation could not continue the cycle forward stepwise. Precisely for this reason, the +G:C- base pair is more likely than the others to become the end point of DNA charge migration along the strands.

Finally, we wish to present a general molecular model for phosphate in DNA backbone based on the electron circulations within oxygen atoms. The central phosphorus atom is bound by four oxygen atoms whose electron circulations are linked together forming four loops like a coiled wire, each loop involving an electron octet (Figure 7.5). An electric current passing through the molecule inevitably incurs considerable induction due to the convoluted electron flow, which explains why phosphate is an inductor and perhaps why ATP can serve as a physiological energy source when broken into ADP +P_i even though the energy of the phosphoanhydride bond is not so large. The convolution disruption is the key factor in releasing a large amount of inductive energy.

In DNA the electric tension of phosphate chains must equilibrate with the mechanical torsion of the strands. It has been pointed out that the intensity of the torsion may influence the charge transport through the double helix [38, 45], but we attribute this effect to inductance change in the circuits rather than the variation of π-stack overlap for hopping.

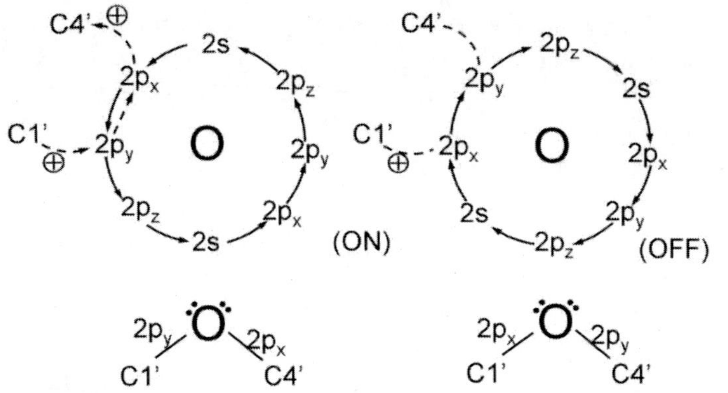

Figure 7.4. Oxygen atomic switch of the ether bond to control the flow of positive charges (dashed arrow) by flipping between two interconvertible states of contrary electron circulation (solid arrows) alignments, along with its chemical structures for on and off at the bottom.

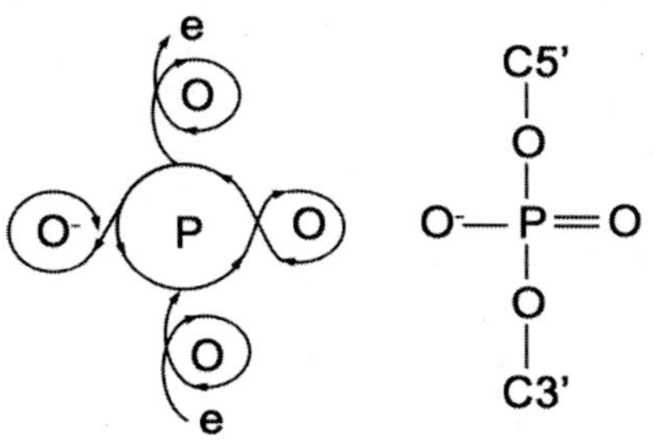

Figure 7.5. Schematic diagram of the electron flow via concatenated electron circulations within a phosphate bridge for electromagnetic induction corresponding to its chemical structure.

To summarize, the electron transport mechanism in DNA is of paramount important to understand the biological function involving the genetic substance. Based on the accurate identification of chemical structures as electronic components, we describe the electron transport mechanism in DNA by a reliable and well-structured oscillatory circuitry that distinguishes a DNA molecule from messy condensed matter. This alternative view is supported by the same established experimental evidence that has supported super-exchange and hopping mechanisms so far. The *LC* oscillatory circuit is robust and classical in physics, yet revolutionary and wonderful in chemistry and biology [50, 51]. This is an interdisciplinary analysis that sparks a general interest in biological physics and may have a potential influence on molecular electronics.

7.3. DNA Wave Functions

The stepwise oscillatory circuits of DNA indicate that the nucleotide sequence is linear in geometric alignment but circular in electrical function. From one nucleotide to another, the electric signals follow the harmonic oscillation principle, which determines that the wave function of each nucleotide transforms by dynamic calculus instead of linear algebra. Upon close examination, we are able to define the mathematical formula of a DNA molecule along with its chemical structure and electronic circuitry.

7.3.1. Harmonic Oscillation between Base Pairs

A DNA double helix consists of two complementary polynucleotide strands wound around each other and crossed by hydrogen bonds. The nucleotides are joined into polynucleotide strands by phosphodiester bond that connects the 3' carbon atom of one deoxyribose sugar to the 5' carbon of the next. Though the polynucleotide seems to be linearly arranged and looks monotonous in structure, DNA has circular elegance and beauty unsurpassed by any other macromolecules.

There are only four kinds of nitrogenous bases in DNA forming two pairs, i.e., guanine (G) matches with cytosine (C), and adenine (A) matches with thymine (T) through hydrogen bonds as were discovered by Watson and Crick. In each pair, a purine base with two rings matches with a pyrimidine base with one ring. If we assign two waveforms to the former, and one waveform to the latter mathematically, then these three waveforms must be coupled together harmonically in a certain way because their mating is so unique and unassailable. We interpret the relationship of these three waveforms as a rotation pathway because it allows alternating current to pass through the hydrogen bonds. Let spherical quantities Γ_{00} and Γ_{01} represent two heteroaromatic rings in adenine, and Γ_{10} represent the ring of thymine tentatively:

$$\begin{pmatrix} \Gamma_{00} \\ \Gamma_{01} \\ \Gamma_{10} \end{pmatrix} = D\begin{pmatrix} \cos\Theta\cos\theta \\ -\dot{\Theta}\sin\Theta\cos\theta \\ -w\sin\Theta\sin\theta \end{pmatrix} \tag{7.12}$$

where D is a dimension factor, w is a velocity dimension, and functions Θ and θ are two radian angles as were analogously described in Chapter 5. There are differential and integral relationships between the three waveforms:

$$-\frac{\partial\Gamma_{00}}{\partial t} = \Gamma_{01}, \tag{7.13}$$

$$\int \Gamma_{01} dl = \Gamma_{10}. \tag{7.14}$$

When describing electronic orbitals, these two equations imply a rotation of one electron into another. Here these similar formulae are used to explain the interaction between a Watson-Crick base pair. Waveforms Γ_{00}, Γ_{01}, and Γ_{10} constitute a rotation pathway that satisfies Faraday's law, which indicates harmonic oscillation between the base pair. The wave function interpretation complies with the DNA oscillatory circuitry.

In a dimension diagram (Figure 7.6), a spherical quantity $D_1 D_2 \cos\Theta \cos\theta$ undergoes a rotatory operation. As radian angle Θ rotates from 0 to π/2, wave function $D_1 \cos\Theta$ transforms into $-D_1\dot{\Theta}\sin\Theta$; and as θ rotates from π/2 to 0, wave function $D_2 \cos\theta$ transforms into $-D_2 \frac{1}{\theta'}\sin\theta$. We may express this rotation by three of the four chords concisely as $\cos\Theta - \dot{\Theta}\sin\Theta .... -\frac{1}{\theta'}\sin\theta$ with the dotted line indicating hydrogen bonds. If we designate this rotation in a DNA base pair as A:T, then the T:A base pair would be $-\dot{\Theta}\sin\Theta .... -\frac{1}{\theta'}\sin\theta + \cos\theta$.

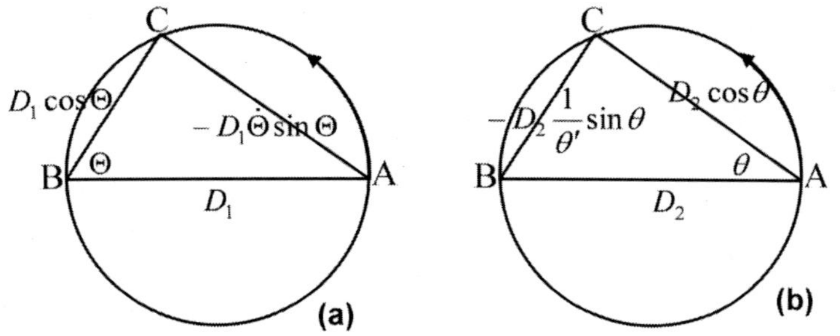

Figure 7.6. Dimension diagram of a rotatory operation to represent the sinusoidal waves of alternating current between the A:T base pair. D_1 and D_2 are dimension factors.

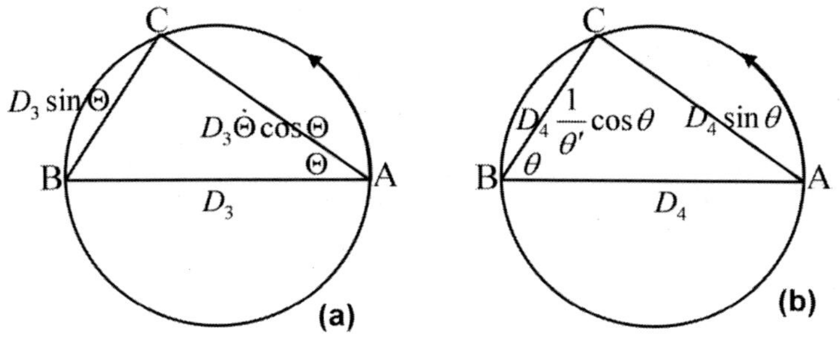

Figure 7.7. Dimension diagram of a rotatory operation to represent the sinusoidal waves of alternating current between the G:C base pair. D_3 and D_4 are dimension factors.

In the same manner, the rotation of a spherical quantity $D_3 D_4 \sin\Theta\sin\theta$ is displayed in Figure 7.7. As radian angle Θ rotates from π/2 to 0, wave function $D_3 \sin\Theta$ transforms into $D_3\dot{\Theta}\cos\Theta$; and as radian angle θ rotates from 0 to π/2, wave function $D_4 \sin\theta$ transforms into $D_4 \frac{1}{\theta'}\cos\theta$ (Figure 7.7). We may express this rotation by three of the four chords simply as $\sin\Theta + \dot{\Theta}\cos\Theta .... \frac{1}{\theta'}\cos\theta$. If we designate this rotation in a DNA base pair as

G:C, then the C:G base pair would be $\dot{\Theta}\cos\Theta....\frac{1}{\theta'}\cos\theta+\sin\theta$. In this manner, we may associate the waveforms with the rings of nitrogenous bases (Figure 7.8).

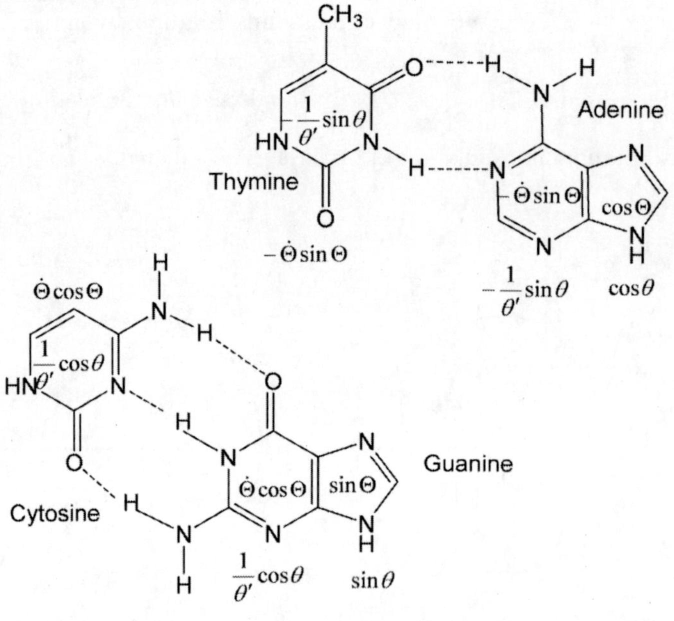

Figure 7.8. Watson-Crick base pairs A:T and G:C with waveforms in heteroaromatic rings for proper rotatory operations. The waveforms labeled outside the rings are for T:A and C:G pairs.

The communication of wave signals between each base pair is through hydrogen bonds, which are implemented by the chain rules of dynamic calculus:

$$\begin{aligned} w &= -\frac{1}{\theta'}\cdot\dot{\Theta} = \frac{dl}{d\theta}\cdot\left(-\frac{d\Theta}{dt}\right) \\ &= \int A_2\cos\psi\ dl\cdot\left(-\frac{dA_1\cos\Psi}{dt}\right) \\ &= \Phi^0 \\ &= u\Phi \\ &= \Omega^0\Phi, \end{aligned} \tag{7.15}$$

where Φ denotes a 2s2p electron within the nitrogen or oxygen atom involved in each hydrogen bond, Ω denotes a 1s electron passing through each adjoining hydrogen atom of the hydrogen bond, and the equal signs indicate the progress of electric pulses between the base pair. Thus the wave signal burrows through the base pair by differential and integral chain rules. The wave signal becomes alternating current passing through the hydrogen bonds. A base pair is an electric capacitor. There are oscillating waves between each pair of nitrogenous bases within DNA nucleotides. Hydrogen bonding is a dynamic electronic action that delivers messages between the two bases, helping to balance the strain in the double helix. Hydrogen bonds serve as the narrowest gateways in delivering messages. In this way, we describe the hydrogen bonds as a communication portal between the pair of bases. This dynamic interpretation is in stark contrast to the traditional static model of hydrogen bonds. Since the expression of any force requires message exchanges, it is the wave information conveyed between the base pairs that makes hydrogen bonds strong enough to hold the double helix together. Without the wave message, the atoms on both sides of hydrogen bonds would be detached and the threads of DNA would start to unravel.

It is interesting to note that the formulae of electronic orbitals can be customized to describe DNA molecular wave functions. Such mathematical uniformity for diverse natural phenomena reflects the universality of the spherical view that has rich connotative meaning as well as general validity.

7.3.2. Mathematical Formula of DNA

We have described the base pairs as electric capacitors and assigned waveforms to the rings in the bases. We have also described the phosphate bridges as electric inductors. In order to express the entire DNA molecule mathematically, we need to express the phosphate bridges in dynamic calculus. As shown in Figure 7.9, we treat both strands of the double helix as time and space components of a DNA molecule. Variables of Z series represent time while variables of ζ series represent the space strand. Both are intertwined together to form a double helix. Here we emphasize the electronic components of base pairs and phosphate bridges disregarding any deoxyribose because the sugar is an electric router only. The base pairs are expressed by trigonometry and the phosphate bridges by dynamic calculus. Both gadgets cooperatively characterize a DNA molecule in an elegant mathematical form as we are endowed with two legs to walk. Please also refer to section 1.3.3 for the description of duality in the *LC* circuit.

A characteristic of an *LC* circuit is electromagnetic oscillation. The charge oscillates back and forth from one plate of the capacitor to the other, and repeats this continuously. Likewise, the current oscillates back and forth as well. When the charge Q reaches a maximum, the energy is all stored in the electric field of the capacitor; but when the charge Q declines to zero, the current is a maximum and all the energy is stored in the magnetic field of the inductor. Thus, the energy oscillates between being stored in the electric field of the capacitor and in the magnetic field of the inductor. The mathematical expression of a DNA molecule must capture this theme of harmonic oscillation, and circular functions in dynamic calculus do just that.

A DNA molecule is a stepwise *LC* oscillatory circuitry. After alternating current passes through a base pair via hydrogen bonds, the electric wave reflects back into the phosphate strands to initiate the next cycle in the stepwise oscillatory circuits. The phosphate bridges in both strands undergo a rotatory operation. The time strand transforms by dynamic differentiation while the space strand by dynamic integration. The rotatory operation produces electric tension on both strands. From a base pair to another, the electric signal proceeds along a closed circuit sinusoidally. The mathematical formula of DNA complies with the *LC* circuitry.

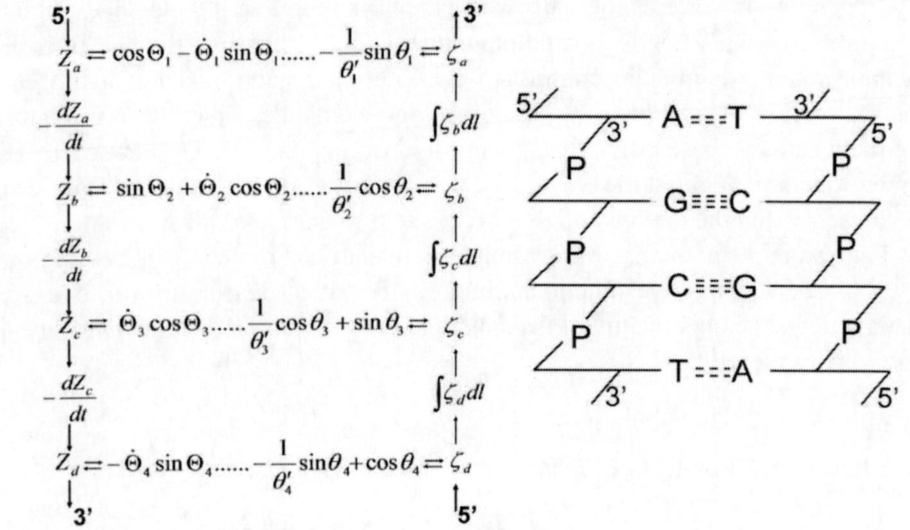

Figure 7.9. The mathematical formula of a DNA molecule along with its abbreviated chemical structure. The arrows indicate the electron flow in rotatory operations, contrary to the electric current direction. Radian angle Θ_i complements radian angle θ_i .

The discovery of DNA oscillatory circuits also reveals an inherent mechanism of genetic mutations. The environmental stress may cause the harmonic oscillation of electron transport out of phase along both strands by affecting the parameters of the LC circuits. The existence of inopportune charges in a nitrogenous base may induce a tautomeric shift (i.e., from keto to enol, or from amino to imino) in the base leading to rare unorthodox pairings. The biologically important unstable tautomers involve keto-enol pairs for thymine and guanine, and amino-imino pairs for cytosine and adenine. The tautomeric shift may result in a transition mutation during DNA replication. In transition mutation, either a pyrimidine is replaced by another pyrimidine, or a purine by another purine in the gene. We ascribe this effect to the congenial spacetime induction for spherical quantity transformation under a specific circumstance rather than a random event. By taking part in inducing the base pair mutation, the environment plays an important role in directing the evolution of species (cf. Section 8.8).

Conclusion

A DNA molecule is insightfully characterized by a stepwise oscillatory circuitry where every base pair is a capacitor, every phosphate bridge is an inductor, and every deoxyribose is a charge router. The circuitry accounts for DNA electric conductivity through both short and long distances in good agreement with experimental evidence that has led to the so-called super-exchange and multiple-step hopping mechanisms to date. However, in contrast to the haphazard hopping and super-exchanging events, the circuitry is a well-defined charge transport mechanism reflecting the great reliability of the genetic substance in delivering electrons. Stepwise oscillatory charge transport through the nucleotide sequence that regulates its frequency along must contain great biological significance.

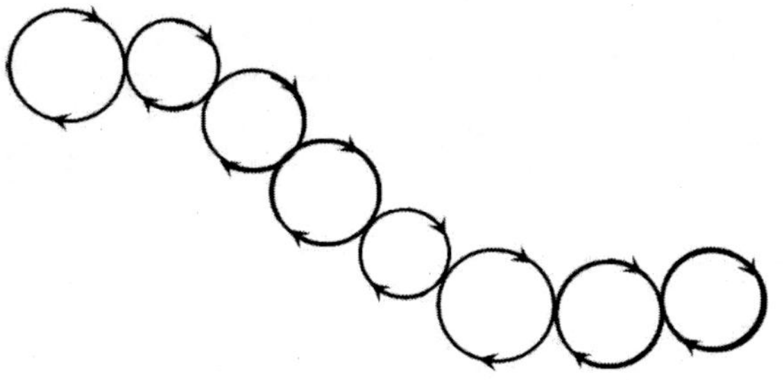

Figure 7.10. Abstraction of electric behavior of a DNA molecule as a stepwise oscillatory circuitry where each ring represents an *LC* oscillatory cycle (i.e., an epitome of Figure 7.1).

Stepwise oscillatory circuits of a DNA molecule indicate that the nucleotide sequence is functioning in a stepwise circular way (Figure 7.10). This permits us to define a DNA molecule by circular functions as well as spherical quantities in dynamic calculus. Three trigonometric waveforms are assigned to each base pair. They constitute a rotatory operation pathway with hydrogen bonds serving as the gateways to deliver electric pulses. Spherical quantities in dynamic differentiation and integration are assigned to phosphate bridges in both time and space strands respectively. Time and space strands are bonded together by hydrogen bonds. The time and space distinction of DNA double strands might contain biological significance that has not yet been discovered (cf. Equation 8.11). For the first time, we have developed the mathematical tools for the characterization of a DNA molecule, parallel to electronic circuits and chemical formulae.

Chapter 8

The Principle of Cells and Life

We have described the organization of various electronic orbitals within atoms in Part I. The order of electrons within the atoms provides the pattern for the order in which a biological cell should arrange its organelles. This is viable because the law of nature is single. In other words, the law that governs the cells is the same as the law that governs the atoms. It is of utmost important to examine the structure of the cells carefully if we are to characterize the various aspects of life phenomena incisively. In this chapter, we demonstrate that mathematical models for atomic shells may be applied to the characterization of biological cells. A helium shell is an electronic rope that may serve as a pattern for other duality systems in nature. The configuration of 1s2s2p electrons in a carbon atom may serve as a pattern for various biochemical reactions within a cell; and the complementary relationship between $3d^{1s}$ and 1s electrons due to an orientational shift within a zinc atom provides a clue for the relationship of endosymbiotic organelles within a cell. A DNA double helix is a stepwise oscillatory circuitry, so is the human lineage. More fundamentally, if dynamic calculus inherent in electronic motion is followed by life phenomena, then the probability density of biological variables may be characterized by circular functions; the mind and body relationship problem can be resolved mathematically by synchronized differential and integral operations; a human body can be regarded as a multiple ring gyroscope always created and governed by the law of nature; and the evolutionary steps leading from low to advanced life forms should observe the same principle of spherical quantities in dynamic calculus.

8.1. Introduction

The elucidation of various atomic structures by spherical quantities provides various models for elucidating the life phenomena because the world is created by the same one natural law. The duality of electrons in the helium shell may serve as a pattern for any dual systems. The carbon model with four active outer electrons may serve as a pattern for gene expression in a cell. The zinc model with complementary electrons may serve as a pattern for endosymbiotic organelles in a cell. The DNA circuitry with stepwise harmonic oscillations may serve as a pattern for the human lineage from one generation to the next. The uniform of wave functions in spherical quantities may serve as a pattern for biological variables. The rotatory operations that characterize electronic orbitals so well may also be used to describe the mind and body relationship in life. The open direction of exponential series with increasing dimensions in the Poisson distribution may point to the direction of increasing biological complexities in evolution forever. The law of nature implemented by dynamic calculus may be used to characterize the human body and will usher an age to break the dawn of calculus biology. If Darwin's evolution unifies all kingdoms of lives, then our concrete dynamic calculus may govern all natural entities, biotic and abiotic.

8.2. The Helium Model for Rope Structure

A helium shell contains both 1s electrons in cyclic transformation: $\Omega_0^0 = \Omega_2$; $\Omega_2^0 = \Omega_0$. Such a mutual rotation relationship was characterized by rope structure in section 5.2.4. The helium shell provides a model for many other systems with dual elements (Table 8.1). For example, light is one of the simplest ropes in nature. Light consists of electric and magnetic fields that interact with each other forming an electromagnetic wave. Electronic motion is closely associated with light. As was explained in Chapter 3, the physical interaction between electrons within an atom is through emitting and receiving virtual photons that observe dynamic calculus. A virtual photon implements a rotatory operation because it carries a time dimension away and brings a space dimension back between two electrons. Cyclic virtual photon transfers make electrons within an inert atom into a rope. In electromagnetic waves, the electric field and the magnetic field are orthogonal and traveling in the same direction; within a helium atom, both electrons are orthogonal and rotating in the opposite directions and manifest as a pair of electromagnetic waves propagating from the nucleus center to the whole sphere at the same phase. They all observe Maxwell's equations, but the former is linear and unbounded whereas the latter is spherical and confined within the atomic shell.

Table 8.1. Rope structure in life phenomena

Scope and entity	Two strands		Interaction or specialization
General sense	Space	Time	Two symmetric dimensions
Light	Electric field	Magnetic Field	Electromagnetic wave
The helium shell	Ω_2	Ω_0	Rotatory operation
DNA	A	T	Hydrogen bonds between nitrogenous bases
	G	C	
Zygote	Ovum	Sperm	Fertilization
Family	Wife	Husband	Marriage and sex
Life	Plant	Animal	Photosynthesis versus Aerobic respiration
The earth's outer stratum	Abiotic	Biotic	Disorder process versus order process

Atoms are the building blocks of various organic molecules such as nucleotides that form DNA. If we regard each atom such as nitrogen and carbon as possessing a space dimension, then macromolecules have multiple dimensions of space, but they are less stable than a neon atom. Thus, from atoms to nucleotides to DNA, the space dimension increases while the time dimension decreases in the biochemical construction. A nucleotide is a spherical layer built with atoms; and a DNA molecule is another outer spherical layer built with nucleotides. In a DNA molecule, both strands of polynucleotide are twisted together and bound through hydrogen bonds between their nitrogenous bases: adenine pairs with thymine and cytosine with guanine. As was formulated in Chapter 7, there is a rotation relationship between each pair of bases.

From a DNA molecule to a cell and to an organism, the space dimension increases while the time dimension decreases. A DNA molecule is more stable than an organism in the sense that an organism is short-lived while DNA, as inheritance substance, passes down from generation to generation, maintaining remarkable stability over a long history. An organism develops from a single zygote that possesses all genetic information to produce the entire body. A zygote results from the fertilization of an ovum with a sperm. There is a symmetric relationship between a sperm and an ovum, which may be mathematically represented by circular complex functions such as Ψ and ψ respectively (see equations 1.2 and 1.5) in two-dimensional spacetime or Θ and θ respectively (see equations 5.43 and 5.44) in four-dimensional setting. A sperm denotes time dimensions while an ovum denotes space dimensions, both forming a rope. A zygote is a rope with great vitality.

The union of a sperm and an ovum requires impregnation that involves sexual intercourse and other human social behavior. Hence we ascend a ladder from an individual to a society. A husband and a wife play the roles of time and space in the family, like two electrons in a helium shell. It takes a village to raise a kid, and families are the basic units of the society. This cultural orientation is similar to the atomic model where each pair of different spin electrons occupy the same type of atomic orbitals (see Figure 2.10).

A human society is only a part of an ecological community in an ecosystem. Evolution creates various lives to colonize various niches on the earth. These include animals and plants that interact with each other. On the surface of the earth, solar energy is captured in organic matter by plants and free oxygen is released as a result of photosynthesis; and animals derive their energy by digesting organic compounds using oxygen as the ultimate oxidant in respiration (equation 8.1). There are a diversity of animal species besides Homo sapiens and a variety of plants including crops. The interaction between these two kingdoms is critical to their survival such as procuring food, fertilization, and pollination. Thus, we may likewise regard animals and plants as forming two functional strands of a twisted rope.

$$CO_2 + H_2O = CH_2O + O_2 \tag{8.1}$$

Furthermore, the earth outer stratum, featuring biological diversity and abundance, is the most sophisticated layer on the planet. Here biosphere and abiotic sphere also constitute two strands of a rope. Biosphere refers to the full array of life, including animals, plants, fungi, prokaryotes, and protoctists whereas abiotic sphere includes lithosphere, hydrosphere, atmosphere, etc. While biological activities transform elements into ordered patterns, abiotic processes go on spontaneously leading the environment toward disorder characterized by entropy increment. Entropy is an indicator of disorder, which either stays constant (reversible processes) or increases (irreversible processes) in an isolated system according to the second thermodynamic law. However, the real world system is always an open system rather than an isolated one. Disorder has various criteria or meanings from different perspectives. For example, on the surface of the earth, wild grasses are elements of disorder because they are creating parts of heterogeneity in the soil, but from the perspective of organisms, the biological bodies are in good order because the molecules are lined up in accordance with genetic information. Thus, the definition of disorder is not an absolute concept. The meaning is relative depending on the perspective and scope of view. Most people don't really appreciate the contents of relativity, but they think that nothing on earth is absolute. This plain impression turns out to be sounder than what relativity actually means. As James Lovelock contemplated whether there is life on Mars and how life can be recognized, he proposed the Gaia hypothesis [52], which is, in brief, that the earth is itself a form of life in the sense that it reduces or reverses the entropy in a planet's system. He claimed that the earth's biosphere, tightly coupled with atmosphere, ocean, and soil, constitutes a feedback or cybernetic system that seeks an optimal physical and chemical environment for life on this planet. Thus, the globe features alternate elements of order and disorder from living organisms to their abiotic environment, and to the Gaia.

Finally, the earth rotates about itself, interlaces with the moon, and revolves around the sun that travels in the Milky Way. How stars form galaxies and move in the universe is still a mystery, but they obviously emit light and rays that travel a long distance to convey messages that could eventually change the behavior of electrons within atoms. As was described in Chapter 4, when an object travels at the speed of light, one of its space dimensions will transform into a time dimension. Hence light realizes space to time dimension transformation, which is equivalent to a clockwise rotation.

As entities rotate counterclockwise, time dimensions decrease while space dimensions increase. From light, electrons, atoms, nucleotides, DNA, cells, humans, animals, biosphere, the earth outer stratum, the earth, the solar system, galaxies, to the universe brimming with light, the stability of the entities decreases as complexity increases. Light joins both ends. It is the infinite enlarging space in the universe that transforms into the eternal time. A space dimension of the universe becomes a time dimension of light, driving the undulation of the receiving particles such as electrons within an atom. This is a large cycle where life plays a key role in the middle rings. From an individual to a population, and from a population to a single species, life features permutations of alternate entities from one to many and many as a whole. The world is continuous in spacetime. Figure 8.1 draws the structure of a large ring that contains multiple rings.

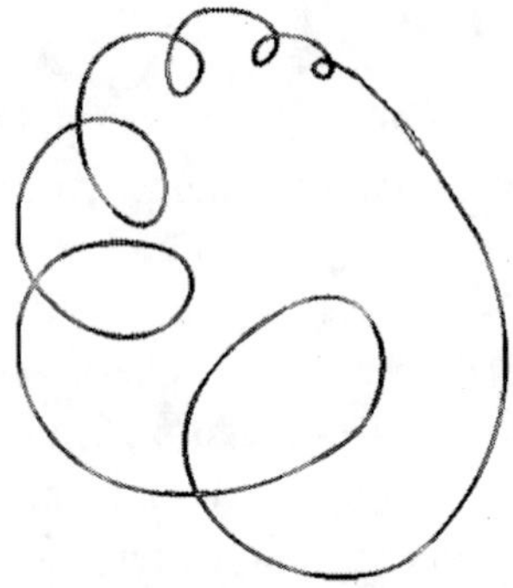

Figure 8.1. The continuity of the world. How many loops are shown in the sketch? Which loop represents properly the whole universe?

Table 8.1 summarizes a variety of ropes made of space and time strands. Space and time dimensions may evolve into two complementary counterparts in the molecular structure, chemical reaction, biological function, etc. As we listed various structures as ropes, one may raise the question why the world pattern must be a rope? The secret lies in the unique property that a rope is able to endure more strength than its two separate strands combined, i.e., "one plus one is bigger than two." Like the win-win situations of a bilateral trade, the harmonic interplay of two strands of a rope is so wonderful. From the perspective of aesthetics, a rope is extremely beautiful besides feasibility. Knitting a rope is the theme of culture as well as nature. No wonder a pithy Chinese idiom runs: "Solidarity as a rope."

8.3. The Carbon Model for Gene Expression

According to the expression of genetic information, biochemical activities within a cell at least include DNA replication, DNA transcription, RNA translation, enzymatic catalyses, and ATP cycle to provide nutrient and energy for the entire cell. These processes correspond to various electronic orbitals in a carbon atom (Figure 8.2). DNA replication keeps the genetic substance as a double helix that resides at the nucleus of a cell, corresponding to the rotation of both 1s electrons in a carbon atom; the other four processes constitute four loops that manifest life phenomena surrounding the nucleus, corresponding to the rotation of $2p_x$, $2p_y$, $2p_z$, and 2s electrons in a carbon atom respectively. We discuss these similarities as follows.

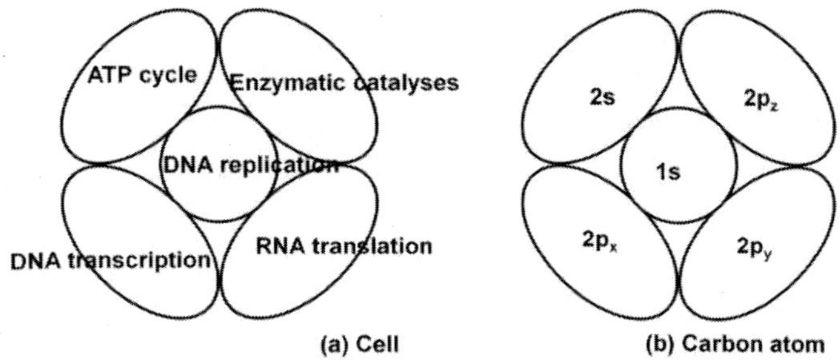

Figure 8.2. Biochemical cycles within a cell in comparison with the electronic orbital cycles within a carbon atom (cf. Figure 11.6). A carbon atom normally uses its $2p_x$, $2p_y$, $2p_z$ and 2s electronic orbitals to form four covalent bonds with other atoms.

8.3.1. DNA Replication (1s to 1s)

All living organisms must possess rapid and accurate DNA synthesis and DNA repair mechanisms. DNA replication must occur before every cell division. The mechanism by which DNA copies are produced is similar in all living organisms. After the two strands of polynucleotide separate, each subsequently serves as a template for the synthesis of a complementary strand. A DNA double helix is composed of two complementary polynucleotide strands like a helium shell composed of two complementary 1s electrons. Because a DNA molecule may be characterized by stepwise oscillatory circuits, the relationship between each base pair observes dynamic calculus (Figure 7.8).

DNA is at the core of life. The relationship between DNA replication and gene expression processes is comparable with that between 1s electrons and 2s2p electrons in a carbon atom (Figure 8.2). Like the inner electrons driving the rotations of the outer electrons, DNA is at the headquarter controlling the construction of the whole cell. DNA replication is not independent of other biochemical processes. On the one hand, DNA replication requires the input of four kinds of nucleoside triphosphate (ATP, TTP, GTP, and CTP) that come from metabolic activities. These required raw materials as well as ATP energy for the synthesis of new polynucleotide strands are represented in Figure 8.2 as ATP cycle. On the other hand, DNA replication produces a DNA double helix, which upon unwinding provides the necessary polynucleotide template for DNA transcription. DNA replication preserves the genetic information for the proper function of the entire cell. Thus, DNA replication is closely coupled with ATP cycle and gene expression processes.

8.3.2. DNA Transcription (2s to $2p_x$)

The ability of living organisms to grow and function properly depends on the precise expression of the genetic information. This information is encoded in DNA, which consists of a linear sequence of deoxyribonucleotides. In the first step in gene expression, information present in one of the two strands of DNA (the template strand) is transferred into an RNA complement through the process of transcription. This RNA complement is called messenger RNA (mRNA) because it carries the message from DNA and is associated with ribosomes for further translation later on.

Modern biochemistry has established that transcription from DNA template strand to mRNA is based on nitrogenous base complements, i.e., C to G, G to C, A to U, and T to A parities. DNA transcripts from the template strand to mRNA base by base. As was shown in Figure 7.9, although DNA looks like a linear stream of nucleotides, the nucleotides are functionally connected through dynamic calculus instead of simple addition mathematically. The coupling of DNA and mRNA during transcription is like the coupling of two cogwheels in motion (Figure 8.3). We have described such a relationship by the differential chain rule in Figures 2.9 and 4.3. The unfolding of a deoxyribonucleotide in DNA involves a differential operation while the elongation of a ribonucleotide in mRNA follows the differential chain. The chain rule of dynamic calculus is observed between both nucleotide sequences as long as the transcription proceeds. The intimate coupling by dynamic calculus ensures the fidelity of message delivery.

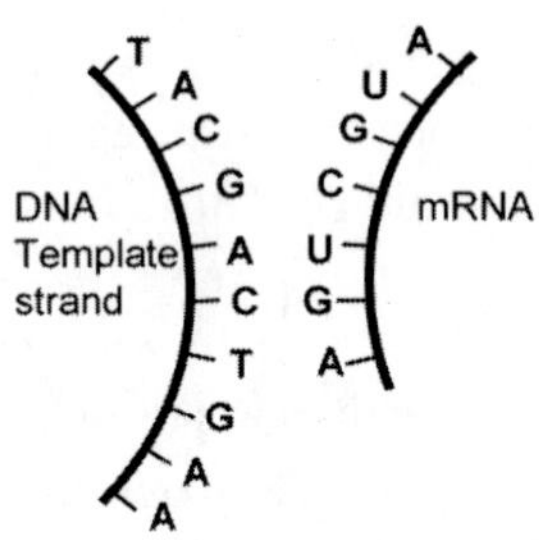

Figure 8.3. Transcription of DNA to mRNA based on the complements of nitrogenous bases. Note that both strands are coupled in a manner like two cogwheels in mesh.

It is also known that some RNA-containing viruses, called retroviruses, may replicate by reverse transcription. Their RNA serves as a template for the synthesis of a complementary DNA strand. In Chapter 5, we demonstrated that the rotation of 1s electrons in a helium shell drives the rotation of 2s2p electrons in a carbon shell, but in a zinc atom, the rotation of $3d^{2s2p}$ electrons drives the rotation of $3d^{1s}$ electrons, which are complementary to 1s electrons. It is interesting to find that DNA may transcript mRNA and some RNAs may reversely transcript DNA, the initiative depending on the circumstances. The virus wakes from dormancy when the condition for its rotatory operation has been met.

8.3.3. RNA Translation ($2p_x$ to $2p_y$)

Once genetic information has been transferred from DNA to mRNA, it is ready to be translated into a polypeptide chain, the end product of most gene expression. Biochemical experiments indicate that a gene's nucleotide sequence specifies in a collinear manner the sequence of amino acids in a polypeptide chain. The ribonucleotide sequence in mRNA is divided into groups of three ribonucleotides, each group, called a codon, specifying one amino acid. The code is thus a triplet. During translation, transfer RNA (tRNA) serves as the adaptor molecule between an mRNA triplet and the appropriate amino acid. Under the direction of mRNA in association with ribosomes, the translation ultimately converts the sequence of codons in mRNA into corresponding sequence of amino acids making up the polypeptide.

Current knowledge of the characteristics of the genetic code is obtained by experimentation. Even though we now know which codon corresponds to which amino acid, the mechanism of their correspondence is hidden in tRNA. The mechanism could be biochemical or electromagnetic, which remains largely unknown to date. However, this magic tRNA relates each codon to an amino acid, translating the mRNA sequence into amino acid sequence in great fidelity. Here we may characterize the coupling of mRNA sequence and amino acid sequence during translation as two cogwheels in mesh. Mathematically, two cogwheels in mesh have been characterized as observing the chain rule of dynamic calculus. A codon is input by a differential operation upon a spherical quantity while an amino acid is output following the differential chain rule. The input of codons drives the production of a polypeptide chain. Like DNA transcription, RNA translation is a process observing dynamic calculus.

8.3.4. Enzymatic Catalyses ($2p_y$ to $2p_z$)

The proteins produced by RNA translation may serve as enzymes to catalyze various biochemical reactions as well as structural materials of a cell. Life is inconceivable without enzymes. Enzymes are enormously powerful catalysts exhibiting high specificity. Each type of enzyme has a unique, intricately shaped binding surface called an active site. An enzyme-catalyzed biochemical reaction achieves significant reaction rate because the active site of the enzyme possesses a structure that is uniquely suited to promote the catalysis. Because of the specificity of an enzyme for a biochemical reaction, we may characterize the coupling of the enzyme and the biochemical reaction as two cogwheels in mesh. The catalysis of an enzyme drives a biochemical reaction. Thus, we may treat the catalysis as a differential operation upon a spherical quantity and the biochemical reaction as a differential chain following the operation. In this manner, the enzymatic catalysis observes dynamic calculus as well. This process corresponds to the rotation of an electronic orbital from $2p_y$ to $2p_z$ that produces necessary chirality to facilitate biochemical reactions around the cell.

8.3.5. Metabolism and ATP Cycle ($2p_z$ to 2s)

Metabolism, the sum of all the enzyme-catalyzed reactions in a living organism, is a dynamic, coordinated activity. Many of these reactions are organized into pathways. There are two major types of biochemical pathways: anabolic and catabolic. In catabolic pathways, large complex molecules are degraded into smaller, simpler products. In the meantime, some of them release free energy by converting ADP into ATP and $NADP^+$ into NADPH. In anabolic pathways, large complex molecules such as polysaccharides and proteins are synthesized from smaller precursors such as sugars and amino acids. The energy and reducing power generated in catabolic pathways are consumed during the biosyntheses. For example, as mentioned above, DNA replication and transcription require the input of free energy. Like 2s orbitals permeating the entire atomic sphere, ADP-ATP cycle is omnipresent within the organism. Metabolism is closely coupled with ADP-ATP and $NADP^+$-NADPH cycles.

To summarize, we have identified five levels of interrelated processes in a biological cell that correspond to $1s2s2p_x2p_y2p_z$ electrons within a carbon atom. DNA replication produces a double helix that resides in the nucleus; DNA transcription produces mRNA that carries the genetic information from the nucleus toward ribosomes, serving as a polor; RNA translation produces proteins that scatter over the cell; enzymatic catalyses dominate various biochemical reactions; and ATP cycle maintains the fine balance and mobility of free energy currency within the whole cell. The dimensions of these activities in a cell correspond to those of 1s, $2p_x$, $2p_y$, $2p_z$, and 2s electrons respectively within the sphere of a carbon atom. Like a tetravalent carbon atom unfolding its chiral wing (section 6.2.1), life is the gradual and continuous expression of the gene. Thus a cell, like a carbon atom, has a four-dimensional chiral structure in its gene expression and physiology. They all proceed by strict dynamic calculus of spherical quantities, the law of nature.

8.4. The Zinc Model for Endosymbiotic Organelles

The degree of mathematical application is a reliable indicator of the maturity of a scientific discipline. Although biology has been developed since antiquity, it remains a young discipline in terms of mathematical involvement. In spite of the widespread biological variable quantification and data analyses nowadays, neither biological statistics nor linear computational models have captured the fundamental aspect of the cells. After the discovery of electronic orbitals within a zinc atom, here we have confidence to address the phenomenon of endosymbiosis in a fundamental way.

As was described in section 5.3.5, the relationship between $3d^{1s}$ and 1s electrons is complementary in geometry. We assume that the relationship between $3d^{2s2p}$ and 2s2p electrons is complementary as well. As outer electrons of a calcium atom, 4s electrons constitute a membrane enclosing the 2s2p shell. In a zinc atom, 3d electrons that attach to this membrane are extroverted as if they came from an extrinsic neon atom engulfed by a calcium atom. We shall not discuss whether it is possible for a calcium atom to fuse with a neon atom under any conditions. The point here is not about the genesis of atoms, but about the complementary relationship between the extroverted 3d electrons and the introverted 1s2s2p electrons (Figure 8.4). Thanks to the extroverted property of 3d electrons, multiple zinc atoms may form a piece of metal with high electric conductivity.

If the law that governs the electronic orbitals is similar to that controls the functions of organelles within a cell, then we may attempt to establish a comparison between the structures of a zinc atom and a eukaryotic cell. According to the endosymbiotic hypothesis by Lynn Margulis [53], a mitochondrium evolves from an aerobic bacterium ingested by a larger anaerobic cell. In exchange for benefits such as protection and a constant nutrient supply, the smaller cell provided its host with energy generated by aerobic respiration. Thus the aerobic bacterium lived inside the larger cell in a harmonious way and gradually became an integral part of it. The striking evidence is that a mitochondrium performs complementary function to the functions of the other parts of the cell. Thanks to the extroverted property of mitochondria and other organelles, many eukaryotic cells may join together to form a multicellular organism.

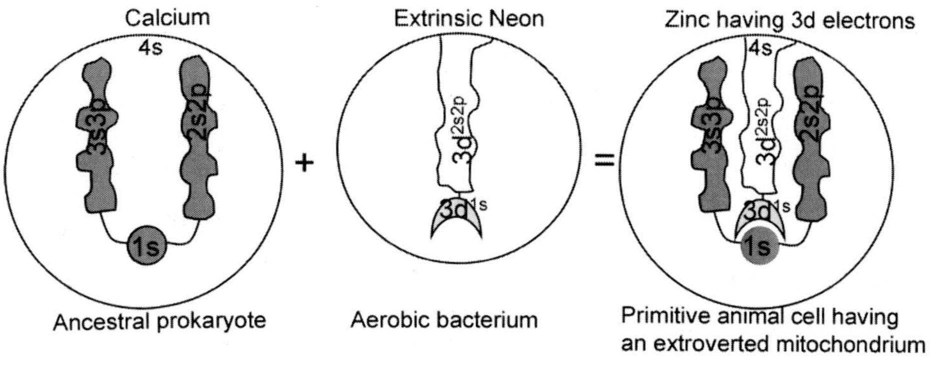

Figure 8.4. The endosymbiotic hypothesis of a mitochondrium evolved from an aerobic bacterium ingested by a larger anaerobic organism, analogous to the complementary relationship between extroverted $3d^{1s2s2p}$ electrons and introverted 1s2s2p electrons in a zinc atom.

The point here is not about whether the endosymbiotic hypothesis is correct or not, but about the similarity between a primitive animal cell having an extroverted mitochondrium and a zinc atom having $3d^{1s}$ electrons. The complementary relationship between $3d^{1s}$ and 1s electrons due to a standpoint shift has been defined mathematically in equations (5.77) to (5.80); and the complementary relationship between an aerobic mitochondrium and the anaerobic host due to the organelle orientation shift follows the same logic. Both are similarly explained by spherical quantities in dynamic calculus. The natural law that governs the atomic structure is indeed the same as that that governs the cellular structure.

We may carry a further step to discuss the functional structure of a human body in a similar way. Here the function of mitochondria corresponds to the function of the lungs that are responsible for aerobic respiration. There is a complementary relationship between the lungs and the heart in a human body (cf. Review question 8.9 in Appendix I). The heart drives blood circulation around the body while the lungs inhale oxygen into the blood and exhale carbon dioxide out of the body. The extroverted lungs and introverted heart are coupled together to realize gas exchanges between the entire body and the environment. We certainly don't suggest that the lungs should come from outside the body abruptly. However, if we regard a human body as a cell, then the origin of the human organs can be traced back to cellular organelles. Like a cell consisting of nucleus, cytoplasm, and various organelles surrounded by a semipermeable cell membrane, a human body consists of a heart and a brain, muscle and fluid, and various organs and structures enclosed and protected by the skin. The extroverted lungs endow humans with language capability, crucial to the social animals for making harmonious communities.

The comparisons among the atomic structure, the cellular model, and the human body are interesting and informative because these three entities have certain similar structures or functional orders. Spherical quantities in dynamic calculus can be applied to the description of organisms. Demonstrating the similar structures or functions affirms that the law of nature is single and universal.

8.5. The DNA Circuitry for the Human Lineage

As was described in Chapter 7, a DNA molecule functions as stepwise *LC* oscillatory circuits. Charges transport from one base pair to another through the double helix constitutes harmonic oscillation. Like a DNA double helix, the human lineage is a stepwise *LC* oscillatory circuitry where the male and female form a capacitor (time end) and the individual development under living stress constitutes an inductor (space end) (see Figures 7.1 and 7.9).

Each person is a multicellular diploid organism that begins life as a fertilized egg called a zygote, which consists of two poles originated from a sperm and an ovum. Like a charged electric capacitor driving electric current along a circuit, the dipolar zygote drives the growth of the individual. Genes contained in DNA within the zygote are the blueprint of our bodies, an initiative that creates a variety of proteins essential to human functioning and survival. During the development process, the body is gaining mass while losing growth momentum. DNA is a spherical quantity with multiple time dimensions, and the entire body has multiple space dimensions in correspondence with DNA time dimensions. Body growth involves the expression of genes through transcription and translation, which observe dynamic calculus. Thus the growth of the zygote observes dynamic calculus as electric current passes through an inductor. During the process, the energy flows from the capacitor to the inductor. Body growth is a process of converting time dimensions into space dimensions.

By the time an organism reaches a mature age, a male produces sperms while a female produces ova, both containing DNA substances that reflect the condition of the subjects. The seeds memorize enough details in the genes to replicate the next generation offspring. The reproduction of the seeds and the mating process can be regarded as a space to time conversion. During the process, the energy transfers from the inductor to the capacitor, a zygote combining a sperm and an ovum. The zygote is capacitor and contains the genes, representing the time end; and the body may be regarded as an inductor, representing the space end. Thus the time and space oscillation carries on from a zygote to a body and from a body to a sperm or an ovum that form both poles of a zygote in a human lineage. A human lineage constitutes stepwise *LC* oscillatory circuits.

The division of humans into both sexes has profound biological significance. The male and female can be expressed mathematically by

$$Man = \Phi_0 = A_1 A_2(\cos\Psi\cos\psi - \dot{\Psi}\sin\Psi\cos\psi), \tag{8.2}$$

$$Woman = \varphi_0 = A_1 A_2(\cos\psi \cos\Psi - \psi' \sin\psi \cos\Psi), \tag{8.3}$$

where *Man* rotates counterclockwise, *Woman* rotates clockwise, A_1 and A_2 are time and space dimension constants with biological significances, and Ψ and ψ denote a sperm and an ovum respectively. Spherical quantities Ψ and ψ have similar forms to time and space components of 1s electrons but have different meanings in their dimension constants and parameters. The dimension encapsulation upon Ψ and ψ produces a zygote, $A_1 \cos\Psi \cdot A_2 \cos\psi$, which begins to rotate in a wider spherical layer. Even though the male and female are rotating in different clockwise directions, they are in a similar chiral pattern (see Figure 5.5). Each human body is a chiral entity oriented by his or her sex hormone. One may express an individual by a spherical quantity such as $A_1 \cos\Psi \cdot A_2 \cos\psi$. However, in order to indicate its sex or its clockwise rotation direction, we had better write out the whole complex number in trigonometric form as in equations (8.2) and (8.3). Gateway functions $\dot{\Psi}$ and ψ' are the male and female sex organs, respectively. The development of a male is driven by the operator $-\partial/\partial t$ whereas the development of a female by the operator $-\partial/\partial l$ through the differential chains of gonads $\dot{\Psi}$ and ψ' respectively that signify the growth of gametes for producing the next generation. Each gonad gains the edge in directing the counterclockwise or clockwise rotation direction in the successive body construction. A good study of the fertilization process may uncover the secret of the biological implementation of sex determination. This sex distinction is in a certain manner similar to the time and space distinction of both strands of a DNA molecule in Figure 7.9 where the operator $-\partial/\partial l$ becomes an integral operator in the 5' → 3' direction. We might expect that the male uses the DNA time strand as a template in transcription whereas the female uses the space strand instead. They are complementary in physiology because time and space complement each other. In other words, they have contrary perspectives in perceiving the time and space strands in DNA. The physiological complement of the male and the female is like the geometric complement of both Φ_0 and φ_0 electrons (Figure 5.1).

By the way, it is impressive to observe that almost every Chinese character is chiral in calligraphy, showing no axis of symmetry. It seems that every entity has a sex or natural tendency to behave in a predisposed manner.

Sex distinction allows males and females to rotate in opposite directions, maximizing human capabilities. He is thinking and driving while she is feeling and expecting. The male organ is basic, and the female organ is acidic. They are complementary in geometry and physiology because of the clockwise direction shift between their rotatory operations (see Figure 5.1). In a human lineage *LC* circuit, both sexes form a capacitor so that electric potential may be built up between them. The division of both sexes normally prohibits electric current or chemical neutralization between them. Only during sexual intercourse do pulses of spermatic fluid pass through the reproductive organs. Thus we may compare this sex capacitor with the capacitor of DNA nitrogenous bases where hydrogen bonds deliver electric pulses between them.

A sperm and an ovum form the two poles of a sex capacitor. They are the haploid gametes resulted from the meiosis of diploid cells. The union of a sperm and an ovum produces a zygote that bears great momentum to grow due to the energy released from fertilization. In a certain sense, every organism is a biochemical battery capable of doing certain kinds of work far more efficient than the finest artificial battery. Man is an energy-storing capacitor or inductor, like a battery, capable of working. The capacity of the energy is related to the capacitance of the initial capacitor derived from the parents. This explains why organisms developed from the zygotes have stronger vitality than organisms developed by artificial cloning technology. After meiosis, a sperm and an ovum possess great mobility. Like the rotation of 1s electrons driving the rotation of 2s2p electrons (Figure 4.3), the mobility of a sperm and an ovum serves as an engine driving the growth of the zygote. An organism developed by artificial cloning technology, without going through the process of meiosis, and with its genetic codes having been transcripted once more, lacks the innate driving power of the sex capacitor so that the rhythms of life are much weaker than normal, remembering that the rotation of spherical quantities in an inner spherical layer drives the rotation of spherical quantities in an outer spherical layer (section 5.4.3). As a result, the cloned organism has a relatively poor immune system.

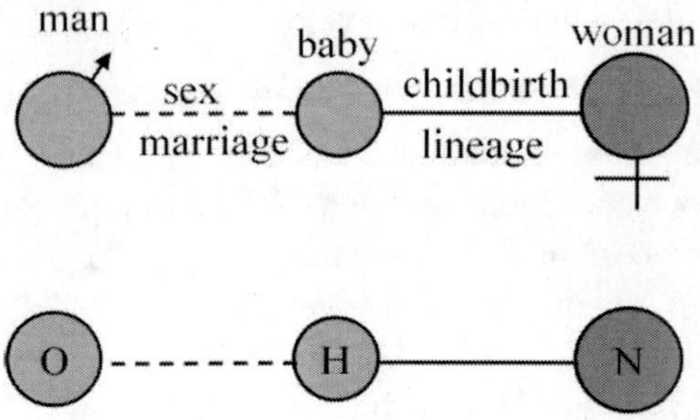

Figure 8.5. The relationship between a man, a woman, and a baby (a zygote) through the reproductive canals (the dotted and solid lines) analogous to a typical hydrogen bond.

Just as two strands of polynucleotide form a DNA molecule, numerous couples of husbands and wives form the human lineage. Like hydrogen bonds between the base pairs, the bond of marriage or love affair plays a key role in social structure (Figure 8.5). The structure of a human society is based on the family unit. At least, the human lineage is like a multiple-strand rope or a web of couples. A DNA double helix is a stepwise *LC* oscillatory circuitry that transfers electric charges from one base pair to another; and the human lineage through time is a stepwise *LC* oscillatory circuitry that passes down genetic information from one generation to the next. The genes-body oscillations of the male and female behave like electromagnetic waves confined within the human lineage where the rotation of the genes indicates growth or body construction and the rotation of the body produces sperms or ova that contain the genes of the species ($Genes^0 = Body,\ Body^0 = Genes$). The union of both gametes produces the next generation through the sexual behavior. The reproductive system and sexual behavior of a biological species implement dimension encapsulation mechanism that is key to the preservation of the species. Thus the human lineage observes dynamic calculus in reproduction that features the calculus chain rules to produce gametes and dimension encapsulation mechanism to produce a zygote for growth. This mathematical characterization is precise and encouraging (cf. Figure 8.9).

8.6. Wave Functions as Biological Variables

As we defined electronic orbitals within atoms, the outer layer electrons were constructed upon the inner layer electrons. The wave functions for electrons are invariantly expressed by sine and cosine functions that curl up multiple dimensions of the inner shell. This provides a clue on how to model physical quantities of life. Should we use sine and cosine functions to account for biological variables? Or should we always stick to data-oriented models such as the normal probability distribution?

Statistics is the scientific study of numerical data based on natural phenomena. Biostatistics is the application of statistical methods to the solution of biological problems. To analyze biological data, it is customary to assume that the data obey the normal distribution. Nowadays, the normal distribution remains a cornerstone of biostatistics. However, is the assumption based on sound biological knowledge? No, the normal distribution is a pure statistical theory based on the central limit theorem. The conditions that tend to produce the normal frequency distributions are all nonbiological: a) that there be many influencing factors; b) that these factors be independent in occurrence; c) that the factors be independent in effect – that is, that their effects be additive; and d) that they make equal contributions to the variance. There are surely many factors that affect the behavior of organisms, but it is hard to tell whether these factors meet the other three conditions. Many factors could be interrelated in occurrence and effect while some factors are doomed to be more important than the others.

Looking at the normal probability density function, there is a loophole in its definition of the variable scope:

$$f(x) = \frac{1}{\sigma\sqrt{2\pi}} e^{-\frac{(x-\mu)^2}{2\sigma^2}}, \ -\infty < x < \infty, \qquad (8.4)$$

where μ is the expectation of variable x, and σ the standard deviation. Although $\mu \pm 3\sigma$ contains 99.73% of the items, the normal distribution predicts that it is possible for x to have a negative number or a huge positive number. If body weight obeys the normal distribution, then can a man have negative mass? Or can he weigh a hundred tons? These are impossible from the viewpoint of biology, but the normal distribution indicates that these are possible even though the probabilities are very low. Thus the application of the normal probability distribution to biological variables encounters insolvable difficulties. At least, the variable must have an upper bound and a lower bound values.

In light of the fact that most biological entities exhibit certain sorts of wave behavior, we hereby introduce a circular function as a probability density function:

$$f(x) = A\sin(ax + b) + C, \qquad (8.5)$$

where A, a, b, and C are parametric constants and x is a variable. The scope of x may be defined within one or preferably half periodic cycle, but in rare circumstances it could cover two periodic cycles with bimodal peaks, which may be alternatively interpreted as belonging to two different characteristics. The suggestion of sine as a probability density function

(Figure 1.18) might not be new, but it deserves to be emphasized when most of contemporary biostatistics textbooks do not recognize the trigonometric function as a probability density function. Mathematically, the theorem for Fourier series indicates that every periodic function satisfying general conditions of continuous and piecewise very smooth can be represented by a trigonometric series. So the trigonometric functions, such as sine and cosine, as probability density functions should have widespread applications. For example, in describing the zonation or cyclic distribution of species in biosphere, each term in the trigonometric series may represent a certain ecological niche for a certain biological species to colonize and thrive.

The circular function agrees with the general style of wave functions. It has clear upper and lower limits for the variable. When one considers the ubiquitous phenomena of harmonic oscillations and waves, the use of trigonometric functions for describing biological data seems justified. However, it is not our intention to replace the normal probability distribution functions with circular functions for all biological variables. We just provide a practical alternative method for analyzing data. The normal probability distribution still has a place in experimental biology when the conditions of collecting data fit the normal distribution conditions as mentioned above.

Although mathematics tries to be exact and accurate, statistics is not an exact science. The choice of using which data analyzing techniques should be based on the specific case of concern, and as always the readers are encouraged to analyze biological data using multiple approaches to derive useful information.

8.7. The Rotatory Operation for the Mind and Body Relationship

The mind and body relationship problem is one of the oldest and most disputed topics in Western philosophy. The early Atomists, such as Democritus (ca. 460 to 360 BC), suggested that the nature of things consists of an infinite number of particles called atoms and ascribed the human mind to the movement of those atoms. By contrast, Plato (ca.428-ca.348 BC) proposed that the world consists of Forms and objects. Forms are changeless, eternal, and nonmaterial essences or patterns; and objects are only concrete instances of Forms. Our souls (and thereby our rational minds) were distinct from our bodies and could not be reduced to material constituents. The dualism of mind and body was further developed by Descartes (1596-1650), who described the interaction of mind and body through the pathway of the pineal gland within our brains. Dualism refers to the view that the world consists of or is explicable as two fundamental entities, such as mind and body. Even though dualism was widely accepted by later philosophers especially religious scholars, contemporary biologists tend to revert to the Atomist's view. According to materialism, our mental experience can be reduced to brain activities, but how are the activities of a physical brain related to the spiritual mind? Without clear definitions of mind and body in mathematics, the question is open to endless disputes.

We have defined spherical quantities in dynamic calculus as the law of nature. Rotatory operations are the standard way of biological activities as well as electronic motion. The probability density functions of biological variables are considered to be circular functions. A rotatory operation includes dynamic differentiation with respect to time and dynamic integration over space, both coupled together inherently. If we define a spherical quantity in dynamic differentiation with respect to time as mind, the experience of time dimension change, then the spherical quantity in dynamic integration over space is body, the construction or destruction of tissue due to space dimension change. To specify, given a spherical quantity with a time component Θ and a space component θ, the mathematical expression $-\partial\Theta/\partial t$ refers to mind while $\int \theta \, dl$ indicates body, both processes synchronized automatically. In other words, mind and body (soul and heart) are represented by time and space components of a spherical quantity in a rotatory operation. They are in dynamic transition as the rotatory operation proceeds.

As a point C traces along the semicircle in a dimension diagram (Figure 2.11), the organism loses a time dimension $\dot{\Psi}$ and gains a space dimension $-1/\psi'$ simultaneously. We assume that each velocity dimension ($u = -\dot{\Psi}/\psi'$) has specific physiological significance memorized through evolution. The lapsing of a time dimension provides the organism with the experience of a spiritual meaning while the acquiring of a new space dimension has a corresponding physical consequence. Mind and body are coupled together because time and space components are in a rotatory operation. Time reflects mind while space connotes body. Any kind of thinking, feeling, or sensing requires a period of time; and any kind of matter occupies a region of space. The dualism of mind and body can be precisely explained by the antiparallel synchrony of time and space components of a spherical quantity (Figure 1.7). They are defined by the differential and integral operations of a rotatory operation. Mind and body are complementary and exchangeable. This indicates that dualism agrees with the worldview of materialism. With Newton's mechanical laws, one has difficulties in explaining the interaction between mind and body; but with dynamic calculus obeying Faraday's law of induction, we have not any obstacle in solving the problem. The mathematical definition is precise and promising, which may push the borderlines of science farther forward into the spiritual realm.

During the daytime, human will consciously drives the motion of the physical body. Time (mind) is the driver. The action is a rotatory operation. At night, human mind is unconsciously driven by the recovery of the body. Space (body) is the driver. Dreaming may be regarded as a clockwise rotatory operation. Thus, the spherical quantity of mind and body (time and space) could rotate counterclockwise or clockwise depending on the consciousness (cf. Review question 5.3 for a simplified dimension diagram). Active doing and passive dreaming constitute harmonic oscillation in synchrony with the human circadian rhythm. Human will may drive the body to work hard; and labor in turn may mold human mind and the brain. The whole human construct, including the mind and body connection, evolves gradually through the continual harmonic oscillations of various biological rhythms between the two poles under the pushing or damping of the environment.

8.8. The Poisson Distribution and Evolution

This section explores the biological significance of the Poisson distribution in statistics. Let x be a constant dimension serving as the parametric mean of the Poisson distribution:

$$1 = e^{-x} + xe^{-x} + \frac{x^2}{2!}e^{-x} + \frac{x^3}{3!}e^{-x} + ... + \frac{x^k}{k!}e^{-x} + ... \quad (8.6)$$

where each term on the right-hand side represents the expected frequency corresponding to the k rare event. Although the Poisson distribution is normally applied to discrete variables, we suggest that its application could be extended to continuous spherical quantities because an exponential function e^x in the polynomial series has the property of a circular complex function (see equation 5.57). If we regard the expected frequency of the k rare event as the difference of counts (ΔM) between two time points delimiting a time interval ($-\Delta t$, the minus sign denotes the countdown direction of time), then we may write the probability of the k rare event as $-\Delta M/\Delta t$ or in differential expression as:

$$-\frac{\partial M}{\partial t} = \frac{x^k}{k!}e^{-x}. \quad (8.7)$$

Let N denotes the probability of the $k+1$ rare event:

$$N = \frac{x^{k+1}}{(k+1)!}e^{-x}, \quad (8.8)$$

then performing differentiation with respect to x dimension on both sides of the equation gives a corollary to the Poisson distribution:

$$\frac{\partial N}{\partial x} = -N - \frac{\partial M}{\partial t}, \quad (8.9)$$

or under the dynamic differential rule that the differentiation is carried out with respect to the most immediate dimension in the x^{k+1} factor only, then instead of equation (8.9), we get

$$\frac{\partial N}{\partial x} = -\frac{\partial M}{\partial t}. \quad (8.10)$$

This indicates that there is a rotation relationship between the probability of the $k+1$ rare event and the accumulative number of counts of the k rare event. There are biological implications associated with this conclusion if we interpret the k or $k+1$ rare event as the emergence of organisms holding k or $k+1$ dimensions in a specific trait. Organisms may evolve from k to $k+1$ dimensions as the parametric mean (or the expectation of the Poisson distribution) shifts under the action of the environment, and the responsive probability drift along the Poisson distribution series represents the evolution direction.

A mutation shifting from M to N in equation (8.10) may be understood as a smooth transformation where dynamic calculus is involved in the implementation of the rotatory operation. Thus, we may say that biological evolution is a process of gradual rotation of a spherical quantity. Biological species are discrete entities along the continuous path of rotatory operations, as if integers are located at the real number axis. The probability of evolving into a $k+1$ dimensional organism from a k dimensional organism is related to the accumulative counts or population of k dimensional organisms. Here population means the totality of individual observations existing anywhere in the world or at least within a definitely specified sampling area limited in space and time. If the intrinsic time signal favors the evolution of organisms toward more dimensions, then given enough time, it is certain that an organism will evolve from k dimensions to $k+1$ dimensions. However, it is not certain whether an organism with higher numbers of dimensions is always more suitable for a given environment. This is where external space signal or natural selection comes into play. The space and time tides of ever-changing nature determine the flourish or recede of a certain species under sustained strain. Whenever a wave of increasing spacetime complexity arrives, organisms response by forking into species of more dimensions accordingly to seize on each place or niche in the economy of nature. The central tendency of the Poisson distribution drifts towards a higher dimension. This probability drift along the Poisson distribution series in response to the parametric mean shifting under the environmental conditions represents the direction of biological evolution. And if any of their progenitor species of k dimensions does not become modified and improved in a corresponding degree to adapt to the new environment, it will be exterminated sooner or later. Hence life evolves one wave forward in harmony with nature following the expectation of the Poisson distribution.

Aside from probability calculation, here we shed light on gene mutation for evolution briefly. Firstly, since DNA sequence contains genetic information, the biological evolution must ultimately reflect in the change of the sequence or in the elongation of the sequence, or affect the structure of DNA between or within chromosomes that may alter an organism's phenotype, giving it greater or lesser advantage in the process of natural selection [54]. While the mechanism of genetic mutations remains largely unknown to biologists, the idea of treating genetic mutations as random and contingent events is never satisfactory. It is hard to imagine how chaotic changes in DNA nitrogenous bases without any direction and order could lead to an upgrade in the genes for better fitness. Genetic mutations must be subject to definite condition changes under the law of nature.

Secondly, if there is any factor that is conducive to gene upgrade, then it must be the environmental change. The environment plays an important role in human development. Organisms are not living within a vacuum. Man is embedded in nature. He needs to interact with the elements to obtain food, clothe, and shelter. Most importantly, he must gain a sense of space and time from the outside world, and pass this reference to the inside body so that genes may transcript accordingly as a C++ program should pass parameters to initialize a class object. The eyes, "the window to soul," serve as the major portal or gateway to human mind and emotion (time) from the environment (space), which marshal the appearance of the external world into the internal brain. Light, which carries images and messages, implements this space to time transformation to a large extent while sound, smell, and touch help realize the transformation in the other manner. Time is the humans' worldview or innate sense of the environment or space. On the earth, the diurnal rhythm and the seasonal rhythm are the most important environmental parameters that would induce the harmonic oscillations of organisms through various complementary relationships as in the case of $3d^{1s}$ and 1s electrons (Figure 8.4).

Furthermore, the interaction between life and the environment is the main cause of evolution, which might involve the elongation of the DNA base sequence or the shifting of the gene expression. As the environmental energy E transforms in space, the biological species evolves in time under the action of momentum p, both being balanced by the natural force F.

$$-\frac{\partial p}{\partial t}=\frac{\partial E}{\partial l}=F. \tag{8.11}$$

Here the environmental gradient includes changing factors of climate, nutrition, work, and society, etc. The environmental variations engendering the natural force would induce mutations in the biological genes. The action and reaction from both the environment and the biological species bind them into a virtual rope. No matter what is the cause, the environmental changes are the main reason for species evolution. A man works at daytime to gain the sense of his living environment while sleeps at night to receive the feedback from the internal biological clock. This rhythm is going on day and night as a diurnal cycle. The interaction between man and the environment is the key to the progress of evolution. Whether a man can live in harmony with nature determines his health condition to a great extent. This is a major foothold of Chinese medicine that has come a long way.

Finally, as humans develop into societies, it is necessary for us to arrive at that degree of sophistication where natural environmental factors refer not only to the physical elements (such as earth, air, fire, and water) but also to the social structures. For examples, politics, economics, and customs have important influences on people's lives; languages, sciences, and technologies play key roles in shaping people's minds. This outer environment (space1) interacts with human mind and emotion (time1) arousing conscious experience that soon permeates the inner body fluid through mind and body oscillation. The body fluid (space2) then produces the seeds (sperms or ova) (time2) for the next generation, and the next generation is destined to change the outer environment as they grow up. Mathematically, space1 $\mapsto$ time1 $\mapsto$ space2 $\mapsto$ time2 $\mapsto$ space1, where the arrows denote rotatory operations. The environment is subject to alteration by humans. Human's hands mold the environment according to his thinking and feeling. The message passings are bidirectional. While changing the world, he changes himself concurrently. Each generation has to meet the challenge of its days or recede. An organism will continue the oscillatory cycles of various rhythms until any cycle is gated. The primal drive in all of nature is to produce life, and the will to live of every organism is to continue the cycle of life. Organisms are coupled with the natural environment as time and space are coupled together. This determines the direction of evolution. Equation 8.11 is a dynamic calculus expression of a base pair in trigonometry in Figure 7.9. The interaction between the environment and mankind may be expressed as a rotatory operation between the base pair. And many base pairs define a gene.

To sum up, genes are the children of the struggle between space and time tides, the ongoing signatures of evolution. The environmental factors, which boil down to space and time effects, may change the genes. The genes so conditioned by the space and time oscillation determine an offspring that in turn may adapt better to the environment in the future. The world is continuous systematically, so are biological species despite their apparent discrete manifestations on the Poisson distribution series that governs the probability of organisms of various dimensions. Any mutation in one part will induce corresponding orientation changes in the other parts of the world. Evolution follows such a continuous route that features interaction and oscillation. It is directed by the law of nature toward the better harmony of the organisms and the environment, reflecting the time and space trends of the general oscillating cycle that encompasses them. These continually conducive environmental conditions and biological vital impulses are faithfully recorded in the genes. DNA is a stepwise LC circuitry with coupled space and time strands. It is an objective history book. This double helix recorded the evolution rhythm of organisms in the time strand along with the fluctuations of the environment in the space strand. The natural script of DNA is not fairly understood yet despite modern biology progress by leaps and bounds. We believe that species are always created and evolved under the continuous guidance of dynamic calculus, the law of nature. By the way, we want to caution you that to manipulate the genes arbitrarily without a thorough understanding of the DNA script is against the will of Nature and seeking self-destruction in the end.

8.9. The Law of Nature and General Gyroscopes

The law of nature is defined as spherical quantities in dynamic calculus. A fundamental law must be simple. If an equation is complex, then it is likely to be a composite of more fundamental ones. The law of nature is fundamental because it only sets the rule of operations for spherical quantities. Throughout this book, the law of nature is repeatedly illustrated by dimension diagrams. The most direct derivation from the natural law is the principle of rotatory operations, which relates a spherical quantity in time to a spherical quantity in space. A rotatory operation describes the motion of an electron in time and space synchronously. Taking the example of equation 8.10, the spherical quantity that is differentiated with respect to a time dimension, M in the equation, can always be regarded as a time component whereas the spherical quantity that is differentiated with respect to a space dimension, N, is its space counterpart. Space and time components undergo harmonic oscillations in various dimensions (see Figure 2.12). The fundamental law of nature governing the oscillations is single even though it was explained in various contexts and formulae (see Appendix III). For example, as will be seen, the famous Pythagorean theorem is only a variant of the natural law (see equations 2.39, 9.2, and 9.3).

Harmonic oscillation is the theme of nature. Each spacetime oscillation is going on within a certain sphere, and various oscillatory cycles may form in parallel or in series in a gyroscope (Figure 8.6). A gyroscope originally refers to a device consisting of a spinning mass, typically a disk or wheel, mounted on a base so that its axis can turn freely in one or more directions and thereby maintain its orientation regardless of any movement of the base. Under the spherical view, we replace each direction of movement with an oscillating entity or a ring. When one of the rings tilts in one direction, the other rings will adapt to it so that the overall orientation of the gyroscope maintains unchanged. For the oscillatory entities within the same spherical layer, the mechanism of coordination between various rings could be Green's theorem, Stokes' theorem, or Gauss's theorem. These theorems relate a spherical quantity in space to another spherical quantity in space. They are faithful interpretations of the law of nature (see section 3.4). For entities in different spherical layers, dimension encapsulation mechanism or calculus chain rules may synchronize their harmonic oscillations as well. The coordination of various spherical layers helps maintaining a constant angular momentum in the overall sphere (see $m\omega r^2 = \hbar/2$ in section 9.2).

If we regard the human body as a multiple-ring gyroscope, each ring representing a physical, physiological, or psychological cycle in a certain manner, then a change in one ring will certainly affect the other rings, but the angular momentum or the net orientation of the organism remains constant. Meanwhile, the change of a specific ring will incur counteractions from the other rings. The capacity to adjust its various biological cycles in various layers to cope with a trouble ring in a body is known as the immunity, which enables the body to repair its injuries and fight against diseases to a great extent.

Furthermore, if we regard the whole universe as a super multiple-ring gyroscope where one spherical layer is tightly coupled with another spherical layer, then a change in the galaxies may affect the solar system. The effect may propagate down to the earth revolution cycle, the biospheric rhythm, and the cycle of a specific biological species by calculus chain rules. The time signal is passed down from light to electrons, atoms, nucleotides, DNA, cells, humans, societies, animals, biosphere, the earth outer stratum, the earth, the solar system, galaxies, and returns to light while space information is transmitted in the reverse order. This bidirectional exchange determines the direction of evolution.

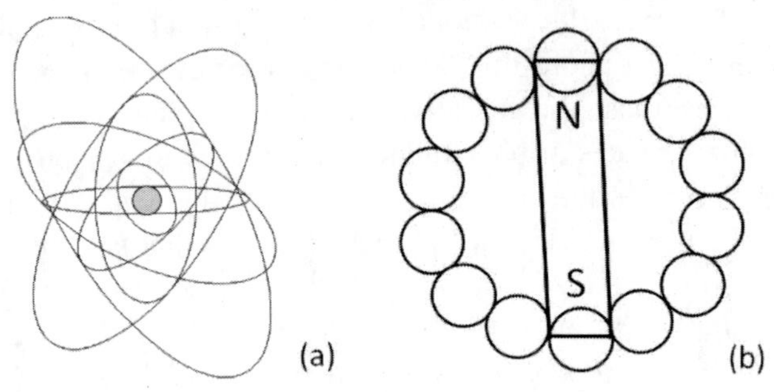

Figure 8.6. Two typical perspectives of an imaginary multiple-ring gyroscope where each coupling ring represents an oscillatory cycle within a certain spacetime sphere. Harmonic oscillations in different spherical layers are connected (a) in parallel and those in the same spherical layer are connected (b) in series.

However, the gyroscopic property must not be construed as fatalism. Although the orbit of biological evolution is mainly determined by prevailing spacetime tides of the biosphere, each individual organism has many degrees of freedom in his or her lifestyle as in a gyroscope even when the net orientation is fixed, each individual ring in the gyroscope could still move actively, inducing the movements of the others, or response passively to the actions of the others, damping the oscillation. Moreover, as we mentioned before, everything is interactive, and none is absolute. The net orientation of a gyroscope is a relative concept depending on the scope of concern, which is never fixed at a direction, so that human creatures can still change their destiny through their prudential collective efforts. For example, reducing the emission of carbon dioxides to mitigate the global warming trend requires our collective efforts currently. We also insist that no any parts of the world are isolated from the others. The second law of thermodynamics that studies isolation systems must be taken with caution when applied to the description of a multiple-layer open system.

In short, space and time oscillations are harmonic and inter-coupled in various layers. In spite of their diverse manifestations, the harmonic oscillations are always governed by the law of nature in certain physical manners. The universe is a natural gyroscope analogous to the atomic structure where each ring is inductively coupled with the others by the process of dynamic calculus, so is a human body.

Naturally, one may raise the question why the human body or the whole universe must be a gyroscope, or why organisms must be evolved from less dimensions to more dimensions. This is not an easy question, but we shall explain it briefly from the origin and creation of the universe. Suppose before the advent of anything, there was nothing, which can be expressed by

$$0^{*}=0 \tag{8.12}$$

where the right-hand side means zero, and the left-hand side means nothing. The difference between nothing and zero is that nothing was soon taken over by space and time vibrations that change the equation into:

$$\frac{\partial W_1}{\partial t}+\frac{\partial W_2}{\partial l}=0\,, \tag{8.13}$$

where W_1 and W_2 are two orthogonal spherical quantities obeying the law of nature. This creation could be the reverse process of annihilation of matter and antimatter such as an electron and a positron. Hence the universe became a system featuring two phases. These two phases were then replaced by more phases in a somewhat recursive way such as

$$\frac{\partial W_3}{\partial t}-\frac{\partial W_4}{\partial l}=W_1\,, \tag{8.14}$$

$$\frac{\partial W_5}{\partial l}-\frac{\partial W_6}{\partial t}=W_2\,, \tag{8.15}$$

where the right-hand sides of the equations can be regarded as quasi-zero or the equilibrium states between W_3 and W_4, and between W_5 and W_6 phases, just as we treat mass as charge neutral. Hence the universe became more enriched and colorful with the emergence of more phases. Equations (8.14) and (8.15) could take the shape of equation (8.9) derived from the Poisson distribution. Bear in mind that equation (8.9) implies the rule of biological evolution from an organism of less dimensions to one of more dimensions. Combining equations (8.13), (8.14), and (8.15) yields

$$\frac{\partial^2 W_5}{\partial l^2}-\frac{\partial^2 W_6}{\partial l \partial t}+\frac{\partial^2 W_3}{\partial t^2}-\frac{\partial^2 W_4}{\partial t \partial l}=0\,. \tag{8.16}$$

This equation in dynamic calculus may be interpreted as a transformation cycle. As shown in Figure 8.7, the four terms on the left-hand side of the equation constitute a two-dimensional oscillatory system, which may be instantiated by both 1s electrons in counterclockwise and clockwise rotations. If the universe has been created and has always evolved according to dynamic calculus, then no matter how many layers of spherical quantities it has now, it must be a natural gyroscope whose various rings are still interrelated by the law of nature. Darwin showed that new species are achieved by natural selection, and absence of plan is an inherent part of the evolutionary process. We believe that no design is the design of the universe. With spherical quantities in dynamic calculus for the creation and revolution of the universe, we have serious reservations about the Big Bang theory even though our description of the origin of the universe is consistent with it roughly.

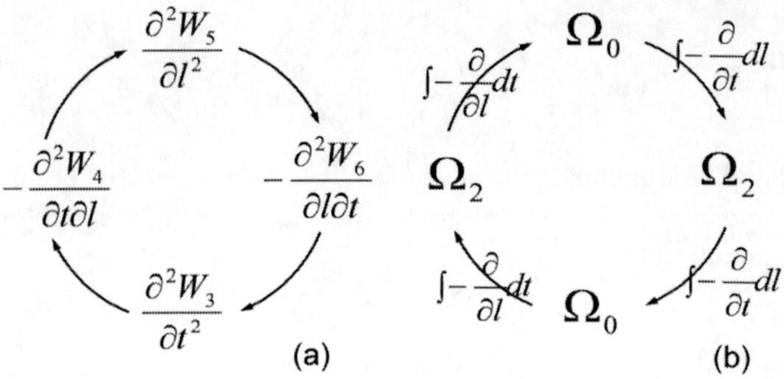

Figure 8.7. Transformation cycle of equation (8.16) in (a) a dynamic calculus interpretation with (b) a concrete instance of wave functions in a helium shell.

By the way, it does not escape my observation that even the social economy complies with the law of nature. If we tactically define money as a time component and commodities as its space counterpart, then the interactions between both components produce diverse economic phenomena. The faster money turns over in business, the more commodities are transacted (see equation 8.10). With the introduction of stock markets and various firms and professions, economic activities become complicated with more dimensions involved. We briefly open this gate for economists to investigate the fine integration of social economics with natural science. We broach the integration because we believe that the social economy is a middle ring of the natural gyroscope, connecting the physical elements with the human behavior. After all, the human society is an inseverable part of nature.

8.10. At the Dawn of Calculus Biology

Life is a kaleidoscope of species, the most splendid part of nature. Life is formed as physical quantities in linear algebra and spherical quantities in dynamic calculus percolate into each other from both ends of gravity and electromagnetism (cf. Review question 5.4). The gravitational field of the earth attracts all weights while the electromagnetic field of the earth induces the motion of electrons within various atoms. As a result, life is corporeally attracted to the ground steadily on the one end and oriented spiritually toward the rhythms of the nature harmonically on the other end. A man on the earth is somewhat similar to a cell within an organism, feeling and hearing the breath of his milieu. This is amazing! While many biological activities and metabolic pathways have been investigated and discovered by biologists, the controlling mechanisms underneath the behavior remain largely an untouched territory, awaiting penetration and colonization by the spherical view. Thus, instead of drawing a conclusion here, this chapter has just raised the various biological issues for further discussion and development, opening the door to a new discipline – Calculus Biology, the mathematical modeling of biological phenomena and the biological interpretation of dynamic calculus, or the principle of the connection and match between dynamic calculus and biology.

It is interesting that organisms can be formulated by precise mathematics. A man and a woman are different in their rotatory tendencies (equations 8.2 and 8.3). The differential chains $\dot{\Psi}$ and ψ' are the gonads of the organisms capable of producing sperms and ova respectively upon sex maturity of the individuals. Both a sperm and an ovum are circular complex functions (equations 5.38 and 5.40) generated by the dynamic calculus chains of gonads – $\partial\Psi/\partial t$ and $\partial 1/\partial\psi$ respectively. As the haploid sperm and the haploid ovum fuse into a diploid zygote (equation 2.12), the cell starts to grow in a larger spherical layer upon dimension encapsulation.

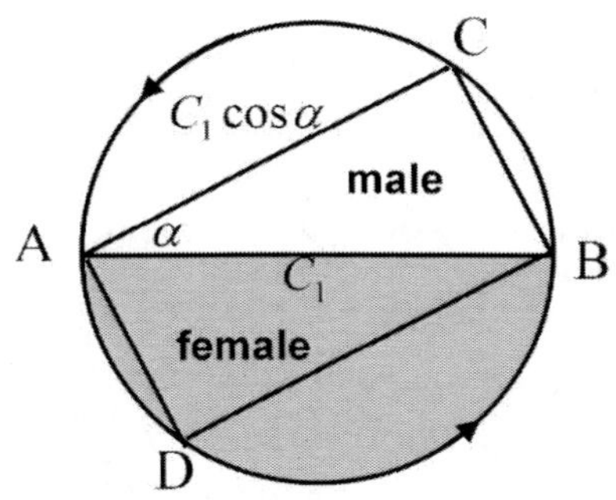

Figure 8.8. Dimension diagram for sex determination of a zygote by the increasing or decreasing state of α in the sperm. As point C traces the semicircular arc from B towards A, α increases from 0 to π/2, driving the zygote to rotate counterclockwise and manifest as a male; and as point D traces the other semicircular arc from A towards B, α decreases from π/2 to 0, driving the zygote to rotate clockwise and manifest as a female.

The sex of a zygote might be determined by the increasing or decreasing tendency of α in the sperm at the moment of sperm and ovum fusion. When radian angle α increases uniformly from 0 to π/2, the zygote undergoes counterclockwise rotation and manifests as a male. When radian angle α decreases from π/2 to 0, the zygote undergoes clockwise rotation and manifests as a female (Figure 8.8). Because in circular complex function Ψ, the probabilities of radian angle α increment and decrement are equal, both sexes have an equal opportunity in the reproduction. We hypothesize that the dominant sex determining factor of a zygote is the random oscillation of radian angle α in the sperm at the cellular level here. How this hypothesis is related to the X and Y chromosomes in sexual differentiation of the embryo remains unclear.

Moreover, the male follows the counterclockwise rotations of equation (2.13), or the carbon model of Figure 11.6, whereas the female the clockwise rotations of equation (5.1). The spherical quantities of a male and a female correspond to the spherical quantities of 2s2p and 3s3p electrons with fresh biological significances in dimension constants of A_1, A_2, C_1, and C_2. Hence the inner spherical layer, upon dimension encapsulation, drives the growth of the outer spherical layer (Figure 8.9).

In a male, by counterclockwise rotations of gene expression, multiple dimensions of time unfold into multiple dimensions of space accordingly as the organism grows. In a female, by clockwise rotations of gene expression in the negative direction of quaternity axes, the growing process also means that multiple dimensions of time are converted into multiple dimensions of space accordingly in the time direction. After the maturity of the organism, the outer spherical layer, like 2s2p or 3s3p, will drive the growth of inner 1s spherical layer through the calculus chain rules, producing the seeds (sperms or ova) for the next generation. This life cycle strictly follows dynamic calculus of spherical quantities, the law of nature.

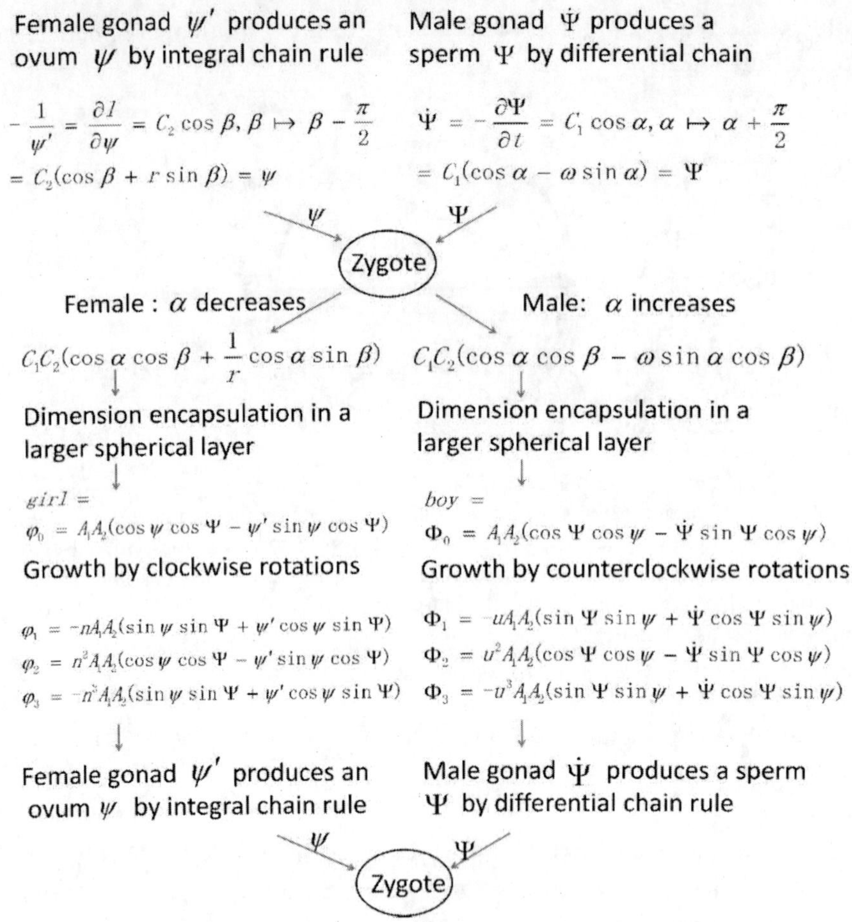

Figure 8.9. A simplified human life cycle in spherical quantities.

At this point, it is revealing to compare electromagnetic waves, stepwise *LC* oscillations in a DNA molecule, and the human lineage where synchronized orthogonal waves traveling in two strands form a rope. In electromagnetic waves, the electric field and the magnetic field are perpendicular to each other and progressing forwards in phase; in a DNA molecule, *LC* oscillations are proceeding from one base pair to another step by step with antiparallel wave signals through the double helix; and in a human lineage, the biological oscillations from a zygote to a body and from bodies to a zygote are propagating from one generation to the next along the complementary roles of both sexes. Even though these three entities are unrelated and with increasing dimensions and complexities, they are undergoing harmonic oscillations similarly on different scales, symbolizing the vital pulses of natural entities required by dynamic calculus.

As we discussed the structure of a helium or a neon atom, the electrons within the atom constitute a complete cycle, rendering the system insulated from the environment, and hence their inert chemical property. An inert atom is a self-contained entity. Conversely, a carbon atom is chemically active because it has only four electrons $2s2p_x2p_y2p_z$ in the outer shell. The half full electron configuration gives it maximum capability in forming chemical bonds with the other atoms. The carbon atom is the central element in organic molecules. This is not by accident but by the rule. Because of its tendency to form four covalent bonds with the other atoms, a carbon atom is not an insulated system. The complete circulation of electron octet involves covalent bonds in the molecule. Although the principle of stable electron octet remains unchanged, the shifting of its location from within an atom to between atoms in a molecule gives it great scalability. For example, a desktop application is self-contained, but an internet application is not. The latter is much more powerful thanks to its extrovert property. Thus, from the topic of inert atoms in Part I to the subject of a carbon atom in Chapter 6, we shifted our focus from atoms to molecules. The evolution of molecules precedes the evolution of life. For example, the advent of extrovert carbon atoms in the environment gives the potential of multilateral chemical bonds between atoms that make organic molecules and then biomolecules. The genesis of life is not without preparation of organic compounds but evolved step by step naturally from inorganic to organic molecules and from abiotic to biotic forms.

The *LC* oscillation system as discussed in Figure 1.5 constitutes an idealized electromagnetic circulation. The electric charges in the capacitor drive the electric current to pass through the inductor. The AC current then induces magnetic flux in the inductor, which provides it with an electromotive force in the circuit, charging the capacitor in the reverse polarity. In the absence of any energy loss, the *LC* oscillation will continue back and forth indefinitely. In Figure 7.1, we introduced an *LC* oscillation system where the *LC* circuits shift their location every cycle due to the electric current routing property of deoxyriboses in a DNA molecule. The stepwise *LC* oscillations provide the DNA molecule as a versatile script for encoding the gene of life. Although the principle of *LC* oscillations remains unchanged, only the extrovert stepwise *LC* oscillations make it possible for DNA to become the blueprint of life.

In section 8.9, we discussed an imaginary multiple-ring gyroscope where each coupling ring represents an oscillatory cycle within a certain spacetime sphere. But here we want to emphasize that life must involve itself in open rings because it needs the nourishment from the environment. Life is an intangible gyroscope with multiple dimensions of cycles, several of which are open ones coupled with the environment (Figure 8.10). Every organism needs air, water, food, etc. from outside the body. In harnessing energy from the environment, homo sapiens becomes social animals and is powerful in shaping its living conditions. The openness of the oscillatory cycles integrates life with the environment intimately and the request of their renewing harmony is the driving force to the continuous evolution of life. However, the principle of dynamic calculus for life is similar to that for an atomic structure such as a zinc atom to a certain extent (Figure 8.4). The elucidation of the isolated atomic structure in Part I is helpful for understanding the multilateral natural pattern in Part II even though the open systems are much more powerful and challenging than the isolated ones due to their extroverted associations.

We shall also take this opportunity to touch on the mechanism of biological gene mutation. As was discussed in section 5.4.3, the 1s electronic cycle drives the 2s2p electronic cycle, and the wave signals then reflect back from 2s2p toward 1s electrons through the calculus chain rules, reinforcing the original 1s waves. Obviously, if both inner and outer waves are out of sync for any reasons, then they may struggle to balance the system. We have assumed in Figure 2.8 that the circles are strictly round. What kind of feedback will happen if one circle is actually oval in shape? When we treat the 1s electronic cycle as the sperm or ovum that carries the gene and the 2s2p electronic cycle as the body construction that expresses the gene, it becomes evident that the rhythm of the body might affect the gene of the next generation during the gene-body oscillation. Electrons in the outer layers affect those in the inner layers, which in turn affect those in the very most inner layer. The external cause will make contribution, through altering the internal cause stepwise, to the determination of the gene mutation for the offspring. This effect boils down to the shifting of space and time dimensions of spherical quantities during their rotatory cycles. Thus, some environmental factors that mold the body might eventually alter the internal gene and hence promote the inheritance of acquired character. When applied to the human lineage, the customization of atomic structure model could be very instrumental in explaining the gene mutation induced by the environment. We believe that the gene should mutate according to the rule of dynamic calculus rather than by a random event. And even random events observe the law of large numbers – when sampling size is sufficiently large, the frequency of the occurrence of an event calculated from repeated observations of a random variable must approach its distribution probability.

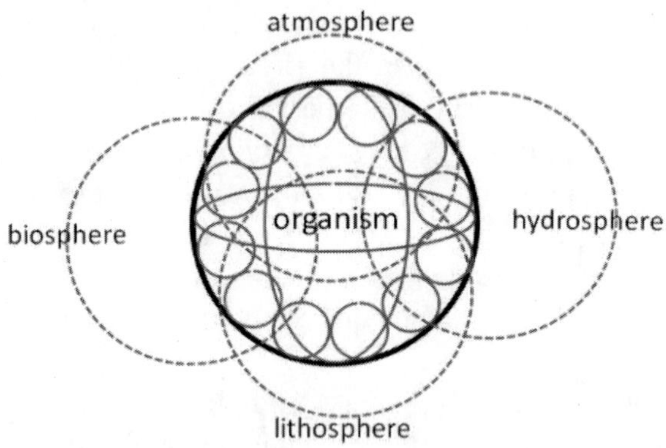

Figure 8.10. An imaginary model of an organism with multiple dimensions of oscillatory cycles, several of which are coupled with the atmosphere, hydrosphere, lithosphere, and biosphere.

Furthermore, the atomic model of transition metals (such as zinc) might be applied to the explanation of substances passing through the cytomembrane of a cell within an organism. The mechanism of dimension encapsulation (equation 2.12) represents the fusion of both haploid circular complex functions. It is the fertilization process of an egg. If we regard the diploid as a cell, then the dimension encapsulation establishes the membrane of the cell. This membrane function has been expressed mathematically by equation (5.86). Substances passing through the cell membrane are subject to regulation by the calculus chain rules. Like 3d or 4f orbitals capable of passing gateways to deliver wave messages, nutritional uptake by the cell is controlled by the same calculus chain rules in passing through the cytomembrane. Only substances that complement the acceptors in someway are selectively absorbed by the cell. We have drawn Figures 5.15 to illustrate the geometric complement of $3d^{1s}$ and 1s electrons; and we have also drawn Figure 8.4 to illustrate the model of endosymbiotic organelles. They describe complementarity of two acute radian angles in a right triangle that observes the principle of the Pythagorean theorem. This principle can be adopted to explain how the substances pass through the cytomembrane and are eventually absorbed by the cell as nutrients. The structural complementarity is the key to the nutrient absorption. On the other hand, the excretion of waste from a cell may be modeled by Figure 5.1 where a 2s orbital displaces into a $2p_x$ orbital, i.e., from the center to the periphery. In this way, we have developed process mathematics to account for physiological phenomena.

Conclusion

This chapter has lightly tapped the great potentialities of the spherical view in a passing review over biology, enough to whet the appetite of researchers for further modeling the organization and evolution of life in a precise way. We have seen how spherical quantities in dynamic calculus may be applied to the formulation of the DNA structure in section 7.3 and to the description of a biological cell in sections 8.2 and 8.3. We have seen in section 8.7 how the mathematical introduction of a rotatory operation to the characterization of the mind and body relationship might be a crucial step to exploring the related fields such as psychology and philosophy. We presented atoms with various electrons of complementary, symmetric, coordinative, harmonic, and interactive relationships instead of discrete and independent ones. These mathematical models developed from the atomic structure can be customized for describing the biological pattern in a coherent manner. We presented a living organism as a general gyroscope with maximum coordination among its various cycles or functions. The existence of open oscillatory cycles in the general gyroscope also means uncertainty, challenging, and blessing to life that potentially determine the direction of evolution. The law of nature is so powerful and fundamental in governing biological kingdoms as well as chemical elements. Indeed, until spherical quantities in dynamic calculus have been fully realized, the full array of life phenomena will remain a mystery.

Chapter 9

Circular Motion and Central Force

We now go back and focus on the fundamental nature of two-dimensional circular motion in terms of spherical quantities as well as physical quantities. Uniform circular motion is the motion of an object in a circle at a steady speed. As an object moves in a circle, it is constantly changing its direction. At any moment, the velocity of the object is tangent to the circle while the acceleration of it is directed toward the center of the circle. In Chapter 1, we characterized general harmonic oscillation in spherical quantities by a rotatory vector with its tip on a circle. Electronic oscillation within a helium atom is simple harmonic in spherical quantities and hence related to circular motion. A planet revolves around the sun in an elliptical orbit. An elliptical orbit is the projection of a circle onto a tilted plane. Thus both electronic motion and planetary movement can be unified in the framework of circular motion. Instead of an elliptical orbit of a planet around the sun, an electronic orbital takes the form of a matter state transformation cycle. While the kinematic movement of a planet is governed by Kepler's first law, electronic transformation obeys the Pythagorean theorem, both being equivalent in their physical principle. Kepler's second law for a planet takes the form of angular momentum conservation for both electronic orbitals within a helium atom. This chapter tries to establish the correspondence between electronic orbitals and planetary orbits through a comparative study. From the energy density function of a 1s electron, we shall also derive the dimension of centripetal force density that the electron incurs along a hypothetical semicircle within a helium atom. The result is exciting and surprising that questions the applicability of venerable Coulomb's law for electrons in atoms.

9.1. Introduction

Circular motion has various manifestations. Electronic transformation and planetary revolution represent two extremes of it. An electron revolves around the nucleus through changing matter state instead of changing position whereas a planet revolves around the sun through kinematic movement without changing matter state in itself. Because these two ideal extremes are orthogonal in their nature of circular motion, they can be distinguished by a hypothetical angle δ in their wave functions. When δ is equal to $\pi/2$, the orbital refers to the pathway of orthogonal state induction; and when δ is zero, the orbit declines into a kinematic circular track. We ordinarily call the former an electronic orbital and the latter a planetary orbit. As shown in Figure 9.1, both electronic orbitals and planetary orbits are in two perpendicular planes with a right angle of δ. Radian angle δ measures the degree of an oscillating entity deviating from kinematic movement toward matter state transformation. Note that we had similar treatment on two orthogonal electronic components in section 1.3.3 where $\cos\beta$ is associated with a capacitor and $\sin\beta$ with an inductor. Both electronic components are orthogonal in their electrical property, which allow the assignments of cosine and sine functions that have a calculus relationship.

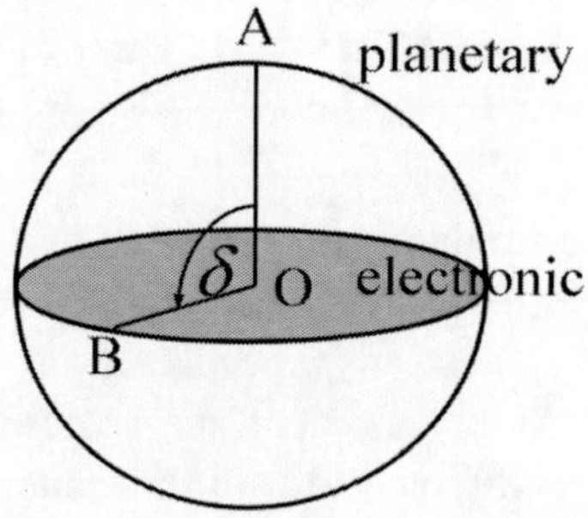

Figure 9.1. The orthogonal relationship of an ideal electronic orbital (shaded circle) and an ideal planetary orbit (blank circle) on two mutually perpendicular planes. For 1s electronic orbitals, $\delta = \pi/2$.

Between electronic orbitals and planetary orbits, the motion of many entities follows circular motion on tilted flat planes, having δ values in the range of (0, π/2). For example, during a year, a flame-tree (*Delonix regia*) in front of my house experiences four seasons. The plant lives out a seasonal cycle similar to circular motion by changing matter state in response to the weather changes. During the same period, a flock of geese migrate from Canada to the United States in the fall, and return to the north in the spring when the weather is getting warmer. The animals are able to lessen the effect of harsh temperature changes upon themselves through the strategy of migration. While the plant has to withstand harsh weather changes by increasing its physical adaptability, the animals choose kinematic movement instead of physical adaptation. To a certain extent, the strategies of the plant and the animals in four seasons are analogous to the circular motion of electrons and planets, respectively. However, both plants and animals have physical adaptation and kinematic movement in the spectrum from electrons to planets, their wave functions lying on the tilted planes of $0<\delta<\pi/2$.

For a real entity whose circular orbit is within a tilted plane ($0<\delta<\pi/2$,), the orbit may manifest as an ellipse on the equatorial plane ($\delta=0$) instead. Here the equatorial plane of the sun refers to the orbital plane of eight major planets including the earth. The equivalence of a circular orbit on the tilted plane and an elliptical orbit on the equatorial plane can be illustrated by the geometrical property of an ellipse as shown in Figure 9.2. In the first form (Figure 9.2a), a circular orbit with a radius r remains its circular shape (shaded) but its orbital plane rotates a certain angle δ away from the equatorial plane whereas in the second form (Figure 9.2b), an orbit keeps in the equatorial plane but deforms into an ellipse whose focus shifts from central point O to O_1. When the vertical projection of the circular orbit onto the equatorial plane in the first form is in congruence with the ellipse in the second, we have a relationship concerning the eccentricity of the ellipse:

$$\cos\theta = \sin\delta, \tag{9.1}$$

where θ is the angle between OO_1 and O_1C in right triangle OO_1C. In the right triangle, the distance of the focus from the geometric center of the ellipse (OO_1 = c) serves as a side, semi-minor axis (OC = b) serves as another side, and semi-major axis (O_1C = a) serves as the hypotenuse. When δ =0, Figure 9.2(a) describes uniform circular motion by kinematic movement in a flat plane, without any induction at all. As δ angle increases, θ decreases, and the eccentricity (c/a) of the corresponding ellipse increases (Figure 9.2b). If δ increases up to the maximum of $\pi/2$ making the oval become a line with two foci at both ends, then the kinematic movement becomes harmonic oscillation of the entity along the one-dimensional line AB. In the real world, planets have not this manifestation; comets approach it but with their planetary planes seriously deviate from the equatorial plane of the sun; only electrons within the atoms display such a phenomenon with matter state induction instead of kinematic movement. We may characterize the electrons with sinusoidal time and space spherical quantities as in equation (1.14). Within the atoms, electrons oscillate around the nucleus under the action of a force density obeying Hooke's law of a spring. We shall demonstrate this feature in section 9.3. Electrons tend to have a δ value at $\pi/2$ whereas planets have a θ value close to $\pi/2$. As shown in Figure 9.2, Both the kinematic movement and matter state transformation are unified by equation (9.1). The relationship is a well-known geometrical property of an ellipse, but it has a physical implication that goes unnoticed.

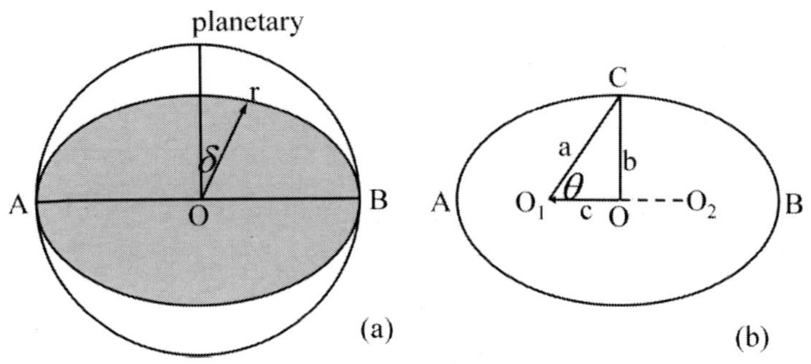

Figure 9.2. Schematic illustration of the equivalence of (a) a circle on a plane tilted an angle δ from the equatorial plane with (b) an elliptical orbit on the equatorial plane.

The equivalence of an elliptical orbit in the equatorial plane and a tilted circular orbital reflects the space and time symmetry of nature. If we regard the focus shift OO_1 in the equatorial plane as in space, then the tilting departure of the shaded plane from the equatorial plane is in the time direction, orthogonal to space. Equation (9.1) can be interpreted as a calculus relationship in a way similar to equation (1.19) that a movement in space is equivalent to a movement in time. The movements in space and in time are a mathematical trade-off with different physical manifestations. We believe that electrons travel in time primarily via state transformation as circular orbitals whereas planets take the route of space movement along elliptical orbits during the early genesis. Thus, both electronic orbitals and planetary orbits are governed by a unified natural principle, and as such Kepler's laws that govern the latter can be logically extended to the former [55, 56].

9.2. Kepler's Laws in Atomic Spacetime

Kepler's first law states that a planetary orbit is an ellipse with the sun at its focus. What is an ellipse? An ellipse is the projection of a circle onto a plane tilted at a certain angle δ as was discussed previously. Mathematically, an ellipse is defined as the locus of points, the sum of whose distances from two fixed points, O_1 and O_2, known as the foci, is a constant (Figure 9.2b). We believe that such an elegant mathematical property must have its physical significance in atomic spacetime. Under atomic spacetime, if the movement of OO_1 corresponds to the departure of the orbital plane from the equatorial plane at an angle of δ, then the movement of OO_2 must be equivalent to the rotation of the orbital plane at the opposite direction at an angle of $-\delta$. Since within a helium shell both 1s electrons transform at opposite directions, they are symmetric about space and time, i.e., the space component of one electron is equivalent to the time component of the other, and the time component of one electron matches the space component of the other. Appling the property of the ellipse to the electronic orbitals, we draw a conclusion that the sum of the time components of both 1s electrons is a constant, i.e., time conserves in a helium sphere. In symmetry with this, space conserves too. However, the sum must be understood as the addition of two orthogonal quantities that observe the Pythagorean theorem as was illustrated by a dimension diagram (Figure 1.7):

$$(C_1 \cos\alpha)^2 + \left|(-C_1\omega \sin\alpha)^2\right| = C_1^2, \tag{9.2}$$

$$(C_2 \cos\beta)^2 + \left|(C_2 r \sin\beta)^2\right| = C_2^2, \tag{9.3}$$

$$\cos\alpha = \sin\beta, \tag{9.4}$$

where the hypotenuse corresponds to O_1O_2 of an ellipse at its maximum eccentricity, which naturally degenerates into a line segment AB (Figure 1.7). At this state, $\delta = \pi/2$ and $\theta = 0$, and electrons transform by electromagnetic induction. For both 1s electrons, equation (9.2) indicates electronic time component conservation whereas equation (9.3) indicates electronic space component conservation, but they are indeed two aspects of the same process. Equation (9.4) synchronizes the space and time components of both 1s electrons in motion, which states that both radian angles α and β are complementary at any moment. Hence Kepler's first law for planetary orbits takes the form of the Pythagorean theorem for electronic orbitals in spherical quantities. Moreover, because the space component of one electron is equivalent to the time component of the other electron, the Pythagorean theorem also means spacetime dimension conservation for each electron. The last conclusion applies to every electron in rotatory operation in any atoms.

How is the Pythagorean theorem in spherical quantities related to the energy and momentum conservations in physical quantities? As shown in Figure 9.3, diameter AB is a time dimension representing $-\omega$ in the direction from A towards B. $AD = AB\sin^2\alpha = -\omega\sin^2\alpha$; and $DB = AB\cos^2\alpha = -\omega\cos^2\alpha$. The energy of an electron is proportional to the projection of its time spherical quantity onto diameter AB.

Because AD plus DB equals AB and the total energy of both electrons is E= $-\omega\hbar$ with the minus sign indicating that the electrons are trapped in the helium shell, time conservation for both electrons corresponds to energy conservation for both electrons. Likewise, diameter AB is also a space dimension representing r dimension. The decrease of a time dimension is accompanied by the increase of a space dimension. $\mathrm{AD} = \mathrm{AB}\cos^2\beta = r\cos^2\beta$; and $\mathrm{DB} = \mathrm{AB}\sin^2\beta = r\sin^2\beta$. The momentum of an electron is inversely proportional to the projection of its space spherical quantity onto diameter AB. Because AD plus DB equals AB and the total momentum of both electrons is $p = \hbar/r$, space conservation for both electrons corresponds to momentum conservation for both electrons. However, the momentum conservation manifests as $1/p_1 + 1/p_2 = 1/p$ where p_1 and p_2 are the momentum of each electron and p is the total momentum of both electrons. In spite of the enormous momentum of each electron inside the atomic sphere, both electrons show momentum conservation as a whole system. As state point C traces along semicircular arc ACB, electron Ω_0 is increasing energy and decreasing momentum; and electron Ω_2 changes the other way around. When point C reaches B, electron Ω_0 attains maximum volume with minimum density and has minimum momentum with maximum energy zero while electron Ω_2 becomes an infinitesimal point at the nucleus with infinitely great momentum and the lowest energy. Thus, within the helium shell both electrons conserve in energy as well as in momentum as a whole system. The energy conservation path is actually the course of dynamic calculus [57].

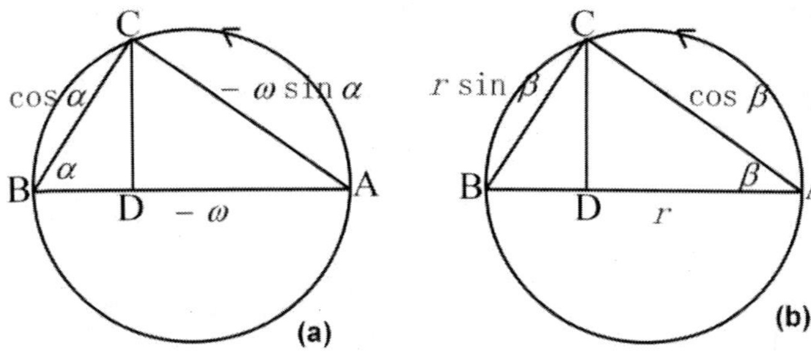

Figure 9.3. Dimension diagram of time conservation for energy conservation and space conservation for momentum conservation in a helium shell.

Kepler's second law is closely associated with the first one. To illustrate its analogy in electronic orbitals, we need to clarify the relationship between angular velocity ω and orbital radius r. As shown in quaternity coordinates (Figure 9.4), orbital radius r is a vector rotating around the origin O from dimension 1 to m while angular velocity ω is a vector tangential to arc ACB at point C, both being orthogonal with their product $v = \omega r$. As a point C moves along the arc, angular velocity ω rotates smoothly from dimension 1 to 1/s. The velocity v indicates the speed of the electromagnetic waves within the atomic sphere. For a planetary orbit, Kepler's second law states that the line connecting the planet and the sun sweeps out equal areas during equal intervals of time. If we interpret that area swept out by the connecting line within a unit time interval in an elliptical orbit as corresponding to vr in atomic spacetime, then because *vrm* is a dimension of angular momentum, Kepler's second law defines a constant angular momentum for both electrons as a whole, i.e., $m\omega r^2 = \hbar/2$, which may easily be derived from equations 1.42 and 1.43.

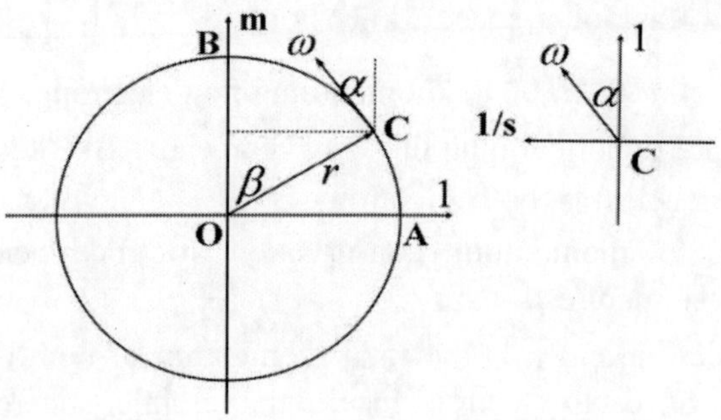

Figure 9.4. The orthogonal relationship between angular velocity ω and orbital radius r in atomic spacetime.

For a planetary orbit, Kepler's first law clarifies the elliptical orbit and the second law further specifies a constant angular momentum. Both laws combine to describe elliptical motion in a definitive way. Because Kepler's two laws may combine to define the solution of two variables in the equation group, a planetary orbit may still be treated as a two-dimensional orbit. For 1s electronic orbitals, Kepler's first law governs the spherical quantities at the dimensional level of α and β (i.e., $\alpha + \beta = \pi/2$) while the second law governs the spherical quantities at the dimensional level of ω and r. The former indicates energy and momentum conservations for both electrons as a whole; and the latter indicates angular momentum conservation for the electron pairs of similar orbital type. When an electronic orbital is specified in two consecutive dimensional levels, it is well defined. Electronic motion is totally deterministic, overthrowing the principle of uncertainty. This is the physical significance of Kepler's two laws in the spherical view.

9.3. The Energy Density and Coulomb's Law

Force is of paramount importance in physics. In classical mechanics, a central force on an object is a force whose magnitude only depends on the distance r of the object from the origin and is directed along the line joining them. Gravitational force and Coulomb (1736-1806) force are two familiar examples. For a planetary orbit, Newton deduced the famous gravitational law as follows:

$$F = G\frac{m_1 m_2}{r^2}, \tag{9.5}$$

where m_1 and m_2 are the masses of the two entities, r is the distance between them, G is a proportionality constant, and F is the gravitational force on either object. For electrostatic interaction, Coulomb's law states:

$$F = k\frac{q_1 q_2}{r^2}, \tag{9.6}$$

where q_1 and q_2 are the charges of the two particles and k is a proportionality constant. For centuries, physicists observing the similarity of the gravitational law and Coulomb's law have tried to generalize them into a unified theory to cover other interactions. However, Coulomb's law is obtained experimentally rather than theoretically. In addition to experimental evidence, the expression of a formula as complex as Coulomb's law must be a natural result of a certain mathematical deduction. If theoretical derivation of its expression has not been made satisfactorily, the law is not completely established and should be regarded as an empirical formula only. In the context of electronic wave functions in atomic spacetime, we shall explore a probable theoretical background of Coulomb's law as follows.

We have characterized both electrons within a helium shell as a pair of electric and magnetic fields. Electromagnetic waves carry energy from one region of space to another. The energy density U of an electromagnetic wave is proportional to the square of the electric field strength or of the magnetic field strength.

$$U = \frac{\mathrm{B}^2}{\mu} = \frac{\mathrm{B}^2}{-j} = \mathrm{B} \times j\mathrm{B} = C_1 \cos\alpha \times C_2 \cos\beta, \tag{9.7}$$

where α and β are complementary so that by a double-angle formula, we have

$$U = \frac{C_1 C_2}{2}\sin 2\beta. \tag{9.8}$$

Here the time and space components $C_1 \cos\alpha$ and $C_2 \cos\beta$ of spherical quantity Ω_0 correspond to physical quantities **B** and $j \cdot \mathbf{B}$ respectively, where j is an imaginary unit converting the magnetic field strength from a time orientation to its orthogonal space orientation. The term $C_1 \cos\alpha \times C_2 \cos\beta$ is treated as a product of the space and time factors of the spherical quantity for a 1s electron, different from Ω_0 as the coordination of the space and time components in equation (1.14). Radian angle β is in the decreasing range of ($\pi/2$, $-\pi/2$) for an electron while α is in the increasing range of (0, π) for another 1s electron (cf. Figure 1.18). Spherical quantity U represents an energy density or a probability density of the electron whereas spherical quantity Ω_0 represents a magnetic field strength in spacetime. Spherical quantity U is the energy density of the electron because an energy density of a magnetic field equals B^2/μ where μ is magnetic permeability of atomic space having a spacetime dimension of $-j$ (cf. Section 10.3). If we only count the energy density of an electron in the scope of $\beta \in (\pi/2, 0)$, then $U = \mathrm{B}^2/2\mu$, which is in agreement with the energy density of the magnetic field in electromagnetic waves.

Radian angle β is a circumferential angle and radian angle 2β is a central angle in Figure 9.5(a). They are always subtended by arc L as a point B traces the dashed semicircle smoothly from the nucleus point O to point A. Given the diameter of the semicircle as r, there is a relation of

$$\beta = \frac{L}{r}. \tag{9.9}$$

Substituting the β value into equation (9.8) yields

$$U = \frac{C_1C_2}{2}\sin\frac{2L}{r}. \tag{9.10}$$

Because energy density U is an inherent characteristic of the electronic orbital, the centripetal force that the electron cloud incurs per unit volume, or the force density, during the harmonic oscillation can be derived from the negative gradient of the energy density. As we consider the source of the reciprocal of distance square in Coulomb's law, one probable introduction mechanism is through the differential operation upon the reciprocal of distance. Hence radian angle $2L/r$ provides a vital link between the wave function and force. If we treat arc L as a fixed constant and regard orbital radius r as a variable, then the force density can be derived by

$$F = -\frac{dU}{dr}, \tag{9.11}$$

whence

$$F = L\frac{C_1C_2}{r^2}\cos\frac{2L}{r} = L\frac{C_1C_2}{r^2}\cos 2\beta, \tag{9.12}$$

or in terms of infinitive progression as

$$F = L\frac{C_1C_2}{r^2}\left(1 - \frac{2^2L^2}{2!\,r^2} + \frac{2^4L^4}{4!\,r^4} - \frac{2^6L^6}{6!\,r^6} + \ldots\right). \tag{9.13}$$

This is the expression of the force density that the electron incurs under the combined actions of the nucleus and the other 1s electron. The force is not equal to the central force or Coulomb force because the center of the electron cloud always stays at the nucleus (Figure 1.16). Instead, this force acts upon the volume of the electron. If the electron does not transform at all, then β is a fixed value, so is $\cos 2\beta$. Thus equation (9.12) has a similar form to Coulomb's inverse-square law. If C_1 represents time, then C_2 represents space; and If C_1 represents positive charge, then C_2 represents negative charge. Factor C_1 is the counterpart of factor C_2 in the two-dimensional spacetime framework. Without the drudgery of measurement, this indirectly proves Coulomb force as a conservative electrostatic force. However, within a helium shell, because the electrons rotate continuously by dynamic calculus, the electrostatic Coulomb's law does not hold for the oscillating electrons within the atom.

The electrostatic interaction is tricky to measure because charges are volatile or dynamical in nature. There have been some experimental reports of the inexactness of Coulomb's inverse-square law, but without theoretical guidance, they are generally considered to be within measurement errors. To think of the trigonometric term in equation (9.12) is out of the question. Theoretical attempts for modifying the inverse-square law in relativistic or quantum context are normally cumbersome and farther from the truth and hence have never been accepted either.

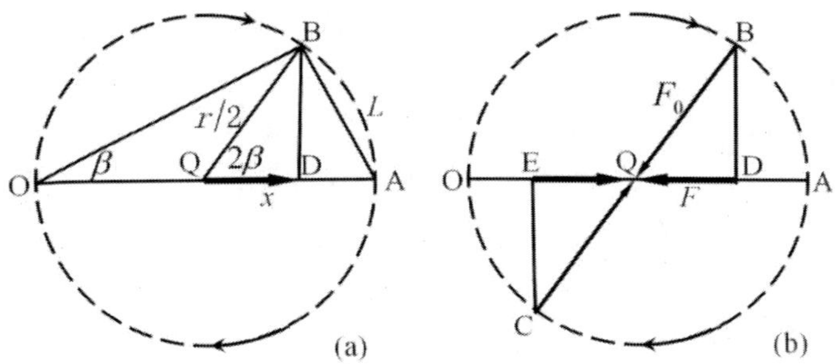

Figure 9.5. Geometric relation of β and 2β as radian angles and their corresponding arc length L in a circle with reference to Figure 1.16. For a point B tracing a dashed semicircle from the nucleus O toward A, $\beta \in (\pi/2, 0)$; and for a point C tracing the other semicircle from A toward O, $\beta \in (0, -\pi/2)$.

On the other hand, we may also regard the arc L as a variable and orbital radius r as a constant parameter. With the smooth movement of arc L or the rotation of 2β, the electronic cloud volume undergoes expansion and contraction under the influence of the centripetal force that balances its centrifugal force in sinusoidal motion.

$$F = \frac{dU}{dL} = \frac{C_1C_2}{r}\cos 2\beta. \tag{9.14}$$

As shown in Figure 9.5, the sinusoidal force F acting on the electron is the projection of a constant quantity F_0 onto the segment OA:

$$F_0 = \frac{C_1C_2}{r}, \tag{9.15}$$

$$F = F_0 \cos 2\beta. \tag{9.16}$$

Moreover, the sinusoidal electronic motion may be regarded as a harmonic oscillator under the action of an invisible spring that obeys Hooke's law. With the equilibrium position at point Q, the displacement of the electron x (segment QD) and the spring constant K can be calculated as follows.

$$x = -\frac{r}{2}\cos 2\beta, \tag{9.17}$$

$$K = \frac{2C_1C_2}{r^2}, \tag{9.18}$$

$$F = -Kx, \tag{9.19}$$

where the direction of the restoring force F is always contrary to the direction of displacement x. Since the restoring force is proportional to the negative of the displacement, the electron undergoes simple harmonic oscillation. As spherical quantity $C_2 \cos \beta$ in Figure 1.7 defines an electron transforming along a hypothetical semicircle, the spring force F likewise guides the electron on harmonic oscillation or in formal circular motion (Figure 9.5). The time dimension C_1 is associated with the property of the particle at the nucleus while the space component C_2 is related to the property of the electron cloud in the atomic sphere. Within the helium shell, when one electron resides at the nucleus and neutralizes one proton, the other electron is totally outside the nucleus. Therefore, the force is the interaction between the other proton and the electron outside. However, instead of acting upon the center of the electron, the electromagnetic force density acts upon the volume of the electron cloud. It is the contraction of the electron rather than the movement of its center that matters. Other than at this moment, Coulomb force is beside the point. We might need a radical hypothesis to circumvent this problem (cf. Review question 11.9). The force densities acting upon both electrons (at circular points B and C in Figure 9.5b) have the same magnitude with contrary directions.

In a helium atom, both 1s electrons are oscillating at a $\pi/2$ phase difference in terms of radian angle β. When one electron particle resides at the nucleus O and begins to expand, the other electron attains the maximum volume and starts to contract (Figure 1.16). They are under the action of the maximum force directed toward the equilibrium position at this moment. At the equilibrium position Q when $\beta = \pm\pi/4$, both 1s electrons have a similar volume and a similar maximum probability density (Figure 1.18), and they do not incur any force but have a strong tendency to expand or contract. To clarify, the equilibrium position Q is not a point or even a spherical surface, but a sphere at $\beta = \pm\pi/4$ (Figure 9.5a). At the equilibrium position, both electrons also have an equal energy and an equal momentum. When one electron is expanding at point B, the other electron is contracting at point C (Figure 9.5b). Both 1s electrons undergo harmonic oscillations with complementary radian angles α and β in the atomic sphere (Figure 1.16). And bearing in mind that general harmonic oscillation by dynamic calculus is a fresh concept as described in section 1.3.1, different from traditional mechanical motion, the force density that governs the electrons within atoms must be adapted to this new perception accordingly.

In a neon atom, because $2p_y$ and $2p_z$ electrons are transforming along a real circle and a spherical surface with a center at the nucleus (Figure 2.12), their centripetal force has the same direction as the central force or Coulomb force, even though these coincident phenomena should be construed as the combined effects of the other electrons and the nucleus, too. This also proves Coulomb's law for central force elegantly because these electrons do not move nearer to or farther from the nucleus during the motion.

In atoms with multiple electrons, one cannot formulate a Schrödinger's equation with Coulomb's law because Coulomb's law is an electrostatic law whereas electrons within atoms are transforming between space and time, not static. Instead, a force density similar to the form of equation (9.14) governs the electron cloud in general harmonic oscillation. Maxwell's equations may be applied to solving the electrodynamic problem. It is recognized that the principle of rotatory operations conforms to Faraday's law; the duality equation conforms to a synthesis of Maxwell's equations in free space; and the quaternity equation contains multiple levels of rotatory operations in conformity with three-dimensional Schrödinger's equation (section 4.5). Since the energy densities for electronic orbitals are all in a form similar to equation (9.8), the centripetal forces are in the form of equation (9.14) invariably. The forces accounting for electrons and nucleus interactions are transmitted sinusoidally outward through the waves of electrons (see section 5.4.3).

Because electronic motion and planetary movement are manifestations of the same physical principle in different directions, the inverse square law expression of Coulomb force can be applied to gravitational force as well. For planets in orbits, the central force must always equal the centripetal force of the planets. The deviation of Newton's gravitational law from equation (9.5) towards equation (9.12) depends on the deviation of the planets from kinematic movement towards matter state transformation. The radian angle δ in Figure 9.1 may be regarded as the upper limit of radian angle β in equation 9.12. Because planets are orbiting around the sun in the equatorial plane ($\delta = 0$), equation (9.5) holds precisely. While the sinusoidal force, which basically extends Coulomb's law to governing the dynamic interaction between the nucleus and the transforming electron cloud within atoms, stands to be verified experimentally, we are happy to see that the inverse square law is the natural outcome of differentiation with respect to the radius upon a circular function, beyond dirty data analysis.

Conclusion

This chapter has provided an excellent example on the unification of physical entities. Even though planets and electrons are remarkably disparate, they obey the same fundamental physical principle in essence. While planets revolve around the sun through kinematic movement, electrons loop around the nucleus via matter state transformation. While planets and orbits belong to different concepts, electrons and electronic orbitals are indistinguishable. While each planet revolves around the sun for incredibly numerous cycles, many electrons, say eight within a neon shell, concatenate into a single complete cycle. Kepler's first law for a planetary orbit takes the form of the Pythagorean theorem for the spacetime dimension conservation of each electronic orbital. Time conservation for both electrons indicates energy conservation for both electrons while space conservation for both electrons indicates momentum conservation for both electrons in the helium shell. Kepler's second law defines a constant angular momentum for both electronic orbitals. This equivalence demonstrates that the spherical view has grasped the fundamentals of nature rather than its diverse superficial manifestations.

In spite of the multifarious dimensions of the earth that supports diverse forms in the biosphere, the earth is a two-dimensional entity in terms of its revolution around the sun. From electrons, atoms, molecules to organisms, the number of entity dimensions increases; but from organisms, life kingdoms, biosphere to the earth, the complexity of entity dimensions can be simplified. If planetary orbits and electronic orbitals can be unified by circular motion, then an array of biological organisms converged from these two ends of waves can also be unified by the same principle. No wonder all entities share the same natural pattern characteristic of harmonic oscillations.

When an object moves in uniform circular motion, it does not do any work so as to maintain a constant speed. When an object undergoes harmonic oscillation, its kinetic energy converts into potential energy and vice versa so as to maintain a constant total energy. When an electromagnetic wave travels through induction, its electric and magnetic fields transform between each other so as to maintain a constant total energy, too. In an ideal *LC* oscillatory circuit, the electric field and magnetic field transform between each other without losing energy to the environment. In a helium shell, both 1s electrons transform between time and space without losing energy to the environment. For the total energy conservation, the law of nature must agree with the principle of uniform circular motion or general harmonic oscillation. The energy conservation path is actually the course of dynamic calculus or the semicircle defined by the Pythagorean theorem. And the energy conservation for the electron pairs of similar orbital type is the reason why the particle energy is quantized within an atom of multiple electrons such as a neon atom.

People have been searching for a grand unified theory in earnest for many centuries, but the Holy Grail of science lies on the level more fundamental than Euclidean space with Newtonian time. One can never manufacture a gigantic ship on board a small boat. It is only on the atomic spacetime platform that various physical entities could be compared and forces and wave functions could be unified in a coherent manner. And who has the temerity to challenge the applicability of Coulomb's law for electrodynamic electrons within atoms?

Chapter 10

The Order of Physical Quantities

After the unification of physical entities in Chapter 9, we now turn to the unification of spherical quantities and physical quantities themselves that characterize the physical entities. A spherical quantity assumes a fixed number with a variable dimension. When the dimension variable takes the value of a whole dimension in space and time, the spherical quantity corresponds to a physical quantity of that dimension; and when the dimension variable takes another whole dimension value, its spherical quantity corresponds to another physical quantity accordingly. The dimensions of physical quantities are discrete points on the continuous dimension curve of spherical quantities as integers are located discretely at the real number axis (Figure 2.13). Thus, a curve of continuously transforming spherical quantities connects multiple physical quantities of a series of spacetime dimensions. This chapter tries to establish such a dimension series of physical quantities and order them into a periodic table. As it turns out, this approach provides sharp insight into the properties of and relationships between various physical quantities.

10.1. Introduction

In Chapter 2, we provided an orbital interpretation of quaternity spacetime in a neon shell. There are four kinds of dimensional combinations termed the scalor, the polor, the metor, and the vitor in a four-dimensional world, each being separated by a velocity dimension (m/s). We normally label the positive direction of four quaternity axes with the 1, u, u^2, and u^3 dimensions and the negative direction of four quaternity axes with the u^4, u^5, u^6, and u^7 dimensions. Between the 1 and u dimensions, there is a $\dot{\Psi}$ or $-1/\psi'$ dimension; and between the other adjacent axes, there is a similar space or time dimension, too. For spherical layer 2, it is necessary to use sixteen physical quantities for describing 2s2p electrons in the neon shell and another sixteen physical quantities for describing 3s3p electrons in the argon shell. While physical quantities of space dimensions higher than three cannot be easily understood in Euclidean geometry, they can be expressed concretely and conceivably in quaternity spacetime that provides a wider scope in interpreting physical phenomena [58, 59].

As shown in Table 10.1, the first two rows correspond to dimensions of the 2s2p electron octet and the last two rows correspond to dimensions of the 3s3p electron octet. Each electron consists of two adjacent dimensions in a vertical column as were defined by wave functions (2.18), (2.25), (5.5), and (5.11). The first row of the table has the u^i dimensions whereas the third row features the n^i dimensions where i takes an integer from 0 to 7. Within spherical layer 2, a spherical quantity completes a cycle after undergoing sixteen consecutive rotations so that

$$u^8 = n^8, \tag{10.1}$$

which means that the unit of s^8/m^8 is equivalent to that of m^8/s^8 in an unperturbed argon atom. Dimensions u^8 and n^8 meet so that all electrons within the argon atom form a spacetime continuum.

Table 10.1. Time and space units of spherical quantities for characterizing electronic orbitals within an argon atom where s represents a time dimension and m represents a space dimension

1	m/s	m^2/s^2	m^3/s^3	m^4/s^4	m^5/s^5	m^6/s^6	m^7/s^7
1/s	m/s^2	m^2/s^3	m^3/s^4	m^4/s^5	m^5/s^6	m^6/s^7	m^7/s^8
1	s/m	s^2/m^2	s^3/m^3	s^4/m^4	s^5/m^5	s^6/m^6	s^7/m^7
1/m	s/m^2	s^2/m^3	s^3/m^4	s^4/m^5	s^5/m^6	s^6/m^7	s^7/m^8

Each cell in the table represents a dimension of a spherical quantity within an argon atom, which can be interpreted as the dimension of a physical quantity as well. If spherical quantities reflect electronic properties correctly, then their whole dimensions must correspond to physical quantities. This means that there are arrays of physical quantities that are entirely composed of space and time dimensions so that we can arrange them into a periodic table according to their dimensions. The unification of physical quantities is an appealing idea.

Given that quaternity spacetime has a wider scope than Euclidean space with Newtonian time, how do current existing physical quantities fit into the quaternity framework? Basically, current existing physical quantities having only space and time dimensions (with units of meter and second only) must fit nicely into Table 10.1 with proper interpretation of their physical significance. Examples are velocity, acceleration, wavelength, frequency, etc. Physical quantities bearing other units such as mass, electric charge, and magnetic dipole moment entail new definitions through rediscovery of their space and time dimensions. These integration processes could be long and painful. With the obstacles to overcome, can we still tabulate all physical quantities in a pattern similar to Table 10.1? We shall approach the question seriously through dimensional analyses of current existing physical quantities in the following sections.

10.2. Dimensions of Physical Quantities

The foundation of physics rests upon physical quantities, in term of which the laws of physics are expressed. A physical quantity in the general sense is a property ascribed to phenomena, bodies, or substances that can be quantified for, or assigned to, a particular phenomenon, body, or substance. It has been well accepted that the dimension of a physical quantity is the qualitative nature of the physical quantity. Dimensional analysis is often used to derive or check formulae by treating dimensions as algebraic quantities. For example, quantities can be added or subtracted only if they have the same dimensions, and quantities on both sides of an equation must have the same dimensions. Based on the status quo of dimensions of physical quantities, we shall organize physical quantities in proper space and time orders to get some clues about their dimensional relationships.

Mechanical quantities include kinematic and dynamic quantities. Kinematics is the study of bodies without reference to mass or forces involved, in which bodies are considered particles so that we can ignore the object's size, orientation, and internal structure. Commonly used kinematic quantities include position, length, time, velocity, acceleration, period, frequency, etc. They serve to tell us how a particle moves in space with time. On the other hand, dynamics describes the motion of objects taking forces, mass, and shapes into considerations. Commonly used dynamic quantities include mass, force, momentum, energy, pressure, etc. Here we arrange some mechanical quantities into a table according to their space and time dimensions (Table 10.2). Three SI base units (second, meter, and kilogram) are used to characterize their dimensions.

Table 10.2. Units of mechanical quantities where s represents a time dimension, m represents a length dimension, and kg represents a mass dimension

Kinematic quantities			
			Area m^2
	Time s	Length m	Kinematic viscosity m^2/s
	Number 1	Velocity m/s	
Wave number 1/m	Frequency 1/s	Acceleration m/s^2	
Dynamic quantities			
	Mass kg	Momentum kg.m/s	Energy, work, heat $kg.m^2/s^2$
		Force $kg.m/s^2$	Power $kg.m^2/s^3$
Dynamic viscosity kg/ms	Surface tension kg/s^2		
Pressure, stress kg/ms^2			

It can be seen from Table 10.2 that the major difference between dynamic and kinematic quantities is an additional kilogram unit associated with dynamic quantities. No direct connection between kinematic and dynamic quantities is implied here. When considering kinematic and dynamic quantities separately, from left to right each successive column increases a velocity dimension, and from top down each successive row reduces a time dimension. A space or time dimension is the smallest possible dimension, which is indivisible. By the way, we never ponder over the atomicity of mass, i.e., whether or not mass could be divided into some space and time dimensions.

Electromagnetic quantities describe various phenomena in regard to electricity and magnetism. In SI unit, they can be expressed in terms of four base quantities namely length, time, mass, and electric current. These four base quantities are assumed to be mutually independent, from which other physical quantities can be derived. Some commonly used electromagnetic quantities are tabulated in two groups reflecting their space and time relationships within each group (Table 10.3).

Table 10.3. Units of some electromagnetic quantities where A denotes an ampere dimension

Quantities with an [A] unit			
			Electric charge sA
			Electric current A
	Electric flux density sA/m^2	Magnetic field strength A/m	
Electric charge density sA/m^3	Electric current density A/m^2		
Quantities with a [kgA^{-1}] unit			
		Magnetic flux kg A^{-1} m^{2}s^{-2}	
		Electric potential kg A^{-1} m^{2}s^{-3}	
Magnetic flux density kg A^{-1}s^{-2}	Electric field strength kg A^{-1}ms^{-3}		

One group of quantities all bear an ampere unit in addition to meter and second, and the other group of quantities contain [kgA^{-1}] units. Other electromagnetic quantities are less regular in units for such a categorization. For each group in the table, from left to right, each successive column increases a velocity dimension whereas from top down, each successive row reduces a time dimension. No relations are implied here between both groups as shown in Table 10.3. After listing physical quantities in this way, it becomes obvious that there are many blank holes in the table, which indicates that many possible physical quantities remain obscure or undefined.

Ampere was defined in SI system as that constant current which, if maintained in two straight parallel conductors of infinite length, of negligible circular cross section, and placed 1 meter apart in vacuum, would produce between these conductors a force equal to 2×10^{-7} newton per meter of length. The effect of this definition was to fix the magnetic constant (permeability of vacuum) at exactly $4\pi \times 10^{-7}$ Hm^{-1}. However, if the permeability of vacuum were actually a non-unit constant, then the unit of ampere would be dependent upon the unit of newton! (Note: Hm^{-1}=kg.m.s^{-2}.A^{-2}=N.A^{-2}). In that case, ampere would become a derived quantity. We explore the possibility of unifying the four base quantities as follows.

10.3. Interpretation of Electromagnetic Quantities

Compared with spherical quantities as were shown in Table 10.1, it is obvious that current existing physical quantities are a hodgepodge of scattered items, not in unified units. They are severely fragmented because we have arbitrarily introduced the units of kilogram and ampere without realizing that both units might be the composites of some space and time dimensions. Since Chapter 1, we have decided that two fundamental spherical quantities, space and time, are enough to express all spherical quantities. Therefore one way to unify current physical quantities is to find spacetime dimensions for kilogram and ampere.

Postulate 1. Every physical quantity is in a certain combination of space and time dimensions. All physical quantity units can be expressed in terms of two base units: meter (m) and second (s).

This postulate assumes that the world is a rendition of space and time. If space refers to the whole universe, then time is only a snapshot of it. If space refers to a particle, then time is the intricate motion of it. For a given entity, the smaller its space is, the denser its time is, and vice versa. In a four-dimensional sphere, it is reasonable to deduce that space and time quantities are enough to describe all phenomena, one serving as a reference while the other as a measuring rod. We have defined four spherical quadrants in various space and time combinations for describing electrons within a neon shell. We have also used wave functions of various spacetime dimensions to account for electrons in larger atoms. Space and time are the ultimate dimensions of spherical quantities. It stands to reason that we may use two base units, meter and second, to express all physical quantities, among which mass and electric current are the most prominent ones.

To rediscover the spacetime dimension for ampere, we note that there is a rotation relationship between electric field strength **E** and magnetic field strength **B** as expressed by Maxwell's equations:

$$\nabla \cdot \mathbf{B} = 0, \tag{10.2}$$

$$\nabla \times \mathbf{E} = -\frac{\partial \mathbf{B}}{\partial t}, \text{ (Faraday's law)} \tag{10.3}$$

$$\nabla \cdot \mathbf{E} = \frac{\rho}{\varepsilon_0}, \text{ (Gauss's law)} \tag{10.4}$$

$$\nabla \times \mathbf{B} = \frac{1}{c^2}\frac{\partial \mathbf{E}}{\partial t} + \mu_0 \mathbf{J}, \text{ (Ampère's law with Maxwell's correction)} \tag{10.5}$$

where ρ and **J** are charge density and current density, respectively. The equations also involve two fundamental constants characterizing the electrical permittivity ε_0 and magnetic permeability μ_0 of free space which are related by

$$\varepsilon_0 \mu_0 c^2 = 1. \tag{10.6}$$

Because operators $\nabla\cdot$ and $\nabla\times$ reduce their operands a space dimension but in different orientations distinguished by l_0 and l_1 as was discussed in section 2.1.2, the results of the operations on a operand produce two spherical quantities with the same number of space dimensions but they must be distinguished by an imaginary unit *j*, which transforms a space dimension to its orthogonal orientation.

$$jl_0 = l_1; \tag{10.7}$$

$$-j\nabla\cdot = \nabla\times . \tag{10.8}$$

If we treat $\nabla\cdot$ as $\partial/\partial l$, and label $\nabla\times$ as $\partial / j\partial l$ to reflect the sequence of space dimension shifting, then Maxwell's equations can be rewritten as

$$\frac{\partial \mathbf{B}}{\partial l} = 0 , \tag{10.9}$$

$$\frac{\partial \mathbf{E}}{j\partial l} = -\frac{\partial \mathbf{B}}{\partial t} , \tag{10.10}$$

$$\frac{\partial \mathbf{E}}{\partial l} = \frac{\rho}{\varepsilon_0} , \tag{10.11}$$

$$\frac{\partial \mathbf{B}}{j\partial l} = \frac{1}{c^2}\frac{\partial \mathbf{E}}{\partial t} + \mu_0 \mathbf{J} , \tag{10.12}$$

where **E** and **B** are physical quantities of electric and magnetic field strengths, and l and t represent a space dimension and a time dimension, respectively. There is a pleasing symmetry about Maxwell's equations, which is particularly striking in free space where ρ and **J** vanish. Under that circumstance, performing the second derivatives upon **B** and **E** in equations (10.10) and (10.12) yields:

$$\begin{aligned} \frac{\partial^2 \mathbf{B}}{\partial t^2} &= -\frac{\partial^2 \mathbf{E}}{j\partial l \partial t} = -c^2 \frac{\partial^2 \mathbf{B}}{j^2 \partial l^2} \\ &= c^2 \frac{\partial^2 \mathbf{B}}{\partial l^2} . \end{aligned} \tag{10.13}$$

Likewise,

$$\frac{\partial^2 \mathbf{E}}{\partial t^2} = c^2 \frac{\partial \mathbf{E}}{\partial l^2} . \tag{10.14}$$

These two equations conform to the duality equation. Hence the duality equation is equivalent to Maxwell's equations in free space. The electric field strength and the magnetic field strength are two perfect solutions to the duality equation. For spherical quantities in atomic spacetime, we discarded dimension shifting indicator j because dynamic differentiation is performed with respect to the most immediate space dimension of the operand by default and the shifting of the space dimension in consecutive differential operations, such as from l_0 to l_1, is implied automatically.

By the way, we have encountered three different meanings of the imaginary units. First, when placed before a trigonometric function, a unit imaginary number indicates a space or time dimension (i.e., radius orbital or angular velocity) as shown in equations (1.2) and (1.5).

It gives the unit of the function. Second, when placed directly before a physical quantity, the imaginary unit is an operator shifting the physical quantity to its orthogonal function. This usage was exemplified in equations (1.46) and (9.7). Third, when placed before a dimension, the imaginary unit shifts the orientation of the dimension to its orthogonal direction such as equation (10.7) without changing its SI unit. In two-dimenional spacetime, the negative imaginary unit *-i* or *-j* converts a time or space diametrical dimension into a radian angle (equations 1.3 and 1.6), which signifies a close relationship between the irrational number π and the negative imaginary unit. This will become apparent in the following discussion that $-j$ equals μ_0 in spherical quantity dimension and μ_0 equals $4\pi\times10^{-7}$ T·m/A in physical quantity value. A radian angle has a physical quantity dimension of $-j$.

Due to their rotation relationship, **E** and **B** must be a velocity dimension apart, i.e., they must be in adjacent positions in the same row if both groups of physical quantities in Table 10.3 are merged together. We label the second group by a negative imaginary unit *-j* so that Table 10.3 becomes Table 10.4 in continuous space and time dimensions. From equation (10.10), we deduce a dimensional relationship between **E** and **B** by their units as

$$j(kg.A^{-1}.m.s^{-3}) = (m/s)(A.m^{-1}), \tag{10.15}$$

where *kg*, *A*, *m*, and *s* represent the units of kilogram, ampere, meter, and second, respectively. This equation can be simplified into

$$kg.m.A^{-2}.s^{-2} = -j \tag{10.16}$$

so that ampere is no longer independent of the other three base units. The left-hand side of equation (10.16) happens to be the unit of vacuum permeability μ_0, which leads to the second postulate:

Postulate 2. The dimension of the magnetic permeability of vacuum is a negative imaginary unit -j.

It can be seen from Table 10.4 that the dimensional interval between electric charge density ρ and ρ/ε_0 is velocity square because they are in the same row and separated by a slot. The position of ρ/ε_0 is defined by equation (10.11) so that ρ/ε_0 has a space dimension less than an electric field strength. If we adopt c^2 to indicate velocity square, then we have the following relationship concerning ρ and ρ/ε_0:

$$j\frac{\rho}{\varepsilon_0} = \rho c^2, \tag{10.17}$$

whence

$$\varepsilon_0(-j)c^2 = 1. \tag{10.18}$$

Comparing equation (10.18) with (10.6), we draw a conclusion that the magnetic permeability of free space μ_0 has a dimension of $-j$ and the electrical permittivity ε_0 has a unit of $j.s^2/m^2$, which agrees with the second postulate.

Table 10.4. Dimensions of some commonly used electromagnetic quantities

			Electric charge sA	Magnetic flux -j.A m
			Electric current A	Electric potential -j.A m/s
	Electric flux density sA/m^2	Magnetic field strength A/m, Magnetic flux density -j.A/m	Electric field strength -j.A/s	
Electric charge density ρ sA/m^3	Electric current density A/m^2	$\frac{\rho}{\varepsilon_0}$, -$j$.A/ms		

The above assignments of μ_0 and ε_0 dimensions clearly honor SI system and abandon CGS system of units because we may easily derive from Postulate 2 that an electric field strength and a magnetic field strength have a velocity dimension interval in SI system and hence are orthogonal, obeying the principle of rotatory operations and Maxwell's equations, whereas in CGS system, both quantities have a same spacetime dimension, leading to the difficulty of explaining their rotation relationship (10.3). Indeed, Postulate 1 implies that the dimension of a physical quantity must reflect its spacetime reality and is not an arbitrary definition. The junction between quaternity spacetime and SI system is coherent with the principle of rotatory operations. The dimensional interpretation of electromagnetic induction brings two subgroups of physical quantities in Table 10.3 together into one group in Table 10.4. The merging of both subgroups complies with Maxwell's equations. By the way, the magnetic field strength and magnetic flux density are in the same cell, both characterizing the strength and direction of a magnetic field with a different factor of –j. The magnetic field strength is orthogonal to the electric field strength whereas the magnetic flux density does not have such an orientation distinction with the electric field strength.

10.4. The Unification of Physical Quantities

A space or time dimension is the smallest dimension in physics. Below a dimension, Postulate 2 suggests that the property of a vacuum such as the unit of vacuum permeability [m.kg.s^{-2}.A^{-2}] is a negative imaginary unit. This means that the unit of ampere is a derived unit rather than a base unit. From equation (10.16), we have an algebraic relationship between four units:

$$A = \sqrt{j\frac{kg \times m}{s^2}}. \tag{10.19}$$

Hence our question becomes how to express mass in terms of space and time dimensions to ensure that ampere and other electromagnetic quantities have meaningful expressions given equation (10.19). At least, the right-hand side of the equation must be able to get rid of the square root.

A significant breakthrough is made as we ponder over the concept of a central force. What kind of spacetime dimension should force be assigned abstractly? Suppose an entity, A, is orbiting around a massive object at point O, which is exerting a centripetal force on the entity. Since time and space are symmetric, we may assume that there is a point ∞ at the very far end imposing a centrifugal force on the entity (Figure 10.1). Because point ∞ is a hypothetical point at the very far end, it is good to treat it as a vacuum property. If we use a dimensionless quantity $-j$ to represent the centrifugal property of the vacuum, then the spacetime dimension of the centripetal force from point O must be in symmetry with it. In a four-dimensional sphere such as an argon atom, an electron rotated sixteen times consecutively returns to its original state. We assign the dimension of ju^8 to the centripetal force because a quantity of $j.\text{m}^8/\text{s}^8$ dimension is in symmetry with a negative imaginary unit in quaternity coordinates.

By the way, since the vacuum permeability is postulated to be of $-j$ dimension, it can of course be treated as a special kind of spacetime force, or a centrifugal force due to spacetime curvature.

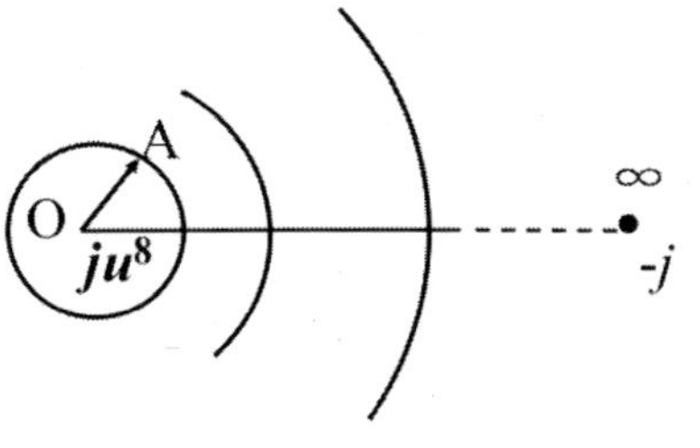

Figure 10.1. Schematic diagram of an entity A orbiting around a massive object O under centripetal and centrifugal forces from points O and ∞, respectively.

Postulate 3. Force has a unit of $j.m^8/s^8$ in symmetry with a negative imaginary unit $-j$.

Based on the above three postulates, it is easy to derive that kilogram has a dimension of $j.\text{m}^7/\text{s}^6$ according to Newton's second law, and ampere has a dimension of $j.\text{m}^4/\text{s}^4$ as calculated from equation (10.19). The dimension of other physical quantities can be deduced accordingly in this four-dimensional spacetime framework. For example, inductance has a dimension unit of (henry = $\text{m}^2\text{kgs}^{-2}\text{A}^{-2} = -j.\text{m}$), which confirms equation (1.37). Some commonly used physical quantities are ordered into Table 10.5. Such unification of physical quantities naturally results from an original exploration of physical quantity dimensions. In this periodic table, each column has the same space dimension. From top down, each successive row reduces a time dimension vertically. From left to right, each successive column increases a velocity dimension horizontally. Obviously, the periodic table is not perfect in its first version. Like a word puzzle, there are many blank holes to be filled and many dimensional relations to be verified in the table. Many existing physical quantities have not been framed in the table yet. Even some mistakes might exist in the table. We broach the methodology for organizing physical quantities with the hope that the readers may study it carefully and improve it in the end.

After organizing various physical quantities into a unified space and time dimension table, we may deduce dimensional relationships between physical quantities. Although the conditions that govern those relationships remain unknown, the periodic table provides useful clues to the possibility of certain relationships between physical quantities. It also indicates that there are a lot of physical quantities and relationships still untouched by physicists. A good periodic arrangement must be able to predict and retrodict important physical relationships. For example, Table 10.5 predicts that there is a rotation relationship between acceleration and electric flux density, and there is also a rotation relationship between ρ/ε_0 and mass density. These predictions indicate strong coupling relations between mechanical properties and electromagnetic properties even though the conditions for producing such inductions remain unspecified. We therefore suggest that careful experiments should be designed and conducted to check the validity of the table.

Table 10.5 also provides a pattern for the physical laws that govern two adjacent physical quantities. For example, although subject to verification, any two adjacent quantities in the same row obey the principle of rotatory operations, i.e., Faraday's law; any two adjacent quantities in the same column have a relation of differential with respect to a time dimension; any two adjacent quantities along the top-right diagonal direction have a relation of integral over a space dimension; and any two quantities having a dimensional gap of m/s^2 obey Newton's second law in the form of $F = Ma$ (Figure 10.2). If the architecture of the periodic table of physical quantities turns out to be correct, then it is a very useful guide to discovering new physical quantities and the laws of physics.

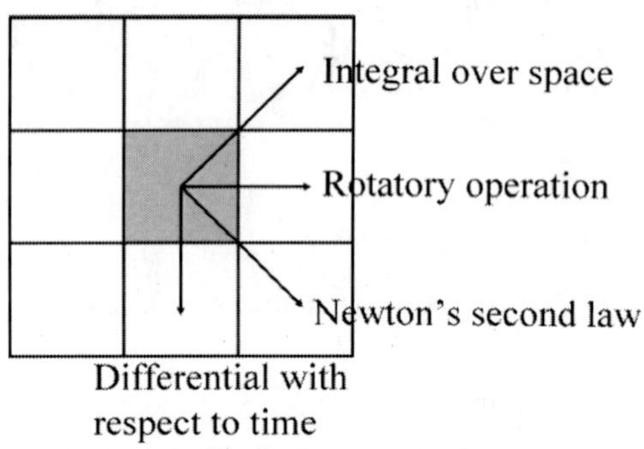

Figure 10.2. Physical laws that govern various adjacent quantities in the periodic table of physical quantities.

The significance of Table 10.5 goes far beyond what have been expected from traditional dimensional analyses. The periodic table of physical quantities is comparable to the Mendeleev (1834-1907) periodic table of elements in the 19th century. The table effectively lays the foundation for the grand unification of physical quantities and physical laws. There are still many physical quantities to be defined. Each physical quantity must have its physical correspondence and reflect the reality of the world. No matter which physical quantity is discovered and filled in the table in the future, it must observe the dimensional rule of the periodic table, obey the theory of quaternity, show its familial characteristics, and reflect physical reality by its unique property and behavior.

Table 10.5. Periodic table of commonly used physical quantities and their units in time and space dimensions. The negative imaginary unit *-j* has a radian angle dimension, which technically shifts the orientation of a dimension to its orthogonal direction

s^3/m	s^2	sm	m^2 area	m^3/s flow	m^4/s^2	$j.m^5/s^3$ electric dipole moment	$j.m^6/s^4$ magnetic dipole moment	m^7/s^5	m^8/s^6	$j.m^9/s^7$ angular momentum
$j.s^2/m$ capacitance (farad)	s time	m, length, $-j.m$ inductance (henry), $j.m$ orbital radius	m^2/s kinematic viscosity	m^3/s^2	$j.m^4/s^3$ electric charge (coulomb)	m^5/s^4 magnetic flux (weber)	m^6/s^5 electric flux	$j.m^7/s^6$ mass (kg)	$j.m^8/s^7$ momentum	$j.m^9/s^8$ energy, heat (joule)
$j.s/m$ electric conductance (siemens)	$-j$ permeability, radian angle, 1 scalar	m/s velocity, $-j.m/s$ electric resistance, (ohm)	m^2/s^2 specific energy, absorbed dose(gray)	m^3/s^3	$j.m^4/s^4$ electric current (ampere)	m^5/s^5 electric potential, electromotive force (volt)	m^6/s^6	m^7/s^7	$j.m^8/s^8$ force (newton)	$j.m^9/s^9$ power (watt)
1/m wave number	1/s frequency (hertz), $-j.1/s$ angular velocity	m/s^2 acceleration	$j.m^2/s^3$ electric flux density	m^3/s^4 magnetic flux density (tesla), $j.m^3/s^4$ magnetic field strength	m^4/s^5 electric field strength	m^5/s^6	$j.m^6/s^7$ dynamic viscosity	$j.m^7/s^8$ surface tension	m^8/s^9	m^9/s^{10}
1/sm	$1/s^2$	$j.m/s^3$ electric charge density ρ	$j.m^2/s^4$ current density	m^3/s^5 ρ/ε_o	$j.m^4/s^6$ mass density	m^5/s^7	$j.m^6/s^8$ pressure (pascal), $-j.m^6/s^8$ energy density,	$j.m^7/s^9$ heat flux density, irradiance	m^9/s^{10}	m^9/s^{11}

10.5. Dimensions of Thermodynamic Quantities

There are still some difficulties in finding all the connections between the dimensions of various physical quantities. It might be easier to define a new set of physical quantities to account for a specific phenomenon than to define it with existing physical quantities. Because the laws of physics have been constructed upon physical quantities, the crucial point in understanding the physical laws lies in the correct definition and rediscovery of the dimensions of physical quantities. Take the unit of joule as an example. It may refer to quantities of energy, work, torque, heat, enthalpy, etc., which share the same spacetime dimension. In the history of science, thermal energy was connected with mechanical energy by James P. Joule (1818-1889). The mechanical equivalent of heat states that motion and heat are mutually interchangeable and that in every case, a given amount of work would generate the same amount of heat, provided the work done is totally converted to heat energy (1 calorie = 4.186 joules). Energy is an important quantity that can be expressed as the product of other quantities with various space factors such as:

$$E = Fl, \tag{10.20}$$

$$E = \tau A, \tag{10.21}$$

$$E = PV, \tag{10.22}$$

where E represents energy or work; F, τ, and P are force, surface tension, and pressure, respectively; and l, A, and V represent one, two, and three space dimensions, respectively. In quaternity spacetime, these three equations can be written as:

$$\frac{\partial E}{\partial l} = F, \tag{10.23}$$

$$\frac{\partial^2 E}{\partial l^2} = \tau, \tag{10.24}$$

$$\frac{\partial^3 E}{\partial l^3} = P, \tag{10.25}$$

which mean that the gradient of energy is force, the flux of energy is surface tension, and the density of energy is pressure where $\partial / \partial l$, $\partial^2 / \partial l^2$, and $\partial^3 / \partial l^3$ are defined as gradient, flux, and density respectively in the spherical view. Note that a review of Table 10.5 reveals some confusion over the meaning of "flux" and "density" in the nomenclature of physical quantities. Since quaternity space consists of four basic space dimensions, we need to search for the four-dimensional space quantity beyond three-dimensional cubage.

One probable candidate for a four-dimensional space is entropy because it is an extensive state variable that measures the disorder of a system. In thermodynamics, entropy is a measure of the number of specific ways in which a thermodynamic system may be arranged, often taken to be a measure of disorder, or a measure of progressing towards thermodynamic equilibrium. We suggest that entropy has one space dimension more than a three-dimensional volume, the fourth space dimension being the order of elements that gives entropy a statistical significance associated with the number of possible microstates. The connection between order and space is not as far-fetched as it might seem at first glance. To illustrate, we all know that a disordered office gives us the feeling of crowding. After properly organizing the items in the office, more working space becomes available. Thus the order of elements occupying a certain volume possesses a space property because the arrangement directly affects the apparent space.

For another example, as shown in Figure 10.3, when four streaky squares are bound properly, they make five squares, the fifth one being composed of the four. If the four squares represent a space dimension, then the fifth square is another space dimension created by ordering. The entropy of the fifth square has to do with the permutation of the four squares without violating the feature of the fifth one as a streaky square. The order of elements makes an additional space dimension.

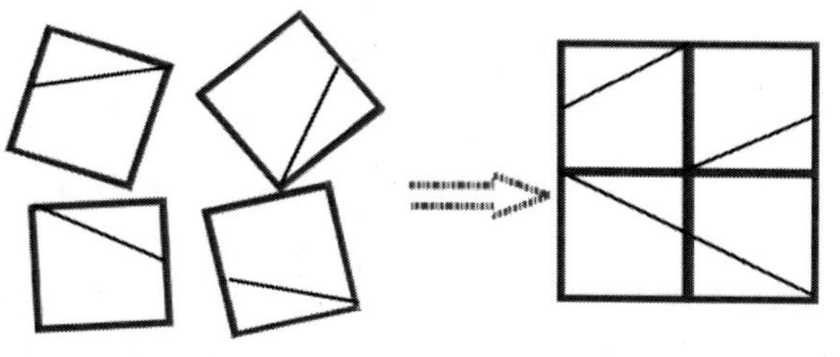

Figure 10.3. When four scattered squares are ordered properly, they form a bigger square unit with a similar streak.

Due to the presence of order, the whole is more than the sum of its parts. Suppose we are offered a choice between the parts of a disassembled watch collected in a heap and the exact number of parts but arranged into a fully functioning timepiece. Obviously, the assembled watch is preferable to a pile of loose pieces. Every organized, functioning whole is worth more than the sum of its parts. Isn't the intangible relationship between all the functioning parts an additional space dimension that must be enumerated as part of the whole?

To further illustrate with a real entity, within a human body, many cells form a liver, an organ that can be regarded as a large cell capable of independent functioning. The fourth dimension of the liver, besides length, width, and height (as measured by the number of liver cells in certain lineups), is the order of liver cells to form the liver while ensuring the proper metabolic function of it. Entropy represents the activity of liver cells.

Conversely, according to the third law of thermodynamics, the entropy of any pure substance in a perfect crystal at absolute zero kelvin is exactly equal to zero. As the temperature increases to disturb the crystal, the entropy increases, leading to disorder in the system. Thus, disorder is a dimension beyond three space dimensions of a crystal. Entropy has four space dimensions.

Postulate 4. The functional order is a space dimension in addition to length, width, and height, and entropy is a four-dimensional space quantity

If this hypothesis proves to be a correct rediscovery, then the differential change of dS in the entropy of a system when differential heat transfer dQ accrues to the system at temperature T can be expressed as:

$$dS = \frac{dQ}{T}, \tag{10.26}$$

$$S = Vl, \tag{10.27}$$

so that

$$\frac{\partial^4 Q}{\partial l^4} = T. \tag{10.28}$$

In this way, we have identified four space dimensions as physical quantities of length, area, volume, and entropy, and effectively integrated thermodynamic quantities into the periodic table of physical quantities where entropy bears a spacetime unit of m^4 and temperature has a dimension of $j.m^5/s^8$. The relationship between temperature and entropy is parallel to that between pressure and volume in the sense that pressure is acting upon three dimensions of the volume as in the case of compressed gas whereas temperature influences four dimensions of the entropy. Both pressure and temperature are intensive state variables, whose values are independent of the size of the system, while volume and entropy are extensive quantities, whose values are proportional to the size of the system.

Table 10.6. Four space dimensions and their combinations for physical quantities. The orthogonality of dimensions warrants their multiplication or division in calculus

Spherical quantities	Integral	Dimension number	Spherical geometry	Linear geometry	Derivative
Polor	$\int \Phi_0 dl_3$	1	Point displacements	Delimited length	$\partial^4\Phi_4/\partial l_0\partial l_1\partial l_2\partial l_3$
Metor	$\iint \Phi_0 dl_3 dl_2$	2	Semicircular arcs	Circumscribed area	$-\partial^3\Phi_4/\partial l_0\partial l_1\partial l_2$
Vitor	$\iiint \Phi_0 dl_3 dl_2 dl_1$	3	Hemispherical surfaces	Surrounded volume	$\partial^2\Phi_4/\partial l_0\partial l_1$
Scalor	$\iiint\int \Phi_0 dl_3 dl_2 dl_1 dl_0$	4	Occupied sphere	Imbued entropy	$-\partial\Phi_4/\partial l_0$

The relationship between various dimensions and their physical significance is illustrated in Table 10.6. For example, a displacement is one-dimensional; a pair of semicircular arcs of spherical quantities define a circumscribed circular area of a physical quantity, which is two-dimensional. The volume is an empty space inside the spherical surface, which is three-dimensional. As long as any items are introduced into the volume, there is some order of the items inside the volume accompanying them so that the system of maximum space dimensions must be characterized by entropy rather than volume. Compared with the traditional three-dimensional space with one-dimensional time world, our four-dimensional space sphere has an entropy variable that corresponds to the time dimension. Because according to the second law of thermodynamics, the entropy of any isolated system always increases with time, the entropy increment may represent the time direction. The entropy change indicates motion of the system, as the time change produces events. The connection between space and time dimensions is through an event in which the time dimension affects all three space dimensions, bearing in mind that length, width, and height have different dimensional levels (Figure 2.5). On the other hand, entropy is a variable of the order of items in the three-dimensional space. Thus, it is rational to regard entropy as a four-dimensional space physical quantity with the introduction of the item order as the fourth space dimension (Figure 10.4).

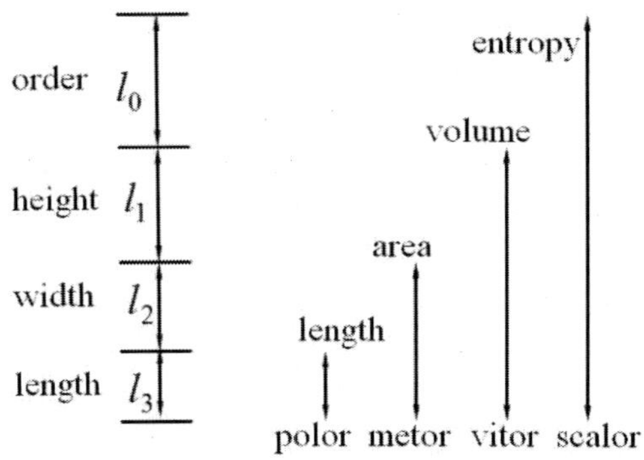

Figure 10.4. Schematic diagram of four quaternity space dimensions l in relation to representative physical quantities.

Why does the functional order qualify as a space dimension? As was described in section 1.5.2, a dimension is defined as an entity undergoing uniform circular motion or general harmonic oscillation. Each electron rotating in a neon atom constitutes a spacetime dimension, i.e., half a space dimension with half a time dimension in its rotating operation $\int (\partial\Phi/\partial t)\ dl$. The simplest pure space dimension can be found in equation (1.5):

$$\begin{aligned}\psi &= \cos\beta + j\sin\beta \\ &= e^{j\beta} = e^{j\frac{l}{r}} = e^{l}.\end{aligned} \tag{10.29}$$

In this way, a space dimension l appears in the exponential index. The imaginary number j is equivalent to orbital radius r except that j has not a unit but r has a length unit so that the entire exponential index is dimensionless after cancelling each other out. Without exchanging with half a time dimension, here e^l indicates two semicircular arcs in a full space dimension. As was shown in Figure 1.7(b), allowing radian angle variable β in equation (10.29) to rotate in the range of (π/2, -π/2), the moving point C traces the circle ACB once, constituting a full space dimension l. Circular complex function ψ is the set of all possible state combinations of chords AC and BC as point C traces the whole circle.

On the other hand, according to Boltzmann (1844-1906), the entropy of a system in a given macro state can be written as

$$S = k \ln W, \tag{10.30}$$

where k is Boltzmann' s constant (k=1.38x10^{-23} J/K) and W is the number of microstates corresponding to the given macro state. W is called the thermodynamic probability, or sometimes, the disorder parameter. This equation can be written in exponential form as

$$W = e^{\frac{S}{k}}, \tag{10.31}$$

which conforms to equation (10.29) in one-dimensional space and equation (5.54) in the four-dimensional space framework if we regard S/k as a space dimension index. This comparison effectively provides sharp insight into the nature of entropy as a physical quantity of four-dimensional space dimension.

If S/k indicates a space dimension index, then how much is the magnitude of S/k to define a physical quantity dimension? The answer depends on the specific system of concern as the length of space dimension l depends on the specific object. For example, in a chemical reaction system, reactant molecules must reach the activation energy for the reaction to progress. The proportion of reactant molecules that reach the activation energy E is determined by exp(-E/RT), or exp(-S/k), where R is gas constant 8.315 J/mol·K. These activated reactant molecules occupy the microstates beyond W, capable of chemical reaction. They may be regarded as the other genus of molecules with a significantly different property from the normal molecules, or the other space dimension beyond the dimension of S/k. The four-dimensional index S/k allows function exp(S/k) to define possible permutations of molecules with energy not greater than E. Thus, the activation energy E or the corresponding entropy value, $S/k = E/(RT)$, determines the upper boundary of the space dimension of the normal molecules. This application example illustrates the qualification of S/k as a dimension index in the exponential function.

From various aspects, this section has successfully established the connection between four dimensions of spherical quantities and four dimensions of physical quantities, giving quaternity space a complete and coherent explanation.

Conclusion

The urge to discover a fundamental theory underlying all natural phenomena has been expressed since the beginning of civilization. However, in seeking the Holy Grail of physics, it was a temptation for physicists to jump to the formulation of certain physical laws without first unifying physical quantities. After all, physical laws are based on physical quantities that describe nature in its crude shape. A successful unification of physical laws should be conducted at the low level physical quantities first. In this chapter, we have approached physics from ground up by delving into very basic concepts of space and time and analyzing dimensions of current physical quantities. Abductive reasoning was wisely applied to defining the spacetime dimensions of physical quantities based on four postulates. Through a periodic organization of physical quantities, we have been paving the way to the unification of physical laws. It is believed that physical quantities have definite patterns like chemical elements in a periodic table, and tabular relationships between any two or more physical quantities constitute the so-called physical laws.

We are confident that a grand unified theory that physicists have long dreamed of does exist and have tried to set up the methodology in a unique manner to accomplish the discovery. A grand unified theory originally refers to a model in particle physics in which at high energy, the three gauge interactions of the Standard Model which define the electromagnetic, weak, and strong interactions or forces, are merged into one single force. We unified forces with wave functions in chapter 9. Here we attempt to unify all physical quantities besides forces. We have actually demonstrated the order of physical quantities in accordance with the four-dimensional spacetime model of an argon atom. However, to discover the grand unified theory, we still need a substantial amount of detailed researches on the junctions of spherical quantities and physical quantities. How to translate physical quantities in Cartesian coordinates into spherical quantities in quaternity coordinates is one major issue; and how to interpret spherical quantities in terms of well-known concepts is another. Considering the fragmentation of current physics in various fields, there is not such a readily available grand unified theory to account for everything. A programmer who wants to develop an all-embracing version of a software application based on existing scattered codes must design unified interfaces and integrate the old and new codes. The grand unified theory will require a consistent manner in describing the world, and quaternity spacetime has provided this new platform for defining spherical quantities in dynamic calculus that bridge the dimensional gaps between physical quantities. Although great barriers exist for many physical theories remain only isolated successes in specific areas, we have shown the shining light of the grand unified theory at the other end of the tunnel.

Although this chapter has put forward a unification theory based on mathematical and logical deductions, it is quite possible that one might probe a certain physical quantity and discover an oversight or conflict in this chapter under conventional paradigms. On the one hand, one must be prepared to change his or her mindset as we migrate from Euclidean space to quaternity spacetime, which might have unexpected properties. On the other hand, this chapter is an initial attempt to suggest a probable way of unifying physical quantities. Appendix III lists some of the fundamental laws that we have unified in the spherical view throughout the book. It is a promising step toward grand unification for physical laws. Mendeleev did not provide the periodic table of elements as it is nowadays, but his original arrangement of chemical elements gave us a sound approach for studying elemental behavior. In the same manner, this chapter furnishes the clues and methodology toward discovering a grand unified theory. A glimpse of Table 10.5 reveals how little we have known about physical quantities and physical laws! Who says physics is a mature discipline?

Chapter 11

The Rational Core of Chinese Medicine

After writing the first ten chapters of this book, I was trying to expound on the implications of the theory of quaternity in philosophy, but ended up with the finding that a good part of what I wanted to say had been stated in traditional Chinese medicine (TCM) and philosophy. This chapter is a bidirectional exploration, extending the theory of quaternity and the law of nature to traditional Chinese medicine on the one hand, and explaining the implications of quaternity by the philosophy of traditional Chinese medicine on the other hand. The merging of the new science with ancient philosophy is a process of mutual validations, which stands out as an emerging enterprise of natural science. The relationship between both electrons in a helium shell can be explained in terms of yin and yang interaction; the five element theory can be illustrated by the geometrical characteristics of electronic orbitals $2s2p_x2p_y2p_z2s$ in a carbon atom; and eight trigrams express consecutive rotatory operations, coinciding with the 2s2p electron octet within a neon shell. Despite its infinite variety, the most amazing thing about nature is its unitary principle. Our rational faculties are capable of conceiving the law of nature for it is by that law that we have evolved from and been an integral part of nature. Rational thinking to reflect truth is universal regardless of the modern or ancient era and of Western or Eastern origin. Here we try to integrate our spherical view with traditional Chinese medicine.

11.1. Introduction to Yin and Yang

Since Chapter 8, we have found that spherical quantities of time and space are so broad in their meanings that they may not refer to the same properties as their corresponding physical quantities. A spherical quantity is a dimensional quantity. A spherical quantity measures a dimensional property in numerical form. In a dimension diagram, a dimension is a diameter for an electron to traverse via its subtending semicircular arc during dynamic differential or integral transformation. For a rotatory operation, the time dimension is the one, with respect to which dynamic differentiation is performed upon a spherical quantity; and the space dimension is the one, over which dynamic integration is performed upon the spherical quantity. In a rotatory operation, time is yang which thins out or vaporizes while space is yin which grows to take shape. Spherical quantities of time and space change sinusoidally rather than uniformly. They transform by dynamic calculus instead of linear algebra. Nevertheless, the abstraction of spherical quantities from diverse entities is comprehensible in terms of harmonic oscillation. Time and space as spherical quantities in harmonic oscillation equations agree with those as physical quantities. And the mathematical formulae for harmonic oscillation equations in both dynamic calculus and infinitesimal calculus are similar as well.

In a two-dimensional world, space and time refer to the two outstanding intertwined dimensions respectively such as Ω_0 and Ω_2, male and female, animal and plant, the biotic and the abiotic, etc. as were shown in Table 8.1. Space means medium while time denotes propensity. They must always be associated with physical entities as two sides of a coin cannot be divorced from the coin itself. In a four-dimensional world, space and time are blended into four spherical quadrants: the scalor (space end), the vitor, the metor, the polor, and the scalor (time end). Space is the whole sphere while time is a black box unit or the core that encapsulates all complexities within itself. If time refers to a seed, then space is the whole plant developing from it. Space and time are not absolute concepts. They are meaningful only within a certain spherical layer, like the relationship between a mother and a daughter in a specific family. Beyond that scope, the mother is a daughter of the grandmother and the daughter may become a mother in the future, too. Whether a spherical quantity is a space or a time quantity depends on its role in the spherical layer of concern.

In a system of more than four dimensions, we may still use the four-dimensional framework to account for most of it by insulating its inner shell. This structural paradigm is scalable. The abstraction of space and time characterizes the main structure of the world and is an important step toward better understanding of the complex world order.

Spherical quantities of space and time correspond to yin and yang concepts in traditional Chinese medicine. Yin and yang are opposite in meaning. Yin refers to feminine, passive, and negative principle in nature; and yang refers to masculine, active, and positive principle in nature. Yin refers to the moon, north or shady side of a hill, and south of a river; and yang refers to the sun, south or sunny side of a hill, and north of a river (on the northern hemisphere). The yin and yang pair is one of the most fundamental concepts in the traditional Chinese medical theory, and it is the foundation of disease diagnosis and treatment [60, 61]. Like space and time concepts, the concepts of yin and yang are relative. They can only be spoken of in a relationship. If yin refers to cool, down, front, and static; then yang refers to warm, up, back, and dynamic. For example, water is yin relative to steam but yang relative to ice. They are a pair of opposites, such as the opposite ends of a cycle like the seasons of a year. If yin refers to spring, then yang refers to autumn. If yin refers to night, then yang refers to day.

Space and time are complementary and constantly transforming between each other, so are yin and yang. We have characterized both 1s electrons within a helium atom in space and time wave functions. Both electrons are complementary within the helium shell. The two-dimensional system is analogous to the Taiji diagram, which features a pair of yin and yang fishes shown in black and white respectively (Figure 11.1). Both fishes are complementary inside a circle. The most interesting point is that the black yin fish has a white eye whereas the white yang fish has a dark point, the meaning of which often escapes public notice. As a matter of fact, the ancient Chinese already knew that yin and yang are transforming between each other. The white eye of the black fish must have indicated its connection with the tail of the white fish; and the tail of the black fish is the eye of the white fish. Nothing remains totally yin or totally yang. Just as a state of total yin is reached, yang begins to grow from the inside. Yin contains the seed of yang and vice versa. Yin and yang correspond to both 1s electrons and should not be confused with the concepts of negative and positive charges (cf. Review question 11.9). We have described the geometric transformation of both 1s electrons in a sphere (Figure 1.16). As an alternative figure, the Taiji diagram is the best way that I can think of to portray the transformation of both 1s electrons on a plane.

Figure 11.1. The Taiji (supreme ultimate) diagram showing the interdependent relationship of yin and yang.

However, as we unify our space and time theory with the yin and yang theory where appropriate, please bear in mind that the transformation between space and time is by strict dynamic calculus while the transformation between yin and yang is undefined mathematically. A pair of space and time spherical quantities can be regarded as a pair of yin and yang safely; but a pair of yin and yang is not necessarily a pair of space and time. Therefore, the extension of the space and time theory to cover yin and yang must be taken with caution.

Relative levels of yin and yang are continuously changing. Normally this is a harmonious change, but when yin and yang are out of balance, they affect each other. Too much of one can eventually weaken or consume the other; and too little of one may be taken over or surpassed by the other. It is suggested that the human body should always maintain the equilibrium between yin and yang. The imbalance of yin and yang elements might result in illness. This is a basic principle of traditional Chinese medicine. Traditional Chinese medicine is a broad range of medicine practices sharing common concepts that have been developed in China over a long history, including various forms of herbal medicine, acupuncture, massage, exercise, and dietary therapy [62]. Although traditional Chinese medicine is overshadowed by Western medicine nowadays, we shall prove that its very common theories are scientifically sound in this chapter.

Whether it is an electromagnetic wave, an atomic shell, a DNA double helix, or a human lineage, the harmonic oscillation between Yin and Yang is fundamental in nature, ensuring the stability of the rope or entity.

11.2. The Common Origin of *"Yi Jing"* and TCM

The yin and yang theory multiplied and flourished in an ancient Chinese scripture, "*Yi Jing*" (or "*I Ching*"). "*Yi Jing*" and traditional Chinese medicine (TCM) share the same origin in the ancient Chinese history. Written more than three thousand years ago, "*Yi Jing*" is considered to be one of the oldest extant scriptures in the world, China's unique heritage, and a great treasure for all mankind. Some diagrams in it were created five thousand years ago, preceding Chinese characters. By the time of Zhou Dynasty (about 1100 BC), the whole scripture of "*Yi Jing*" had been completed after experiencing three dynasties and over the hands of three generations of saints. This classic text describes an ancient system of philosophy that is intrinsic to Chinese cultural beliefs, and its influence on Chinese people is far-reaching and persistent.

When the ancient clan Bao-Xi ruled the country, they observed upwards the images in the sky, watched downwards the patterns of earth, paid attentions to the characteristics of beasts and birds in harmony with their native habitats, abstracted bodies nearby and multifarious objects far away, and started to draw eight trigrams so as to understand the revelation of the divine power and categorize the temperaments of all lives. It is evident that the quest for truth is a noble cause of intellectuals both ancient and modern. We don't know exactly how much the ancient Chinese learnt about the law of nature, but their unmistakable diagrams indicated that they had mastered the law in reasonable detail, about which even modern scientists do not know. "*Yi Jing*" and "*Yellow Emperor's Inner Canon*" are the very sources of traditional Chinese philosophy and medicine, both originating from the same theory of yin and yang. However, since Confucius's humanistic interpretation of "*Yi Jing*" in ten commentaries, known as ten wings, the study of "*Yi Jing*" has been focused on the ethics, getting farther and farther irrelevant to science over time. Thus the best way for me to do here is to borrow the terms and diagrams from "*Yi Jing*" and vest them with my own views, ignoring their original denotations. What I am trying to accomplish here is to establish the proper connections between my fresh discovery in mathematical physics and the ancient philosophy to avoid reinventing the wheel.

To bridge the philosophical gap between the East and the West is one of my life dreams. While taking a biochemistry course at the University of Delaware in 1996, I discussed with Professor J. Herbert Waite about my initiative of integrating traditional Chinese medicine with the Western sciences, and his opinion was that that isn't easy though he believes in the pharmaceutical effects of Chinese herbs. Humbled by the tremendous task ahead of me, I devoted my time to it assiduously. There are reasons why the West and the East differ from each other. The Occident developed linear mind of thinking for it seemed that there were inexhaustible resources across the Atlantic Ocean and farther in the American West to support economic expansion whereas the Orient was inclined toward spherical mind of thinking by the evidence that limited resources had been recycled once and again over thousands of years of civilizations. Apart from the different mindsets, English is a phonetic language, its fluidity allowing it to become a global vehicle for communication, whereas Chinese began as a pictographic language, the solidity of its compact square characters making it efficient and durable in recording history. When we consider the contrary frames of mind and dissimilar phyla of languages, the barrier to the integration is as formidable as the ten thousand mile long Great Wall, which insulated the Chinese from Hsiung-Nu nomads for peace over a long history. And I could not expect to make great progress in the integration but start with a trivial step as follows.

Although not any part of my spherical view has been proposed based on the knowledge of "*Yi Jing,*" some rudimentary thoughts of "*Yi Jing*" coincide with my ideas. This is where the West meets the East. The very West is the East. So I find myself actually having coded the program prototyped by my ancestors. This section accounts for three correspondences between the spherical view and "*Yi Jing*" philosophy, leaving more for later sections to discuss. First, the law of nature is simple to govern the world robustly; the law of nature is invariable no matter how weird somethings may appear to be; and the law of nature characterizes the continuous changing of everything in the universe. The title of book "*Yi Jing*" expresses these three properties of the law. In Chinese, *jing* means scripture, and *yi* has three basic meanings: simple and invariable as adjectives, and to change or exchange as a verb. Thus, *yi* accurately characterizes the properties of natural law, which are simple, invariable, and about changing. Among them, the simplicity is the synthesis of the pair of thesis and antithesis, namely the invariable rest and changing motion, or yin and yang. "*Yi Jing*" may be literally translated as "The book of changes."

Second, as was discussed in section 11.1, our definitions of space and time dimensions correspond to the core concepts of yin and yang in "*Yi Jing*." The spherical view deals with space and time dimensions whereas the entire scripture of "*Yi Jing*" is about yin and yang interaction. *Yi* has Taiji; Taiji produces two poles of yin and yang; two poles produce four quadrants; four quadrants produces eight trigrams; and eight trigrams determine good or ill luck, which in turn determines the prosper or decline of an enterprise. The interaction of yin and yang is called *yi*. If yin and yang refer to both 1s electrons, then Taiji can be understood as a helium shell. When one electron occupies the entire helium sphere representing a space dimension, the other is an infinitesimal point at the center representing a time dimension. If yin and yang refer to an ovum and a sperm respectively, then Taiji is a zygote at the moment when the sperm enters the ovum, a fertilized egg ready to multiply.

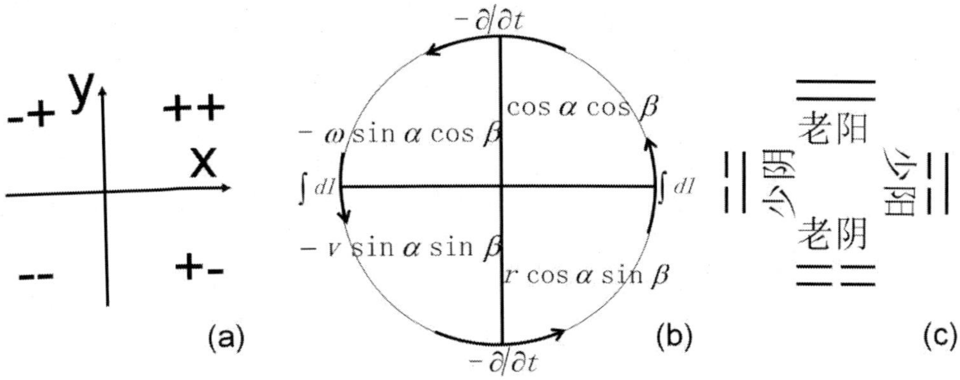

Figure 11.2. Comparison of signs, sines, and lines in four quadrants.

Third, four quadrants produced by both yin and yang poles is analogous to four quadrants on the Cartesian plane. Yin and yang are drawn as a divided line and a whole line to symbolize female and male respectively, and two pairs of yin and yang produce four quadrants. As shown in Figure 11.2, the left plot indicates positive and negative signs of the X- and Y-coordinates in four quadrants in a Cartesian coordinate system; the middle plot lists sine and cosine functions of matrix (1.14), four roots to the duality equation; and the right one shows four quadrants from "*Yi Jing,*" termed old yang, young yin, old yin, and young yang counting counterclockwise from the top. Note that the standpoint of the observer is at the center of the plot in Figure 11.2c. The plus sign corresponds to cosine and to yang; and the minus sign corresponds to sine and to yin. The plots speak for themselves the unity of three systems even though Cartesian coordinates are related to linear algebra, circular functions observe dynamic calculus, and the yin and yang theory indicates changes by its connotative binary expression. A truth can be approached in diverse but unified manners.

11.3. Literal Meanings of Four Spherical Quadrants

In analytic geometry, four quadrants normally refers to the four areas into which a plane is divided by the X and Y axes in a Cartesian coordinate system, designated first, second, third, and fourth, counting counterclockwise from the area in which both coordinates are positive (see Figure 11.2a). After introduction to the quaternity theory, four spherical quadrants refers to the four regions into which a sphere is divided by four quaternity axes, designated first, second, third, and fourth, counting counterclockwise from the region bounded by the *1* and *u* axes (see Figure 2.4). Spherical quantities in four spherical quadrants are termed distinctly: a polor is a spherical quantity of the first spherical quadrant, a metor of the second, a vitor of the third, and a scalor of the fourth. Geometrically, a polor is a point moving along a line segment; a metor is an arc stretching along a semicircle; a vitor is a curved surface extending over a hemisphere; and a scalor is a crescent solid waxing into a sphere. Four kinds of spherical quantities may be vividly expressed by various arrows as shown in Figure 11.3.

For translation, the proper Chinese characters matching four spherical quantities are readily available from "*Yi Jing,*" namely, Yuan for polor, Heng for metor, Li for vitor, and Zhen for scalor, respectively. Regarded as the very classic of Chinese culture, "*Yi Jing*" expresses our ancestor's good understanding of the natural law in pithy vocabulary and vivid diagrams. Besides defining four spherical quadrants in physics and geometry, it is worthwhile to dig out their literal meanings from both English words and Chinese characters. As we'll see in the next paragraph, the evidence for a perfect match between both languages for four varying spherical quantities is overwhelming.

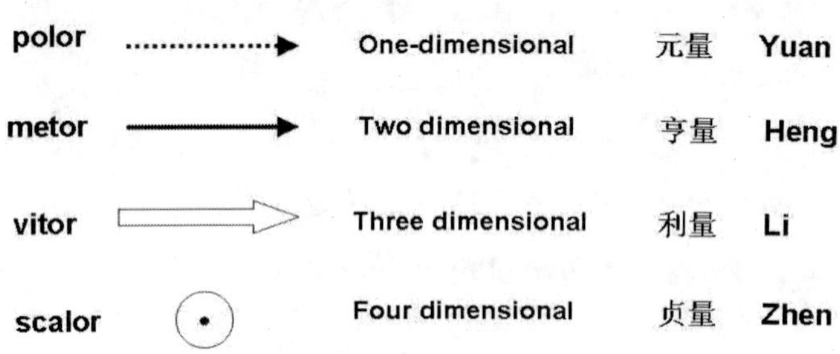

Figure 11.3. Symbols and dimensions of four spherical quantities.

Yuan is in the first spherical quadrant, referring to an entity that is polarizing as "polor" implies. Yuan means inchoation, agglomeration, origination, debut, and walking out of fog with great ambition. Heng is in the second spherical quadrant that connotes metamorphosis after the polor as prefix "meta" implies. As the second developmental stage after Yuan, Heng means progression, penetration, traverse, expedition, and marching circuitously without hesitation. Li is the third developmental stage of the affair that manifests vitality and maturity, which is the meaning of the coined word "vitor" from prefix "vita." Li indicates enclosure, sincerity, well-being, benefit, and steering around aptly and harmonically. Zhen is in the last spherical quadrant that completes a developmental cycle fairly. With a certain scale that delimits a spherical layer, as the meaning of "scalor," Zhen indicates richness, soundness, purity, perseverance, and awaiting orders humbly and instinctively for the next cycle. A new cycle initiates after Zhen grows into a full sphere. Four spherical quadrants also refer to four seasons in a year, forming a full cycle (Table 11.1).

Table 11.1. Chinese and English bidirectional translations of four spherical quantities

Spherical quadrant	Spherical quantity	English meaning	Chinese character	Chinese meaning	Geometric meaning	Season
I	Polor	Polarizing	Yuan	Birth	Moving point	Spring
II	Metor	Metamorphosis	Heng	Growth	Extending arc	Summer
III	Vitor	Vitality	Li	Maturity	Spreading surface	Fall
IV	Scalor	Scale	Zhen	Chastity	Waxing crescent	Winter

A polor initiates another cycle after reaching a full scalor.

In addition, according to "*Wenyan Zhuan*" that literally means "*Classic commentaries on texts and characters*" by Confucius (551-479 BC), Yuan is the growth of kindness because man is innately benevolent at the outset; Heng is the gather of goodness or politeness after education; Li is the sum of justice in conduct; and Zhen is the undertaking of work firmly on principle. Such an ethical interpretation identifies the four characters as gentlemen's morality in four developmental stages of virtue cultivation and reflects the philosophy and value orientation of ancient Chinese culture.

11.4. Geometrical Interpretation of Five Elements

Based on the ancient Chinese philosophy, there are five elements in the natural environment: metal, water, wood, fire, and earth. Everything is made of these five basic elements. The relationships between five elements include generation and overcoming cycles. Each elemental force generates or creates the next element in a creative cycle. Metal generates water because water usually condenses on metal and also because a metal shovel digs out a water spring; water generates wood because rain nourishes trees; wood generates fire because wood is easy to catch fire; fire generates earth because ash is created from fire; and earth generates metal because metal is mined from earth (Figure 11.4).

According to the five element theory, each elemental force also controls or overcomes another element in an overcoming cycle. Metal overcomes wood because an ax cuts wood; wood overcomes earth because trees root into the earth; earth overcomes water because earth absorbs and conceals water; water overcomes fire because water puts fire out; and fire overcomes metal because fire melts metal (Figure 11.5).

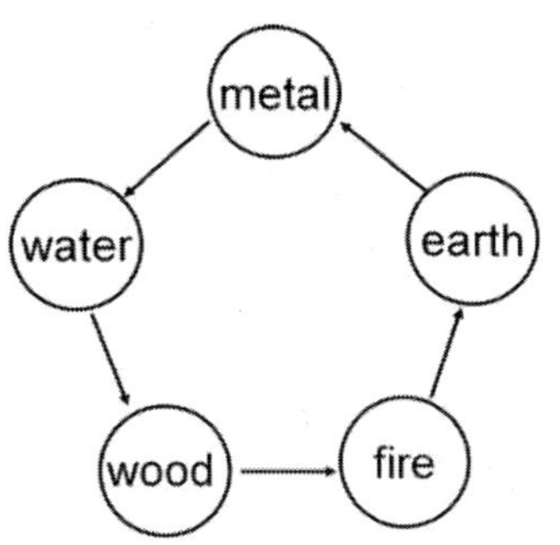

Figure 11.4. Cycle of five element generation.

In traditional Chinese medicine, five elements are abstracted as five properties or five characteristics that support or control one another [63, 64]. Metal behaves like leather. Those having properties of clearing away, quelling down, and convergence belong to metal. Water slicks down. Those having properties of coldness, lubrication, moving downwards belong to water. Wood is straight and resists bending. Those having properties of growth, development, extending, and resilience belong to wood. Fire is hot and oriented upward. Those having properties of warmness and rising up belong to fire. Earth supports crops. Those having properties of production, supporting, and reception belong to earth. These generalizations of five characteristics have widespread applications in nature.

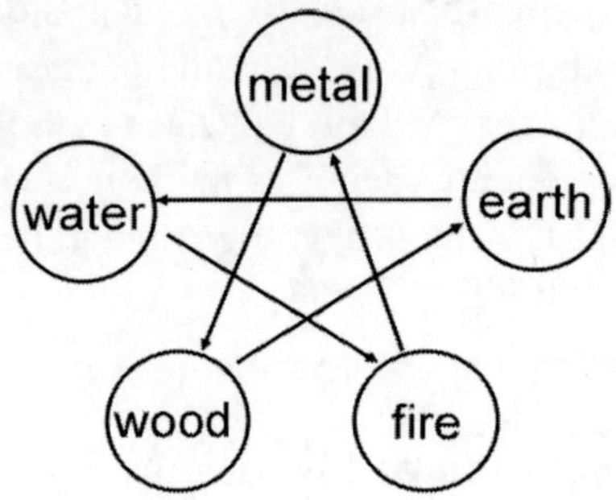

Figure 11.5. Cycle of five element overcoming.

In a human body, every organ may be categorized according to its roles and properties. For example, the lung has the functions of removing carbon dioxide from the blood and providing it with oxygen so that the lung belong to metal; the kidney has the functions of maintaining proper water and electrolyte balance, regulating acid-base concentration, and filtering the blood of metabolic wastes, which are then excreted as urine, so that the kidney belong to water; the liver has the function of discharging bile and tends to be elastic and not to be depressed so that the liver belongs to wood; the heart has the function of warmness so that the heart belongs to fire; and the spleen has the functions of storing blood, disintegrating old blood cells, filtering foreign substances from the blood, producing lymphocytes, and nourishing various organs so that the spleen belongs to earth. Strictly speaking, even though five organs correspond to five characteristics, the concept of an organ in traditional Chinese medicine is a functional abstraction, not the exact concept of a physical organ in anatomy.

Because of their belongings, these five organs have the relationships of generation and overcoming cycles as illustrated in Figures 11.4 and 11.5. For instance, the liver (wood) is said to be the mother of the heart (fire), and the kidney (water) the mother of the liver. One observation is that kidney deficiency affects the function of the liver. In this case, the mother is weak, and cannot support the child. Many of these relationships can be linked to known physiological pathways now. Thus, the five element theory in traditional Chinese medicine is about five characteristics or five phases and their relationships in vivo instead of five substances in the environment.

The foregoing introduction to the five element theory is an excerpt of traditional Chinese medicine. Even though the five element theory is subjected to endless disputes and challenges, we believe that the core concepts of five characteristics are correct. Under the context of quaternity spacetime, here we define five characteristics in rigorous geometry.

As shown in Figure 11.6, five elements are represented by the 2s, $2p_x$, $2p_y$, $2p_z$, and 2s electronic orbitals within a carbon atom. First, the 2s orbital at the bottom is a sphere of four-dimensional time with zero-dimensional space, representing metal. Through a rotatory operation, this orbital gradually polarizes into a $2p_x$ orbital, which is a point of three-dimensional time with one-dimensional space. The polarization process corresponds to metal producing water. Water is drawn as a drop in the figure. Second, by a rotatory operation, the $2p_x$ orbital transforms into a $2p_y$ orbital, which is a two-dimensional semicircle in shape. The stretching process corresponds to water producing wood. Wood is drawn as an arc.

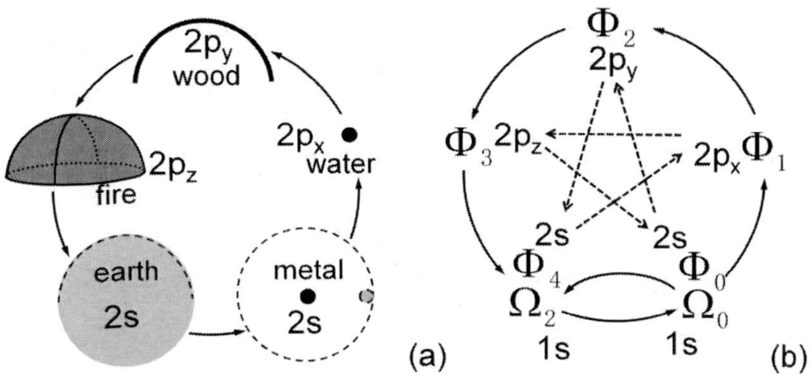

Figure 11.6. Cycle of five element generation (black arrows) and cycle of five element overcoming (dashed arrows) with geometrical shapes, electronic orbital types, and spherical quantity definitions of six electrons within a carbon atom.

Third, by a rotatory operation, the $2p_y$ orbital transforms into a $2p_z$ orbital, which is a hemispherical surface of one-dimensional time with three-dimensional space. The distribution of the electron is dense outside and empty inside, like a typical characteristic of fire. Fourth, by a rotatory operation, the $2p_z$ orbital transforms into a 2s orbital, which is a full sphere of zero-dimensional time with four-dimensional space. The full sphere defines and supports a spherical layer, which is a characteristic of earth. For detailed description of the shapes and transformation processes, please see Figures 2.14 and 2.12. The carbon atom with electron configuration of Figure 11.6(a) has a potential chirality of dextrorotatory tendency. Finally, from earth to metal, the sphere of zero-dimensional time with four-dimensional space condenses into a sphere of four-dimensional time with zero-dimensional space. This condensation process could be realized through the rotations of $2s \mapsto 2p_z \mapsto 2p_y \mapsto 2p_z \mapsto 2s$ (treated as a single step $2s \mapsto 2s$) in the negative direction of quaternity axes or through a $1s \mapsto 1s$ rotation due to the spacetime continuity between the 1s and 2s orbitals. Five element cycle corresponds to five spherical quantities of the solution to three-dimensional Schrödinger's equation (Figure 4.3). Table 11.2 lists five elements in geometrical shapes along with their characteristics. A carbon atom with four unpaired electrons $2s2p_x2p_y2p_z$ is chemically active to form a compound with other atoms.

We have established the correspondences between the cycle of five element generation and the cycle of electrons within a carbon atom in Figure 11.6. Under the octet configuration, each electron is rotating counterclockwise as was shown in Figure 3.5. However, when there are vacancies in an atomic shell such as a carbon atom, some electrons may rotate back and forth (clockwise and counterclockwise) so as to fill the vacancies. To avoid colliding, the existence of a $2p_x$ electron in a counterclockwise rotation may push a $2p_z$ electron to rotate counterclockwise as well because the transformation of $2p_x \mapsto 2p_y$ cannot coexist with the transformation of $2p_z \mapsto 2p_y$. A carbon atom has a vacancy in the electron configuration of $1s^2 2s^1 2p_x{}^1 2p_y{}^1 2p_z{}^1 2s^1$ (i.e., deducting one electron from any orbital type here) as shown in Figure 11.6. This vacancy induces the transformation cycling of elections within a carbon atom in a consistent rotation direction. Thus there are overcoming relationships between electrons at every two intervals in a carbon atom, similar to the cycle of five element overcoming in the environment. There are a cycle of five electron overcoming and a cycle of five electron generation within a carbon atom.

Table 11.2. Five elements with biological and geometrical significance

Element	Metal	Water	Wood	Fire	Earth
General Characteristic	Convergence	Fluid	Extending	Warmness	Reception
Organ	Lung	Kidney	Liver	Heart	Spleen
Electron	2s	$2p_x$	$2p_y$	$2p_z$	2s
Geometrical shape	Inner sphere	Polarized Point	Semicircular arc	Hemispherical surface	Outer sphere
Space dimension	0	1	2	3	4
Spherical quantity	Scalor	Polor	Metor	Vitor	Scalor

Because carbon is a fundamental building block of life, the five phase cycling property is somehow carried over to the human body from the carbon atom, as it is carried over from metal, water, wood, fire, and earth in the environment. To specify, the chirality of carbon atoms in deoxyriboses direct the electric current in the 3' → 5' direction in a DNA molecule, giving the double helix to function as stepwise *LC* oscillatory circuits (Figure 7.1). Gene expression in a cell follows the carbon model of electronic orbitals (Figure 8.2). Thus both the genetic substance and gene expression, which are critical to life, bear the trace of five elements. The cycle of five element generation and the cycle of five element overcoming are corollaries of quaternity theory for chiral entities. The five element cycling exists in multifarious entities inherently owing to the ubiquitous applicability of quaternity spacetime to the entity structure. The five element cycling waves from the environment and from the human gene converge to produce various organs that fall into the categories of five elements (Table 11.2). This establishes the scientific background of five element theory for the proper functioning of life. It is amazing that the five element theory and three-dimensional Schrödinger's equation spell out the same principle of electronic motion within a carbon atom, the most significant element for life on earth. Fundamental theories always agree with each other.

It is important for us to realize that the five element theory is an abstraction of properties and relationships, no more about the composition of matter. Only when five elements satisfy the cycle of generation and the cycle of overcoming relationships do they qualify as metal, water, wood, fire, and earth in the five element theory. This is remarkably different from the fundamental Greek theory of matter (credited to Empedocles, 490-435 BC) that everything in nature is made up of four basic elements (fire, air, water, and earth). While the four element matter theory is untenable, the five element abstraction theory survives up to now.

11.5. Quaternity Interpretation of Eight Trigrams

In Figure 11.2c, we saw that two pairs of yin and yang produce four quadrants. As shown in Figure 11.7, three pairs of yin and yang would produce eight trigrams, forming an octagon. Relative to the observer sitting at the center, eight trigrams are located at eight orientations. Note that in ancient Chinese literature, south was at the top and east was at the left, contrary to the orientation of a conventional map nowadays. From the top trigram counting counterclockwise, they are Qian, Dui, Li, Zhen, Kun, Gen, Kan, and Xun, literally meaning sky, marsh, fire, thunder, ground, mountain, water, and wind, respectively. We may tentatively interpret these eight factors as heaven, biological activities, chemical reactions, electromagnetic phenomena, earth, solid, liquid, and gas in sequence.

Furthermore, the overlay of two various trigrams produces 64 hexagrams, each having specific divine meanings. With six divided or whole lines stacked from bottom to top there are 2^6 or 64 possible combinations, and hence 64 hexagrams. Based on the 64 hexagrams, "*Yi Jing*" consists of sixty-four short essays, enigmatically and symbolically expressed, on important themes, mostly of a moral, social, or political character.

People looking at the trigrams and hexagrams usually arrived at the conclusion that "*Yi Jing*" is the mother of binary numbers. In western nations, the binary system was first studied in detail by Leibniz in 1678 and forms the basis for all computer and digital manipulations. Whether or not Leibniz was inspired by the trigrams and hexagrams from "*Yi Jing*" in the binary number invention is moot. The significance of "*Yi Jing*" is beyond the binary system. It is about change, the law of nature. Here we give Fuxi trigrams a fresh interpretation in terms of quaternity spacetime.

Each trigram consists of three lines: the top line, the middle line, and the bottom line. Each line is either a divided line or a whole line to represent yin or yang. The relative position and number of yin and yang lines within a trigram determine the meaning of that trigram. For example, Qian is a trigram with all yang; and Kun is a trigram with all yin. In the sense of yin and yang, Qian is active, dominating, and creative while Kun is receptive, yielding, and nurturing. The great attribute of heaven (Qian) and earth (Kun) is giving and maintaining life.

As was described in section 11.1, a yin line indicates a space dimension and a yang line indicates a time dimension within each trigram. The middle line of a trigram always has extraordinary importance in determining the meaning of the trigram in "*Yi Jing*." If we regard the middle line as double lines, then Qian contains four time dimensions; Dui contains three time dimensions with one space dimension; Li contains two time dimensions with two space dimensions; Zhen contains one time dimension with three space dimensions; and Kun contains four space dimensions. The transformation from Qian to Dui is a rotatory operation, so are transformations from Dui to Li, from Li to Zhen, and from Zhen to Kun. The relationships between the trigrams are similar to those of $2s \mapsto 2p_x \mapsto 2p_y \mapsto 2p_z \mapsto 2s$ transformations (Figure 6.3). After maximum space has been reached, spherical quantities in the negative direction of quaternity axes undergo clockwise rotations. In the transformations of Kun $\mapsto$ Gen $\mapsto$ Kan $\mapsto$ Xun $\mapsto$ Qian, each step increases a time dimension while reducing a space dimension. Thus eight trigrams correspond to the electron octet within a neon shell. This is our stereochemical interpretation of eight trigrams. The arrangement of eight trigrams implies changes of spherical quantities by rotatory operations. One has to combine logical deduction with imaginal thinking to fully comprehend the ancient script as he or she matches Figure 3.5 with Figure 11.7 compellingly.

Moreover, the Taiji yin and yang fishes diagram was usually placed at the center of eight trigrams (Figure 11.7). Yin and yang fishes denote both 1s electrons in a helium shell; and eight trigrams denote the 2s2p electron octet in a neon shell. From the theory of quaternity, the rotatory motion of both 1s electrons drives the motion of 2s2p electrons. They are coupled together by dimension encapsulation mechanism outwards and by calculus chain rules inwards. Taiji yin and yang fishes are encapsulated within eight trigrams to indicate the engine for rotation. Even though the ancient Chinese didn't know anything about calculus and trigonometry, the unmistakable diagram indicated that they had mastered the gist of natural law before the invention of eight trigrams. It is a miracle! At least, I still cannot imagine a better diagram than Figure 11.7 to express a continuous ten-dimensional structure such as a neon atom with ten electrons.

Figure 11.7. Fuxi trigrams surrounding Taiji yin and yang fishes.

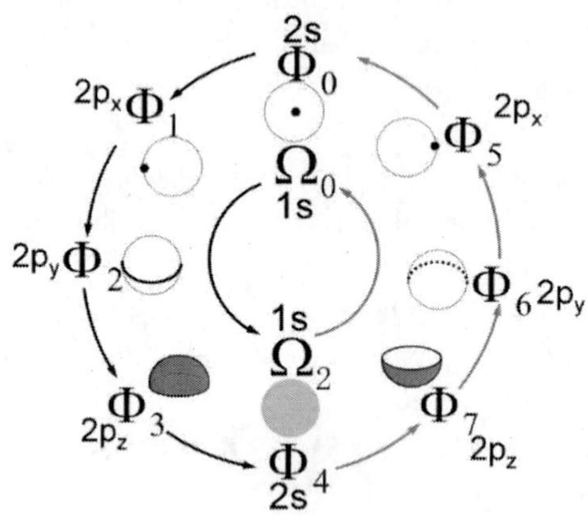

Figure 11.8. The electron octet in a neon shell channeled by both 1s electrons in a helium shell in quaternity space in correspondence with Figure 11.7. Black arrows indicate counterclockwise rotations and gray arrows indicate clockwise rotations.

By the way, the electronic orbitals in a neon shell are four-dimensional in quaternity space (Figure 11.8); the traditional eight trigrams are two-dimensional in a Cartesian plane (Figure 11.7); and if one adopts binary numbers to characterize the eight trigrams, he or she will unavoidably miss some important information on the changing pattern of the trigrams because the number series is one-dimensional only. In a significant way, the four-dimensional geometry connotes the possibility of molecular chirality around a carbon stereocenter with the electron quartet of $2p_x2p_y2p_z2s$ orbitals, which the other figures in lower dimensions cannot foresee. Thus, from four dimensions to two dimensions, and to one dimension, each step loses vital information on the space configuration notwithstanding the great capacity of binary numbers in modern technology.

Conclusion

China has witnessed five thousand years of human civilization. The very headstream of this oriental civilization was documented in "*Yi Jing*" and "*Yellow Emperor's Inner Canon,*" which give us a valuable cultural inheritance. Through the above discussion, we have found that the ancient Chinese already developed a sound philosophy in medicine. The traditional Chinese medical theory and the philosophy of "*Yi Jing*" shared the same origin, which left many loose ends to be tied up and many avenues to be explored. In this chapter, we meticulously reinforce the medical philosophy in coherent mathematical forms through the new interpretation of geometry, trigonometry, and calculus. We advocate calculus biology for studying the intricate life phenomena (section 8.10).

The yin and yang theory corresponds to the space and time model of a helium shell, and the five element theory corresponds to the electronic orbital model of $2s2p_x2p_y2p_z2s$ within a carbon atom, both forming the foundation of traditional Chinese medicine. Fuxi eight trigrams delineate a dynamic four-dimensional spacetime structure visually; and four Chinese characters of Yuan, Heng, Li, and Zhen characterize four spherical quadrants literally. The atomic shells and the human body share the same organized structure with law from the quaternity perspective. In this chapter, we have established the coarse connections between the theory of quaternity and traditional Chinese medicine, leaving the details for medical experts to refine discreetly. In section 8.9, we characterized an organic being with a general gyroscope in which each ring of function is intimately coupled with the other rings. Studies at this high level generalization of vital organs in a human body may help us master the controlling mechanisms underneath the diverse life phenomena.

The yin and yang theory is a general description of the space and time relationship; and the five element theory is more specific on the five characteristics or five phases of electrons in the four-dimensional spacetime continuum. Both theories are consistent with each other under the framework of quaternity spacetime. They describe entities in transformation that can be formulated by spherical quantities in rotatory operations. Five elements in traditional Chinese medicine (metal, water, wood, fire, and earth) correspond to the scalor (time end), polor, metor, vitor, and scalor (space end) respectively in a carbon atom (Figure 11.6). The yin and yang cycle and five element cycle of spherical quantities are exactly the quaternity solutions to three-dimensional Schrödinger's equation (section 4.5). It is amazing that the ancient philosophy with proper interpretation could amount to the Schrödinger's equation. No one relates traditional Chinese medicine to quantum mechanics so adequately as we do.

There is a large amount of research literature on "*Yi Jing,*" "*Yellow Emperor's Inner Canon,*" and traditional Chinese medicine accumulated over a long history. Of course, I cannot expect that all of the ancient knowledge reflected truth, but I do realize that the foundation underlying traditional Chinese medicine is firm. To be an exact science, fundamental theories must be and have been expressed in mathematics instead of vague, general statements. This mathematical confirmation assures us that traditional Chinese medicine is a decent scientific discipline. It is an important step toward serious integration of Eastern and Western sciences.

We have driven home the bright idea of spherical quantities in dynamic calculus through a multidisciplinary approach, touching on mathematics, physics, chemistry, biology, and philosophy. Out of the very fundamentals of all these disciplines I distill a unified scheme of thought that forms the foundation of science. In retrospect, the focus of discussion was fierce, as was my burning desire to achieve the revolutionary goal. As a result, this monograph is so fundamental in the subject and so original in the thought that the whole edifice of quantum physics after this work will require a critical reappraisal of the solidity of its basis. We have demonstrated that the spherical view of unified order, symmetry, and harmony does conform with and even crystallize both Eastern and Western philosophies in many key aspects, signifying an auspice of good science.

Even though the goal of this book is extraordinarily ambitious, the law of nature is conservative. Careful readers surely found that the main theme that has run through the entire book is unchanged. Spherical quantities in dynamic calculus make an excellent complement to physical quantities with linear algebra. The duality equation and the quaternity equation parallel Schrödinger's equations in one and three space dimensions respectively. We have built the comprehensive theory step by step from a variety of visions. In some cases, we exposed the unknown and made intellectual speculations or hypotheses, lending perceptual imagination in a coherent manner. We hope that this inferential composition was helpful for understanding the law of nature and would not dilute our great discovery of the 2s2p electronic orbital geometries.

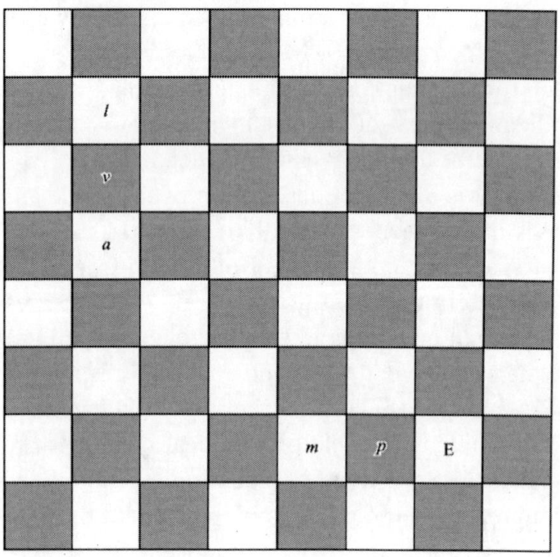

Motif 3. A periodic table of elements or physical quantities, a word puzzle with letters and vacancies, and a chessboard with law and order.

Appendix I. Review Questions

The following review questions aim at reinforcing your knowledge base of the spherical view, deepening your insights into quaternity, and broadening your scientific horizons. They are further discussion of contents in the text.

1.1. Please formulate wave equations for the cyclic motion of a free pendulum (Figure 1.4) and for the harmonic vibration of a ball under the action of springs (Figure 1.12). Are both equivalent mathematically in any way?

1.2. A circle may be regarded as a one-dimensional line such as a geodesic line around the earth (Figure Q1a), but a circle plus a center is regarded as a two-dimensional system (Figure Q1b, Q1c, and Q1d). Please explain this difference physically by looking at the following illustrations.

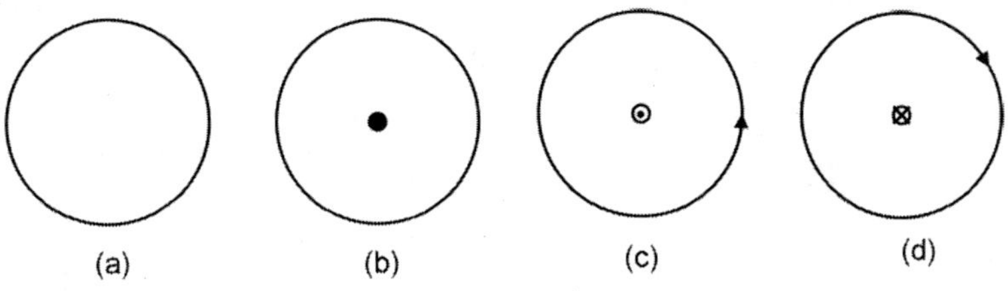

Figure Q1. Comparison of a circle without a center and three circles, each with a center.

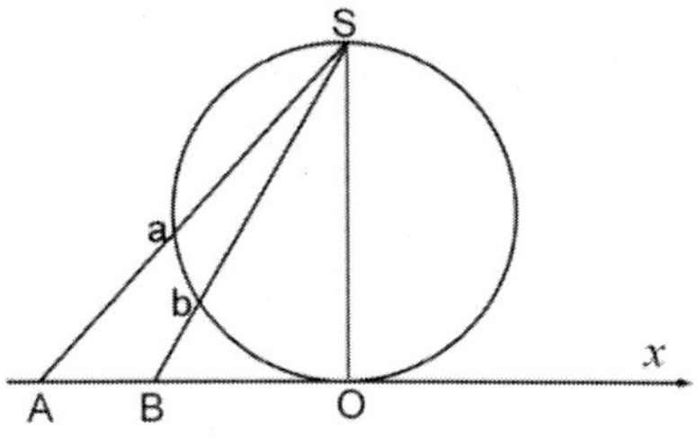

Figure Q2. The one-to-one point correspondence between a circle without point S and a real number axis. For every point b except the standpoint S at the circle, there is a point B at the axis corresponding to it; and for every point A at the line, there is a point a at the circle corresponding to it. The standpoint S corresponds to ∞ or $-\infty$ at the real number axis, both being connected.

(Hint: In Figure Q1a, sitting at any point of the circle, one sees the other points as a line or a real number axis as demonstrated in Figure Q2; In Figure Q1b, sitting at the center of the circle, one sees the circle as two lines and two points as demonstrated in Figure 2.1, in addition, section 9.2 describes a two-dimensional orbit of a planet around the sun governed by

Kepler's laws; In Figure Q1c and Q1d, imagine a magnetic field line around an electric current along a straight wire. The electric and magnetic fields constitute two dimensions.)

1.3. A sphere may be regarded as a one-dimensional radius, but a sphere plus a center is regarded as a two-dimensional system. Please explain this difference physically and biologically by looking at Figure 1.16. (Hint: A sphere indicates a 1s electron, but a sphere plus a center may refer to both 1s electrons oriented in the radial and spherical surface directions, respectively. Biologically, a sphere may refer to an ovum, a haploid cell, but a sphere with a seed may refer to a fertilized ovum, a diploid zygote).

1.4. According to $E = mc^2$ by Einstein, mass may change into energy. Mass and energy are two different manifestations of the same substance. Does the assumption of a physical quantity (a continuous variable for magnitude with fixed units) agree with the fact that mass and energy are interchangeable in their dimensions?

1.5. In parallel with equation

$$\cos x = 1 - \frac{x^2}{2!} + \frac{x^4}{4!} - \frac{x^6}{6!} + \ldots,$$

please express $\sin x$ in a power series and discuss whether a smooth trigonometric function can be expressed by finite polynomial terms.

1.6. Please plot the curve of a density distribution function as $P = -\ln(\cos\beta)$, $\beta \in (\pi/2, 0)$ using a spreadsheet application. (Hint: The trend of $P = -\ln(\cos\beta)$ curve is almost similar to that of Figure 1.17, but the area under $P = -\ln(\cos\beta)$ is about 1.088, not normalized. The normalized function is P versus $\cos\beta$)

1.7. Are there any differences between the dimension concept developed in Chapter 1 and the traditional dimension concept? (Hint: A traditional dimension refers to an independent vector in linear algebra, or a real number axis such as the X-axis. In Chapter 1, a dimension refers to the spacetime span of an entity in circular motion or simple harmonic oscillation where the circle corresponds to the real number axis point to point as demonstrated in Figure Q2. The traditional dimension tends to be linear in scale whereas in Chapter 1, a dimension is sinusoidally attained by a spherical quantity in dynamic calculus.)

1.8. There is a subtle but crucial argument that an angular velocity may be regarded as an imaginary unit. Why? Look at Figure Q3, be sure to pause and think it through.

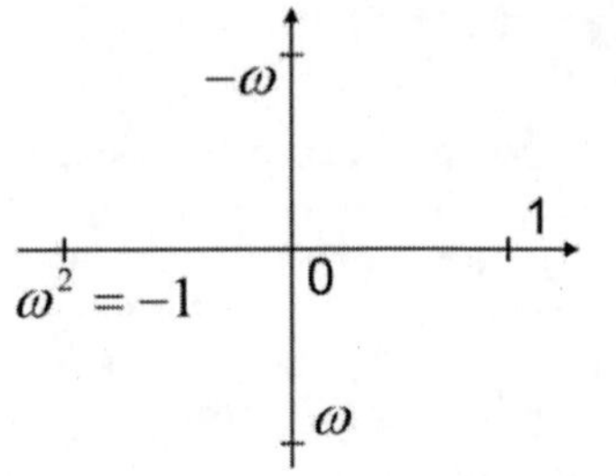

Figure Q3. On the Cartesian plane with a scalar axis and a frequency axis, the number 1 at the abscissa axis multiplied by a unit reciprocal time dimension ω transforms into a number along the ordinate axis clockwise, and multiplied by ω twice returns to the abscissa axis but in the negative direction. The transformations are clockwise because of the minus sign in Equation (1.3).

1.9. Please compare sine/cosine functions of the four roots to the duality equation with +/- signs in four Cartesian quadrants (Figure 11.2). How is continuous two-dimensional spacetime related to the Cartesian plane?

1.10. Prove that f_1 transforms into f_2 (see equations 1.35 and 1.36) after performing $\int -\frac{\partial}{\partial t} dl$ operation once, and returns to its original value after performing $\int -\frac{\partial}{\partial t} dl$ operation twice assuming $QL^2\varpi^2$ as electric charge Q under the two-dimensional scenario.

1.11. Under what conditions can we use a radius instead of the X, Y, and Z axes to characterize a solid ball? (Hint: When the ball is round and isotropic, the distinction of X, Y, and Z directions is artificial and arbitrary, which does not provide additional information to a radius for the characterization of the ball. The ball is one-dimensional in space.)

1.12. Considering the general solution to equation $\frac{d^2\Psi}{dt^2} = \Psi$ as $\Psi = B_1e^t + B_2e^{-t}$ in physical quantities, please prove that the solution is compatible with equation (1.2) in spherical quantities. (Hint: From equation 1.3, $t = -\alpha / \omega = \omega\alpha = i\alpha$ where $\omega = i$. This imaginary number expression transforms both exponential functions into circular complex functions:

$$B_1e^t + B_2e^{-t} = B_1e^{i\alpha} + B_2e^{-i\alpha} = B_1(\cos\alpha + i\sin\alpha) + B_2(\cos\alpha - i\sin\alpha).$$

Since an entity is normally undergoing an oscillatory cycle, either clockwise or counterclockwise, we may keep only one term, either B_1e^t or B_2e^{-t}, in the solution without losing generality. Thus, the equation $\frac{d^2\Psi}{dt^2} = \Psi$ is a general harmonic oscillation equation in spherical quantities. Please refer to Figure Q3 where from axis 1 to the negative direction of the ordinate axis is defined as clockwise transformation).

2.1. Flatland: In 1884, Edwin A. Abbott wrote a book titled "Flatland, a romance of many dimensions" describing the situation of two-dimensional creatures on a spherical surface. From the standpoint of the flatland inhabitants, the country is two-dimensional without the movements of rising up and sinking down. Straight lines, triangles, squares, pentagons, hexagons, and circles all appear as straight lines of the edges, but from the standpoint of the spherical center, those shapes appear as they are honestly. The quaternity theory regards the spherical surface as a three-dimensional geometrical shape since all shapes can move about the spherical surface, left and right, forward and backward, and up and down relative to the spherical center. How does a perspective shift alter geometrical shapes and the number of dimensions in this case? Moreover, traditional geometry treats a point as an infinitesimal small point, a line as a very fine line, and a surface as one without thickness. This treatment deviates from real objects. In contrast, quaternity points have a certain size, and lines and surfaces possess proper thickness. Humans colonize the spherical surface of the earth. From the standpoint of humans, do we live in an environment like the flatland? Does the imaginary flatland have any correspondence in the real world? If not, why not?

2.2. Given a circular ring on a plane whose outer diameter is three times as large as its inner diameter (Figure Q4), how much is its area by the area of the inner hole?

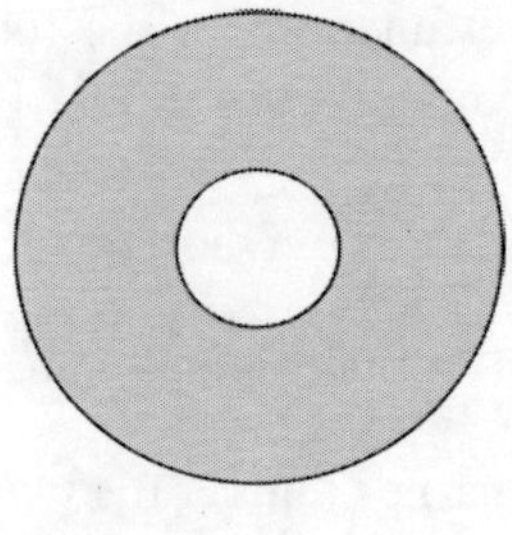

Figure Q4. A planar ring.

2.3. Dimension curling up mechanism: Trigonometric functions such as sine and cosine cast a radian angle into a real value (Figure Q5); a compiler transforms a program of a high-level programming language (such as C++ or Visual Basic) into a machine language so that the program can be executed. How does the compiler insulate the dimension of a program? Please compare the compiler insulation of a high-level language program with trigonometric insulation of a radian angle.

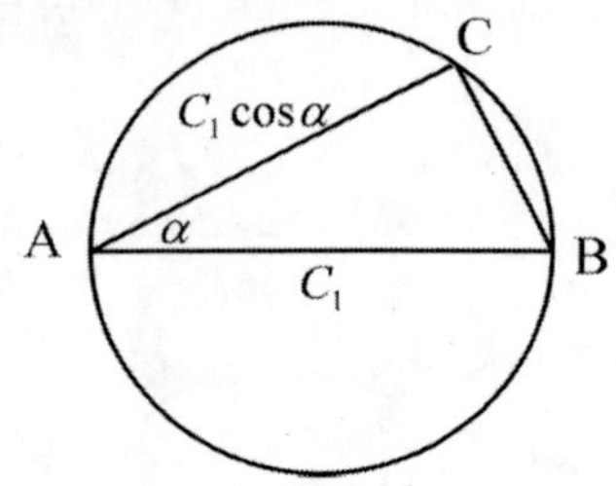

Figure Q5. The rotation of radian angle α determines the time component ($C_1 \cos\alpha$) of an electron within a helium shell.

2.4. Please study the working principle of a four stroke engine and find out how to convert linear reciprocating motion into rotary motion by means of a crankshaft. Please study the mechanical function of a cam, a lever, and a spring and find out how to convert rotary motion into linear reciprocating motion. Please look at Figure Q5 and find out how the rotation of radian angle α around point A brings the motion of chord $C_1 \cos\alpha$ as a point C orbits along the circle, like a crank driving the wheel. History tells us that the extent of exploitation of the wheels in any era is an unbiased indicator of a worker's productivity of that era.

2.5. Spacetime continuity: If a wave function is in the rotation pathway of another wave function, then they are continuous in spacetime. Please demonstrate that wave functions $\cos\Psi$ and $\cos\alpha$ are spacetime continuous if $\Psi = \cos\alpha$ by the differential chain rule, but they are not continuous if $\Psi = 2\alpha$ instead. The spacetime continuity requires that there is not any hole or break in a rotation pathway. For instance, both 1s electronic orbitals are continuous. Please demonstrate graphically that function $\cos(\Psi + \pi/4)$ is continuous to function $\cos\Psi$ via dynamic differentiation.

2.6. Mathematics is the abstraction of objective laws rather than the idealistic logic for applying indiscriminately, and each mathematical operation has its special physical meaning. The plus sign, +, may be used in two different circumstances. It means addition when inserted between two physical quantities of similar dimensions; but it gives a complex number, or a vector when inserted between two physical quantities of orthogonal dimensions. In quaternity, a complex number is a wave function that undergoes a calculus operation so that the plus sign indicates continuity of the wave function transforming from the first term into the second term by dynamic calculus. The plus sign in $\frac{d}{d\alpha}\cos\alpha = \cos(\alpha + \frac{\pi}{2})$ means a smooth functional shift from $\cos\alpha$ to $\cos(\alpha + \frac{\pi}{2})$ through the rotation of radian angle α, equivalent to dynamic differentiation upon the wave function. A plus sign in dynamic calculus signifies an action. For example, a man plus a man equals two men; a man plus a woman signifies marriage and sex, resulting in three people or more in the end; and a teacher plus a student means education. Could you give more vivid examples on both conventional and quaternity usages of a plus sign?

2.7. Unity but diversity: a pair of $2p_x$ electrons correspond to two points displaced from a center symmetrically; a pair of $2p_y$ electrons correspond to two unbound semicircular arcs connected by the two points into a circle; a pair of $2p_z$ electrons correspond to two unbound hemispherical surfaces connected by the circle into a spherical surface; and a pair of 2s electrons correspond to the central point and an open solid sphere bounded by the spherical surface (Figure 2.14). These four symmetric pairs constitute four basic elements in a neon shell and have interesting analogs in biological anatomy: a pair of eyes are symmetrical and synchronized; a pair of ears are separated but interconnected; a nose has two symmetrical nostrils; and the mouth and the anus are channeled by the alimentary canal. They function properly in response to four orthogonal vehicles of light, sound, smell, and fluid, respectively. Is the analogy far-fetched or relevant to the biological expression of quaternity space? Could you list more symmetrical and interrelated pairs of organs in a human body?

2.8. After reading Part I, think about the question why the theory of quaternity must be developed by a single mind instead of a crowd. (Hint: Any non-trivial theoretical work must be finished by an individual such as Newton's mechanics, Maxwell's electrodynamics, and Einstein's relativity. A crowd is chaotic, only capable of suggesting conflicting ideas)

2.9. Chord AC and arc AC are two parameters determining the radius of a circle (Figure Q6). If you stand inside the circle, you see the chord AC=AB $\sin\alpha$; but if you stand at point A on the circle, the intercepted arc AC or angle $\angle DAC = \Psi$ is what you see. In order to convert a circular complex function Ψ into a radian angle, we cast function Ψ by a cosine operator into $A_1 \cos\Psi$. How does the standpoint shift change the role of function Ψ adopted in dimension encapsulation? (Hint: The cosine operator is a differential operator, which has different consequences from the perspective of helium shell and the perspective of neon shell.

The helium shell perspective:

$$\begin{aligned}
&\cos(\mathrm{AB}\sin\alpha) = \cos\overline{\mathrm{AC}} \\
&= -\frac{d(-C_1\omega\sin\alpha)}{dt} \\
&= C_1\cos\alpha.
\end{aligned}$$

The neon shell perspective:

$$\cos(AB\sin\alpha) = A_1 \cos \overset{\frown}{AC}$$
$$= A_1 \cos \angle DAC$$
$$= A_1 \cos \Psi.$$

Both $C_1 \cos\alpha$ and $A_1 \cos\Psi$ coincide and connect at point A at the moment $\Psi = \alpha = 0$ when $C_1 \cos\alpha$ is about to rotate in the inner helium cycle and $A_1 \cos\Psi$ in the outer neon cycle. They belong to the helium and neon shells respectively. The perspective shift results in radian angle shift from α to Ψ in the cosine functions, the latter fulfilling the dimension encapsulation. Please also refer to Figure 2.8).

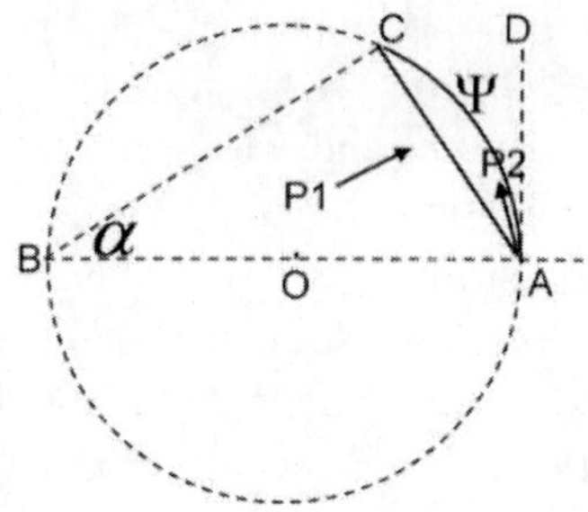

Figure Q6. Schematic diagram of a segment as a chord from perspective P1 at the initial point B and as an angle $\angle DAC$ or the measure of its corresponding arc AC from perspective P2 at the terminal point A on the diameter with reference to Figure 2.6(a).

2.10. A magic number: 1.618*1.618≈2.618; 2.618*6/5≈Π. Π is the ratio of the circumference to the diameter of a circle. In a dimension diagram, the diameter or the chord is represented by a spherical quantity whereas the circumference or the arc may be represented by a radian angle. From equations (1.3) and (1.10), $\alpha = -\omega t = -i \times t$. A diameter remains a diameter when viewed from the inner shell; but a diameter is a circumference when viewed from the outer shell. Thus Π is the ratio of a diameter viewed from the outer shell to that viewed from the inner shell. As shown in Figure 2.1, a circle with a diameter in a flat plane has two orthogonal dimensions: a diameter (two radii) and a circumference (two semicircles). It is interesting to note that Π symbolizes a door and was assumed to be the ratio of the height to the width of the door (say, 22/7) in ancient Greek buildings. Can you remember the value of Π to ten significant digits?

2.11. Scientific progress in every step will have a great impact on philosophy. Are there any philosophical significance for dimension encapsulation mechanism? (Hint: In an effort to explain the significance of such an important mathematical innovation, we find that the differential process from $\cos\alpha$ to $-\omega\sin\alpha$, and to $\cos\Psi$ corresponds exactly to Hegel's triadic movement from thesis to antithesis, and to synthesis. This philosophical point of view is verified mathematically by dynamic calculus. Whether it is Hegel's dialectic idealism or Marx's dialectic materialism, dimension encapsulation mechanism is the precise mathematical expression for synthesis.)

3.1. As infinitesimal calculus and complex number notation drive mathematics toward idealism farther and farther, the quaternity definition of dynamic calculus and dimensional interpretation of complex numbers bring it back to reality. Please describe the differences and connections between physical quantities and spherical quantities. Why can they express harmonic oscillation in a similar mathematical formula (see equations 1.1 and 1.67)?

3.2 Demonstrate mathematically that the integral chain rule in equations (2.13) and (2.16) agrees with Green's theorem (equation 3.32) so that 1s electrons synchronize with 2s2p electrons in a neon atom.

4.1. There are two fictitious methods for characterizing the property of a man. The first one is to investigate his social relationships by asking who his parents are, who his wife is, who his children are, and what his job is. The second one is to detect his physical response upon an electric shock, say, how much the probability of his jumping over a preset hurdle is at the moment of the dangerous shock. Which approach, do you think, is more pertinent and relevant to the purpose of knowing the man? How are quaternity and quantum mechanics different in characterizing an electron? (Hint: Both characterizations are complementary and compatible.)

4.2. Imagine the different pictures of a forest viewed from the perspective of an eagle in the sky and from the perspective of an ant residing in one of the trees!

4.3. What might be the different descriptions of today's events from the standpoint of yesterday and the standpoint of tomorrow? Do both descriptions deviate from today's reality as we are experiencing now? How does your answer help in evaluating the perspective and objectivity of historical records? (Hint: Regarding an event as a rotatory operation, how to describe it at an upstream standpoint and at a downstream standpoint?)

4.4. When David was a kid, his father expected him to become a CEO. David was indeed the CEO of a software company in the later part of his career, of which his son is very proud. What might be the different images of David from the expectation of his father and from the impression of his son? Do both images have any kinds of symmetry? (Hint: Is there any symmetry or complementarity between the images of a coffee mug viewed from the left and right directions?)

4.5. Standpoint shift: The earth is only one of many planets in the solar system, yet it is the only one that is known to bear life in the universe. From electrons, atoms, molecules, cells, lives, moons, planets, stars, to galaxies, human race is at the center of the sequence and is the most intricate entity of space and time convergence from the microcosm and macrocosm. Human race is at the center of the universe in that we always examine the environment from the human perspective; the earth that we colonize is the center of the celestial sphere; and the sun and the moon revolve around the earth as we can see. Can you list some descriptions where human-centric conception is assumed, geocentric conception is assumed, and heliocentric conception is assumed, respectively? (Hint: Paradoxically, modern science is human-centric, but it suggests heliocentric viewpoint in astronomy).

5.1. By the theory of quaternity, space and time are complementary. Thus man is opposite to woman; forward is contrary to backward; and a brave man is in contrast to a coward. How to define the complement of two spherical quantities mathematically? (Hint: see equations 5.77 and 5.78 in set theory and equations 9.2 and 9.3 in the Pythagorean theorem.)

5.2. A rotatory operation in reality may refer to the functional or material connection between two things such as a chisel and marble, a plough and soil, a man and food, a bird and worms, a pen and ink, cheese and milk. A rotatory operation makes a spacetime connection. With your eyes wide open, please examine how things on the earth constitute a harmonious spacetime web instead of a hodgepodge of items.

5.3. Dream: If a rotation pathway represents the course of a physiological activity, then physiological activities must at least include a pair of semicircles in a diurnal cycle, say, doing consciously in daytime and dreaming passively at night. What dreaming in sleep tries to accomplish is to recover the body to its fresh state on the next morning under the direction of the natural law and through a complementary biochemical channel to that of the daytime activity. Please compare the pair of doing and dreaming with both electrons in a helium shell that are symmetric in spacetime (Figure Q7), i.e., the time component of one electron is equivalent to the space component of another, and vice versa. Doing is a rotatory operation and dreaming is a clockwise rotatory operation, both being in time and space reversal. Please also refer to Figure 3.5 for counterclockwise rotations and clockwise rotations in a four-dimensional case.

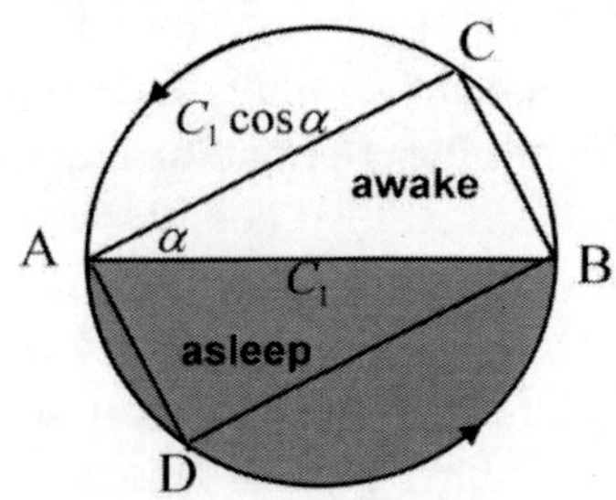

Figure Q7. The complementary relationship between doing and dreaming analogous to that between both electrons in a helium shell.

5.4. Please look at the electronic configuration in an oganesson atom (Figure Q8) and explain the mechanisms of electronic waves converged from the inner and outer spherical layers. (Hint: They are dimension encapsulation mechanism propagating from the inner spherical layer outwards and calculus chain rules reflecting back from the outer spherical layer inwards.)

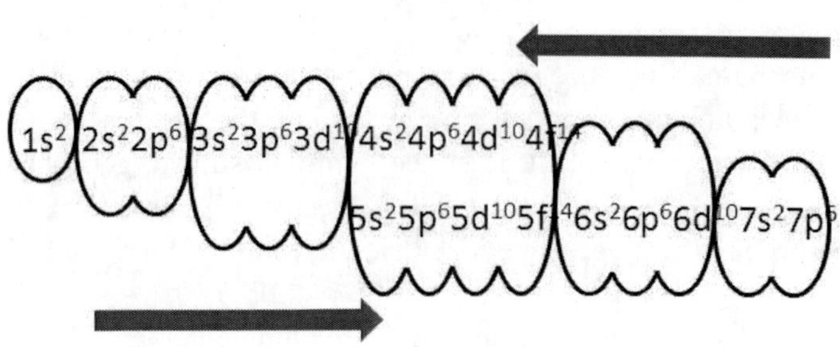

Figure Q8. The electronic configuration in an oganesson atom. It is the middle spherical layers converged from both ends that are the most complex in orbital types.

5.5. Please compare the number of space dimensions in a neon shell, which rotates from 0, 1, 2, 3, to 4 and then starts over from 0, 1, 2, 3, to 4 in a cyclic sequence (Figure 5.2), with the count of hours in a clock, which begins from 0, 1, 2, …, 12 and then starts over from 0, 1, 2, …, 12 again in a daily cycle.

6.1. By the principle of dynamic calculus, two electrons within a helium shell satisfy harmonic oscillation equation $\Omega = \Omega^{02}$, and eight electrons within a neon shell obey harmonic oscillation equation $\Phi = \Phi^{08}$ or $\Phi^{04} = u^4\Phi$ where the times of rotatory operations (and hence electron numbers) must be an even number for the wave equations to have proper solutions because a cosine function returns to the same cosine form only after being differentiated twice, and so does a sine function. Consider a possible solution to harmonic oscillation equation $\Phi = \Phi^{06}$ and its implication in chemistry? (Hint: Six-electron structure is not a stable configuration for an atom, but it could be a perfect pattern for a molecule such as benzene as shown in Figure Q9).

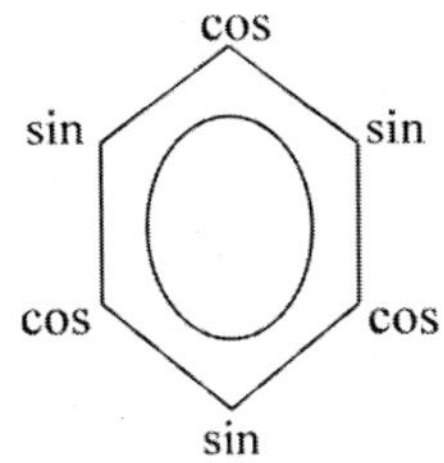

Figure Q9. The structure and carbon waveforms of a benzene molecule.

6.2. Collect all kinds of snails and examine whether their spiral curves are directed in the same chiral orientation (Figure Q10).

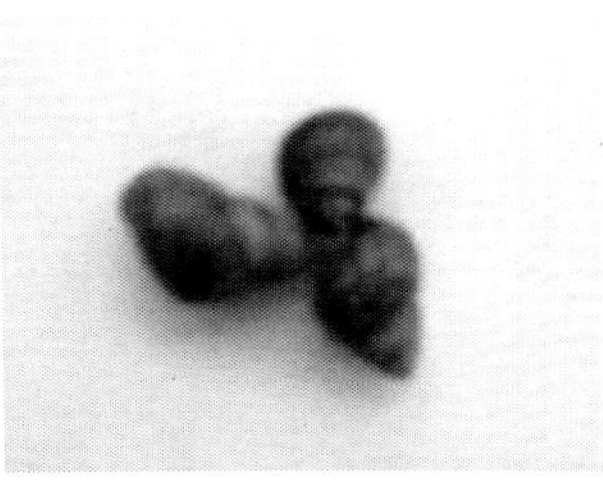

Figure Q10. Three field snails.

7.1. Please visit a rope-building factory to learn various braiding patterns of ropes. What are the basic properties of a rope? How ropes are used in a boat and how to tie various kinds of knots for those usages? Moreover, it is well believed that humans recorded things and events by tying knots in ropes long before the advent of characters. Please list various usages of ropes in business and imagine how the world would be without ropes.

7.2. Please write down the mathematical definitions of curvature and torsion of a three-dimensional curve in Cartesian coordinates and calculate how much they are for a strand of a B-DNA double helix.

7.3. Please calculate the length of a curve $y = \cos x, x \in [0, \pi/2]$. (Answer: 1.91).

7.4. Please calculate the length of a helix around a cylinder with radius R and height H (Figure Q11).

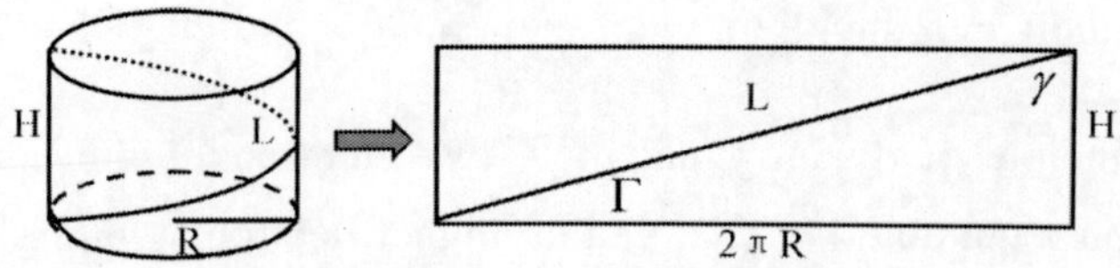

Figure Q11. The relationship between a helix around a cylinder and the hypotenuse of a right triangle.

7.5. Descartes (1596-1650), Spinoza (1632-1677), and Leibniz (1646-1716) founded continental rationalism in the 17th century. The rationalists assumed that what they could think clearly with their minds did in fact exist in the world outside their minds. They interpreted the natural world after the mechanical model of physics. In this monograph, we interpret the natural world after the electromagnetic model of physics. Electrodynamics is about induction and oscillation rather than linear motion. Dynamic calculus of spherical quantities is a revolutionary idea in electrodynamics. Why should electronic orbitals behave according to electrodynamics instead of mechanics? (Hint: In philosophy, mechanics is over-exploited, but electrodynamics is under-exploited. The reason is simple. Even though mechanics is sound and popular, electrodynamics, still based on the same premises of physical quantities with linear algebra, misses something fundamental.)

7.6. Hegel (1770-1931) set forth the general proposition that "what is rational is real and what is real is rational." After reading Chapters 2, 6, and 7, think about the questions whether the theory of quaternity is rational and whether it is real. (Hint: Space and time are symmetric and complementary, so are what is real and what is rational. We propose a double test principle to prove or disprove a scientific hypothesis. Besides experimental tests, a theory must be able to withstand mathematical scrutiny. For example, people are less cautious in accepting the hypothesis of orbital hybridization. How are three 2p orbitals and one 2s orbital "mixed" to form four sp^3-hybridized orbitals? The hypothesis has never passed the mathematical or rational test. Time (mind) and space (body), like DNA double strands, are symmetric and complementary. They are double checked. This redundancy proves biologically necessary. Any successful theories must be spiritually logical as well as materially demonstrated. Modern scientific methods emphasize the latter and miss the former. We don't think Hegel was correct, but we suggest that the rational test should be a part of the standard scientific methods. If a theory is not coherent by itself, then it should be rejected rationally in the first place!)

7.7. Reliable programming: In situations when the reliability of a program is the quality of main concern, speed and resources being trivial compared with the reliability, the programming methodology may simulate the redundancy of a DNA double helix. One thread of the program runs in time, the other thread proceeds in space, both being complementary and mutually checked. Or the software application runs in time, and the hardware memory shifts in space. The discovery of the mathematical formula of a DNA molecule (see Figure 7.9) opens up endless possibilities in software engineering. Is it feasible to design memory drifting hardware in accordance with a certain algorithm? (Hint: For example, see equations 1.35 and 1.36 where a typical capacitor is associated with $\cos\beta$ and a typical inductor with $\sin\beta$.)

8.1. The 80/20 rule: Italian economist Vilfredo Pareto proposed the 80/20 rule for wealth distribution after discovering that, in Italy, 80 percent of the country's wealth was held by 20 percent of the country's population. Thus it was inferred that for the typical product category, eighty percent of the products sold will be consumed by twenty percent of the customers. The 80/20 rule was later adopted for measuring progress. For example, after spending about 20% of the time, many programmers claim that they have completed 80% of the work. Later reality strikes from the back. Upon having spent another 80% of the time, the same programmers come to realize that they have resolved only 20% of the remaining problem.

a) Suppose in a more egalitarian community, wealth distribution follows a probability density function

$$f(x)=\frac{1}{2}\sin x, x\in[0,\pi],$$

where x refers to relative wealth value with π as the highest, and $f(x)$ is the probability density of the population at x value. What percentage of the wealthiest people own half of total community wealth? (Answer: 33%)

b) Supposing the learning progress versus time consumption follows the curve of sine

$$f(x)=\sin x, x\in[0,\pi/2]$$

where $f(x)$ reaches 100% after one spends $\pi/2$ unit time interval, how much percent of time must be spent to finish the first and last 20% workload respectively? (Answer: 13% and 41%)

8.2. The variable of a certain biological property observes trigonometric function with its probability density as $f(x)=A\cos x$, where A is a constant. Supposing x value is confined to the scope of [$-\pi/2$, $\pi/2$], based on the nature of a density function, please calculate the value of A, expectation $E(x)$, and variance $D(x)$. Please draw the graph of the trigonometric function and the graph of a normal probability density function with equal expectation and variance on the same worksheet and explain their differences.

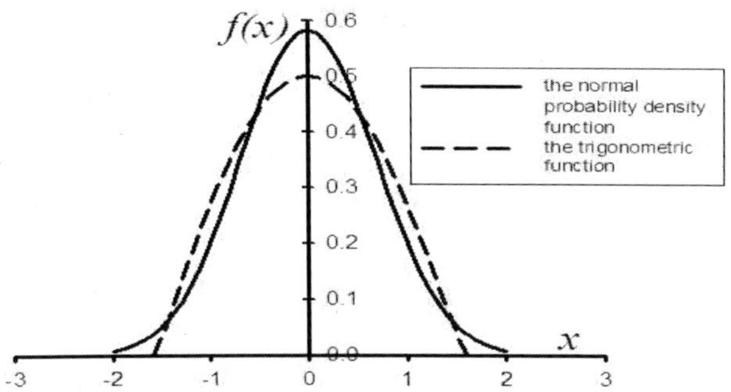

Figure Q12. The comparison between $f(x)=\frac{1}{2}\cos x$ and the normal probability density function of equal expectation and variance.

(Answer: A=1/2, $E(x)$=0, $D(x)$= $\pi^2/4-2$. As shown in Figure Q12, the normal probability density function has a higher peak, longer tails at both ends, and lower shoulders than the cosine function. This graphical feature is similar to the relationship between t-distribution density function and the normal probability density function!)

8.3. Are there any structural similarities between an apple and the earth?

8.4. A seed grows into a tree via a rotatory operation $\int(-\partial/\partial t)dl$ in the sense that the height of the tree increases as time elapses. Thus the seed represents a dense time and the tree is the unfolding space, both being continuous and symmetric in spacetime. The grown-up tree bears another seed through a mechanism other than the reversal of growth, thereby completing a life cycle. Could you give more examples of continuity, symmetry, and cycle like a tree?

8.5. Please compare the blood circulation in a human body with electronic orbitals within a neon atom. Note that the heart pulse from right atrium to right ventricle may be characterized by wave function ($\Omega_0+\Omega_1$) while the heart pulse from left atrium to left ventricle may be characterized by wave function ($\Omega_2+\Omega_3$). Four chambers of a heart correspond to four terms in equation (1.14) for 1s electrons. The systemic circulation corresponds to $2s2p_x2p_y2p_z$ electron quartet in counterclockwise rotations while the pulmonary circulation corresponds to another quartet in the negative direction of quaternity axes. Do cardiograms help in establishing the analogy?

8.6. How many loops are shown in Figure 8.1? Which loop represents properly the whole universe? If you walk along the loop, can you experience ever larger cycles all the time without an end? Which part of the loops represents traveling light?

8.7. Please open an old-fashioned mechanical watch and study the working principle of cog wheels in mesh. How do cog wheels in mesh ensure the accuracy of time counting? Can you mention some other devices or machines where cog wheels in mesh play a crucial role?

8.8. Mosquito's transmission of a disease to organisms can be characterized by calculus chain rules (differential and integral chain rules) in mathematics. Please read Figure 4.3 and section 8.3 and explain cog wheels in mesh in biological propagation of the disease.

8.9. Nature needs no design: As long as nature creates and maintains entities according to dynamic calculus of spherical quantities, all entities will live in harmony, the world on time. Please compare the orbital complementarity of 1s electrons and $3d^{1s}$ electrons in Figure 5.15 with the organelle complementarity in the endosymbiotic hypothesis in Figure 8.4, and with the role complementarity of bees and plants in the process of pollination. While feeding on nectar and pollen, bees make a flower or plant produce seeds by giving it pollen carried from another flower. Making honey and pollination are two orthogonal and complementary processes involving two organisms in mutual benefits. How wonderful the law of nature is!

8.10. Please compare the electronic cycle of Figures 3.5 and 4.3 with the biochemical cycle of Figure 8.2 and tell their similarities and differences. Please compare the electronic cycle of Figures 3.5 and 4.3 and eight trigrams of Figure 11.7 with citric acid cycle in biochemistry. Considering an electron movement or a proton transfer as a redox reaction, how do Eh and pH changes influence the chemical transformation and equilibrium constants within the biochemical cycles?

9.1. The earth revolves around the sun making a year in each circle. During the same period, a flame-tree in front of my house experiences four seasons. Does the flame-tree undergo a cycle in any sense similar to the circular motion of the earth? Are both cyclic phenomena tightly coupled together in any mechanism? During the same period, a flock of geese migrate from Canada to the United States in the fall and return to the north in the spring when the weather is getting warmer. How to characterize their migration cycle in comparison with the earth's orbit and the flame-tree's experience? (Hint: A circle is a cycle and vice versa. Please refers to section 9.1)

9.2. Prove geometrically and physically that a circle has the shortest circumference to confine an equal area; and a sphere has the least surface area to confine an equal volume.

9.3. Prove algebraically that the projection of a circle onto a plane tilted at a certain angle is an ellipse.

9.4. Angular momentum: The earth has orbital angular momentum attributable to its annual revolution around the sun, and spin angular momentum to its daily rotation about the north-south axis. Supposing the earth does not rotate at all, there would be only one day in a year resulting from revolution, and the revolving speed and the circumference of the earth's elliptical orbit would determine the length of the day. Can the earth's angular momentum be converted into spin momentum? How are they coupled together? (Hint: Look at Figure Q13 for a visual aid on the relationship between a circle and a twisted band and tell how the length of the circumference influences the torsion of the twisted band? The shorter the circumference is, the severer the twisting tension is.)

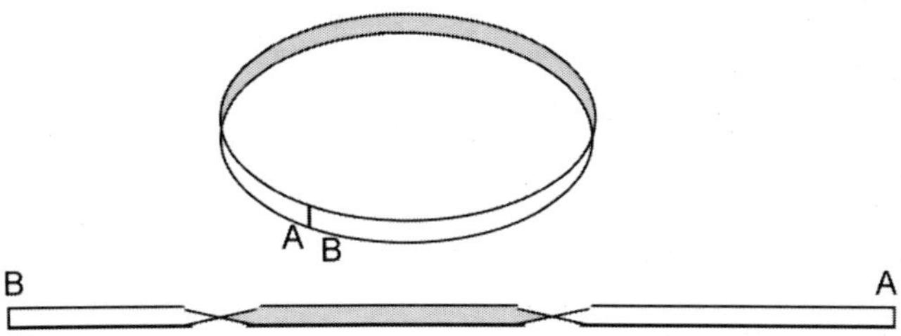

Figure Q13. The topological transformation of a circular band into a 360° twisted linear band by cutting at A-B and translating both ends in opposite directions.

9.5. Given the following equations for electrons within a helium atom that correspond to Kepler's first and second laws, please calculate α and β functions in terms of *t*, *l*, and *K*.

$$\begin{cases} \alpha + \beta = \pi/2, \\ r^2\omega = K. \end{cases}$$

(Hint: $r\omega = v$, $vt = l$, $\alpha = -(v^2/K)t$, $\beta = (v/K)l + \pi/2$, $K = \hbar/2m$).

10.1. Assuming the pattern of nature repeats in various levels to a certain extent as exemplified by the periodic table of elements where the electronic configuration somewhat repeats in every row, please compare the structure of a human society with the anatomy of a human body. (Hint: A society is a living body with various functional departments.)

10.2. Dimension algebra: Ignoring the positive and negative signs of wave functions, the spacetime dimensions of eight electrons in a neon shell are in the sequence of

$$1+\dot{\Psi}+u(1+\dot{\Psi})+u^2(1+\dot{\Psi})+u^3(1+\dot{\Psi})+$$
$$u^4(1+\dot{\Psi})+u^5(1+\dot{\Psi})+u^6(1+\dot{\Psi})+u^7(1+\dot{\Psi}),$$

which equals the product of four factors $(1+\dot{\Psi})(1+u)(1+u^2)(1+u^4)$. But how does dynamic calculus modify the sign of each dimension in the power series? (Hint: see equations 2.18 and 2.25)

10.3. An ideal gas obeys the law $PV = nRT$ where P denotes pressure, V volume, T absolute temperature, n number of moles of gas molecules, and R universal gas constant. We are exploring the possibility of extending its application to the other fields boldly.

a) Regarding each human individual as a free molecule, we might consider a human society to be an ideal gas where P denotes working pressure, V surviving roomage or job variety, T knowledge and technology (or the capability of an average man), n number of working people, and R human activity constant. In a society, is working pressure alleviated when job variety increases? Does technology progress intensify the working pressure of workers? Does technology application reduce the number of working people? If yes, then please evaluate seriously the applicability of the ideal gas law to a human society with the conceptual substitutions.

b) Regarding each star as a free molecule, can the whole universe or a galaxy be treated as a star gas that obeys the ideal gas law?

c) Regarding an electron cloud as an ideal gas, does electronic behavior obey the ideal gas law? Can nR be treated as the entropy of the electron so that $PV = ST$?

10.4. Assuming that the dimension of force equals $j.s^8/m^8$ instead of Postulate 3 in Chapter 10, would you please derive the spacetime dimensions of the other physical quantities. Is this assumption viable in defining the spacetime dimensions of physical quantities? (Hint: We may list all physical quantities in a table according to the alternative hypothesis. The table will indicate that there is a rotation relationship between electric potential and permittivity, which requires experimental verification.)

11.1. According to Chinese folklore, if a toxic herb grows in a region, then it is likely to find an antidotal herb against it in the nearby area. Does the ancient empirical medicine agree with the principle of rotatory operations? (Hint: see equation 8.13 where W_1 and W_2 are orthogonal spherical quantities created from nothing by a rotatory operation. A pair of orthogonal spherical quantities like these might transform in opposite directions and neutralize each other. If W_1 denotes a toxic herb, then W_2 might have an antidotal effect.)

11.2. Chinese ancestors had created eight trigrams (Figure 11.7) by observing celestial bodies in the sky and studying entities on the ground before the advent of characters four thousand years ago. Please read the trigrams carefully and assign each trigram a binary number in a logical way. Furthermore, how to relate eight trigrams to the electron octet in a neon shell? Please refer to the vast Chinese literature of "*Yi Jing*" (or "*I Ching*") and try to explain the philosophy of quaternity, keeping new terminology to a minimum.

11.3. Quantitative to qualitative change: History shows that social and economic orders are in a process of change. Marx's dialectical materialism maintains that what causes change is simply alteration in the quantity of things, which eventually leads to something qualitative new. For example, as I increase the temperature of water, it not only becomes warmer, but finally reaches the point at which this quantitative change changes it from a liquid into a vapor. Reversing the process, by gradually decreasing the temperature of water, I finally freeze it from a liquid to solid ice.

a) Look at Figure 1.6 and examine quantitative to qualitative change in cosine function algebraically. Does the value of $\cos x$ change sign when variable x gradually increases from 0 to over $\pi/2$? Does the slope of the curve change sign when x gradually increases from $\pi/2$ to over π?

b) Look at Figure 2.13 and examine quantitative to qualitative change in the orbital types. Does a $2p_x$ orbital change into a $2p_y$ orbital when radian variable Ψ increases from 0 to over $\pi/2$? Does the $2p_y$ orbital change into a $2p_z$ orbital when Ψ increases from $\pi/2$ to over π? Please describe these changes in geometry.

11.4. Please study the curve for space dimension increment of an electron in consecutive rotations (Figure 2.13) and tell why the future of a youngster is generally bright but the course of a career is usually sinusoidal.

11.5. Please compare the trend of function $y = \arctan x$ with that of a logistic curve function in biometry (Figure Q14). (Hint: When the probability density function is $f(x) = \dfrac{1}{\pi(1+x^2)}$, the distribution function is $F(x) = \dfrac{1}{2} + \dfrac{\arctan x}{\pi}$.)

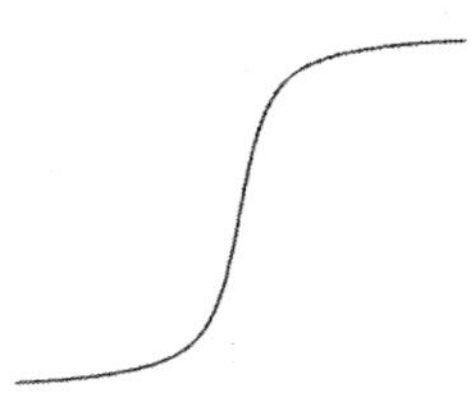

Figure Q14. The Sigmoidal curve of $y = \arctan x$ as a logistic curve in biology.

11.6. He who would behold trigonometry in the greatest glory could not find its application better than in the implementation of dynamic calculus. In a helium shell, both 1s electrons transform between each other:

$$\cos\alpha\cos\beta \Leftrightarrow -v\sin\alpha\sin\beta\,,$$

$$K = \frac{-v\sin\alpha\sin\beta}{\cos\alpha\cos\beta} = -v.$$

The value of K is independent of phases α and β, which are complementary at any moment. Please compare the property of K with that of a chemical equilibrium constant.

11.7. Circular argument: We began the discussion of entities with the assumption of harmonic oscillation. Our formulation of the law of nature by spherical quantities in dynamic calculus led to harmonic oscillation of entities. The theory is a circular argument. In contrast, the conflicting ideas developed in quantum mechanics based on linear algebra of physical quantities are themselves incompatible. For example, I don't like the fancy idea of confining a Schrödinger's cat, being dead and alive, in a wooden box. Why are good arguments always coherent and circular? (Hint: From Figure 1.3 or 8.1, the world is circular; and from Figure Q1a, the real number axis is circular, too. Therefore, the law of nature that describes the world must be circular as well.)

11.8. Review Part II of this monograph, paying attention to the scientific methods that have been applied to the subjects of discussion. In particular, Chapter 6 is an extrapolation of the theory of quaternity from atoms to molecules; Chapter 7 suggests a logical explanation of experimental results; Chapter 8 focuses on model building to define the principle of cells and life; Chapter 9 constitutes a comparative study on electronic orbitals and planetary orbits; Chapter 10 employs abductive reasoning to define the spacetime dimensions of physical quantities based on four postulates; and Chapter 11 is an integration of the new and old sciences. It is the application of diverse scientific methods that makes the theory of quaternity so powerful in elucidating the natural pattern as well as the atomic structure. Since this treatise has elucidated the atomic structure and law of electronic motion from a fresh perspective, it is of real worth and immediate interest to all men who by nature desire to know.

11.9 An audacious hypothesis about the existence of positrons in the atomic shells: According to Stueckelberg (1905-1984) and Feynman (1918-1988), an electron moving backwards in time can be interpreted as a positron. In a neon shell, there are four electrons undergoing counterclockwise rotation $\int(-\partial\Phi/\partial t)dl$ as well as four electrons undergoing clockwise rotation $\int(-\partial\Phi/\partial l)dt$. If dynamic differentiation with respect to time is regarded as the time direction, then dynamic integration over time must proceed against time. This leads to the hypothesis that those electrons in clockwise rotation are indeed positrons. Such a radical idea breaks the traditional conception on the atomic structure that negatively charged electrons are moving around the positively charged nucleus comprising protons and neutrons. Based on this hypothesis, what consequences can you predict on the theoretical construct of an atom? (Hint: In a helium atom, the nucleus will set the ground of spacetime dimensions for the interplay between a pair of electric and magnetic fields confined within the atomic shell. The pair of electric and magnetic fields are a certain manifestation of a pair of positron and electron. This hypothesis, awaiting to be tested, is the last point but not the least important one I try to convey to the readers.)

Appendix II. Glossary of Key Concepts

The invention of a fundamental theory from scratch invariably requires large numbers of new terminologies. In order to avoid misunderstanding, it is important to be aware of the fresh, non-classical meanings of the following words and phrases that we have deliberately adopted in this book.

π-bond – In the quaternity theory, a π-bond is the partial overlap of two $2p_y$ flat sectors in one atom with two $2p_y$ flat sectors in another atom, or the partial overlap of two $2p_z$ hemispheres in one atom with two $2p_z$ hemispheres in another atom at the edges. Contrary to a σ-bond, a π-bond is a directionally oriented and unsaturated covalent bond.

Anisotropic orbitals – Electronic orbitals unequally distributed along the X, Y, and Z directions geometrically. Specifically, a $2p_x$ orbital is a point displacement along the X-axis; a $2p_y$ orbital is a semicircular arc on the X-Y plane; a $2p_z$ orbital is a hemispherical surface with a vertex on the Z-axis; and a 2s orbital is a solid sphere.

Asymmetric carbon – A carbon atom bonded to four different substituent groups giving the molecule optical activity.

Atomic orbital – An unoccupied spacetime dimension in an atomic sphere. When occupied by an electron, it is called an electronic orbital.

Atomic shell – The continuous spacetime occupied by the characteristic electrons in the same period of the traditional periodic table of elements. There are seven atomic shells, namely the helium, neon, argon, krypton, xenon, radon, and oganesson shells.

Atomic shell octet configuration – The configuration of eight atomic shells including hydrogen, helium, neon, argon, krypton, xenon, radon, and oganesson shells in an oganesson atom analogous to the electron octet in a neon shell. Each atomic shell corresponds to an inert atom except hydrogen.

Atomic spacetime – Spacetime within an atomic sphere where time and space components of electrons obey dynamic calculus instead of linear algebra. For example, a helium shell is two-dimensional atomic spacetime, and a neon shell is four-dimensional atomic spacetime.

Atomic structure – The dimensional composition of time and space within an atomic shell, or the organization and relationship of electronic orbitals within an atomic shell based on spherical quantities in dynamic calculus.

Central force – In classical mechanics, a central force on an object is a force whose magnitude only depends on the distance r of the object from the origin and is directed along the line joining them. Gravitational force and Coulomb force are two familiar examples.

Centripetal force – The force that guides the surrounding bodies along circular or elliptical orbits; or the force that attracts electrons in various harmonic oscillations in the atomic sphere. Centripetal force might and might not be equal to the central force depending on circumstances.

Centromere – The state point that connects with four chords (two time and two space components) and moves from one end to the other along a semicircular arc in a dimension diagram during the rotation of a spherical quantity.

Chirality – The structural characteristic of a molecule that makes it impossible to superimpose it on its mirror image. A chiral carbon center results from the anisotropic orbitals of $2p_x2p_y2p_z2s$ in a carbon atom bonded to four different substituents.

Circular complex function – A complex function comprising a complete set of either time or space dimensions within an inert atomic shell. Every two adjacent terms of a circular complex function bear a differential or integral relationship.

Clockwise rotation – A differential operation with respect to a space dimension along with an integral operation over a time dimension upon a spherical quantity in the meantime.

Complementarity of orbitals – The relationship between two electronic orbitals sharing contiguously a full spatial sphere but mutually exclusive at any moment during the whole period of their transformations, such as the relationship between $2p_x$ and $3p_z$ electrons and that between 1s and $3d^{1s}$ electrons.

Complementarity of space and time – A spacetime principle in which the number of space dimensions plus that of time dimensions equals four in a four-dimensional electronic orbital. Reducing a time dimension inevitably increases a space dimension during the spacetime trade-off, and vice versa.

Complex function – A wave function that spans two space or time dimensions with a dimension factor as an imaginary unit.

Continuous variable – A variable that at least theoretically can assume an infinite number of values between any two fixed points in its valid range.

Coordination operation – A dot operation between both time and space circular complex functions to produce a complete set of spherical quantities in an inert atomic shell. It is different from multiplication. When both circular complex functions are placed parallel, only adjacent terms between both functions are associated in the coordination operation (see Figure 5.8).

Course of dynamic calculus – The semicircular arc in a dimension diagram, along which dynamic calculus is performed upon a spherical quantity. The course of an electron in quaternity coordinates has diverse manifestations in geometric paths.

Density gradient of an electron – An electronic orbital in rotatory operation from one state to another resulting in the ratio change of its time component to space component during the process.

Differential chain rule – A differential rule that involves wave functions in two spherical layers geared together. For example, $-\frac{d\Theta}{dt} = -\frac{d\Theta}{d\Psi} \cdot \frac{d\Psi}{dt}$ where $-\frac{d\Theta}{dt}$ initiates wave functions for the electron octet in a neon shell while $-\frac{d\Psi}{dt}$ indicates wave functions in a helium shell (see Equation 2.34).

Dimension – A diameter in a dimension diagram for an electron to traverse via its subtending semicircular arc during a rotatory operation, or the semicircular arc that corresponds to the diameter; the space and time span of an entity in simple harmonic oscillation; or a circle except a standpoint that corresponds to a real number axis.

Dimension diagram – A diagram showing a time and/or a space dimension as a diameter of a semicircle and two orthogonal time or space components as two chords subtended by the semicircle, both chords and the diameter always forming a right triangle that observes the Pythagorean theorem. A dimension diagram indicates a spherical quantity in dynamic calculus.

Dimension encapsulation, dimension curling up – The mechanism by which dimensions of an inner shell are hidden in radian angles from an outer spherical layer. For example, in a neon shell, a spherical quantity $\Phi_{00} = A_1 A_2 \cos\Psi \cos\psi$ encapsulates time and space dimensions of a helium shell in radian angles Ψ and ψ respectively.

Dimension factor – A gateway function serving as a dimension identifier or an imaginary unit when placed before a trigonometric function. But it may be regarded as a differential or integral operator when associated with a spherical quantity (a Greek letter) directly.

Discrete variable – A variable that has only certain fixed numerical values, with no intermediate values possible in between, such as the dimensions of physical quantities.

Double test principle – The suggestion that a scientific hypothesis to become a theory should pass rational or mathematical proof as well as experimental verification. This redundancy check is necessary to ensure the accuracy of a theory.

Dualism – The view that the world consists of or is explicable as two fundamental entities, such as mind and body. Under the spherical view, mind and body are represented by time and space components of a spherical quantity respectively, which are coordinated together in a rotatory operation.

Duality equation – A partial differential equation describing the spherical quantity of electrons in two dimensions within a helium shell: $\frac{\partial^2\Omega}{\partial t^2} = v^2 \frac{\partial^2\Omega}{\partial l^2}$.

Dynamic calculus – A differential or integral operation on a spherical quantity implemented by smooth rotation of a radian angle in sine or cosine function for $\pi/2$ displacement, which is conceptually different from infinitesimal calculus; a ratio whose value changes sinusoidally. Dynamic calculus stands for a physical process of harmonic oscillation rather than a mathematical limit.

Electron cloud – The vibrating medium of an electron in the atomic sphere. An electron within an atom is normally distributed over a large region of space as an electron cloud even though it is always a particle when actually detected by instruments.

Electronic orbital – An electron in harmonic oscillation within an atom (such as helium and neon) that can be characterized by a spherical quantity in dynamic calculus.

Electronic rope – Two or more electrons in a spherical layer twisted circularly by rotatory operations. For example, if we regard rotatory operator $\int(-\partial/\partial t)dl$ upon an electron as a kind of twisting tension, then electronic transformations of $\Omega_0^0 = \Omega_2$ and $\Omega_2^0 = \Omega_0$ constitute the art of making a rope with Ω_0 and Ω_2 as two partaking strands.

Electronic wave – Although an electron is traditionally regarded as a particle, it is a wave of an electron cloud characterized by a wave function in the atomic sphere.

Endosymbiosis – According to endosymbiotic hypothesis by Lynn Margulis, self-replicating cellular organelles such as mitochondria and chloroplasts were originally free-living organisms that entered into a symbiotic relationship with nucleated anaerobic cells. Mitochondria are believed to descend from aerobic bacteria, while chloroplasts derive from cells that were similar to modern cyanobacteria. Endosymbiosis can be mathematically expressed by the model of the complementary relationship between 1s and $3d^{1s}$ electrons in a zinc atom.

Entity – An object or a living organism that exists as a single and complete unit such as a photon, an electron, an atom, a DNA molecule, a cell, a bacterium, an organ, a plant, an animal, the earth, the solar system, and the universe. Harmonic oscillation is the theme of the behavior of an entity.

Euclidean space – A linear vector space that satisfies certain properties of the scalar product operation on pairs of tangent vectors.

Feynman diagram – A diagram illustrating the indirect interaction of electrons through emitting and receiving a virtual photon in between.

Five element theory – An ancient Chinese theory regarding five elements (metal, water, wood, fire, and earth) as the basic constituents of matter. In traditional Chinese medicine, five characteristics or five phases are derived from five elements to explain the interactions and relationships between biological phenomena.

Gateway function – An imaginary unit serving as an intermediate variable in calculus chain rules and as a connector between spherical quantities of an inner sphere and an outer sphere within an atom. For example, function $\dot{\Psi}$ in spherical quantity $A_1 A_2(\cos\Psi\cos\psi - \dot{\Psi}\sin\Psi\cos\psi)$ is a gateway function, a portal from the neon shell to the helium shell.

Gateways combination operator – A subscript symbol "$_\otimes$" placed between two gateway function variables such as $\overline{\Gamma}_{i\otimes}\overline{\Phi}_j$ to denote valid d-orbitals and f-orbitals where $\overline{\Phi}_j$ attaches to wave function $\overline{\Gamma}_i$.

Geodesic curve – A straight line in Euclidean space, which can be characterized as the shortest curve between two points. The whole geodesic curve on a spherical surface is a circle dividing the globe into two equal hemispheres.

Geometric elements – The basic elements of geometry such as points, lines, planes, and solids. In four-dimensional space, geometric elements refer to points, semicircular arcs, hemispherical surfaces, and solid spheres in a spherical layer.

Gonad – See sex organ.

Gyroscope – A device consisting of a spinning mass, typically a disk or wheel, mounted on a base so that its axis can turn freely in one or more directions and thereby maintain its orientation regardless of any movement of the base. Under the spherical view, we replace each direction of movement with an oscillatory cycle or a ring. A human may be regarded as a multiple-ring gyroscope with each ring representing a physical, physiological, or psychological cycle in a certain manner.

Harmonic oscillation, simple harmonic oscillation – Two-dimensional uniform circular motion or its diverse projections in one, two, three, and four space dimensions; physical motion that can be characterized by a spherical quantity in dynamic calculus; or the motion that is purely sinusoidal.

Helium shell – The spacetime sphere containing both 1s electrons.

Imaginary unit – A dimension factor that functions as the imaginary time unit i or the imaginary space unit j in a complex function.

Integral chain rule – The counterpart of the differential chain rule in dynamic calculus that involves integral operations in two spherical layers. For example, $\int A_2 \cos\psi dl = \frac{dl}{d\psi}\int A_2 \cos\psi d\psi = -\frac{1}{\psi'}(A_2 \sin\psi) = A_2 \sin\psi \cdot \int C_2 \cos\beta dl$ where $\int A_2 \cos\psi dl$ is an integral operation over a space dimension in a neon shell and $\int C_2 \cos\beta dl$ is an integral operation over a space dimension in a helium shell triggered by $-\frac{1}{\psi'}$ after the fullness of $\int A_2 \cos\psi d\psi$ dimension integration, both forming a chain.

***LC* circuit** The oscillatory circuit composed of a capacitor and an inductor disregarding electric resistance along the wire. A DNA molecule is a stepwise *LC* circuitry.

Meiosis – The division of a complete set of spherical quantities in an inert atomic shell into two haploid circular complex functions. It is somewhat analogous to the meiosis of cells because time and space variables are separated during the process and hence the number of variables is reduced by half as a result.

Membrane function – The wave function constructed by encapsulating an atomic shell, functioning as a membrane of the shell as plasma membrane encloses the cytoplasm in a cell.

Metor – A spherical quantity in the second spherical quadrant.

Mind and body relationship – The relationship between time and space components of a spherical quantity in a rotatory operation. Mind is the time component in dynamic differentiation with respect to a time dimension while body is the space component in dynamic integration over a space dimension, both being coordinated together inherently.

Natural law, the law of nature – The rules and properties of spherical quantities in dynamic calculus, in stark contrast to physical laws based on physical quantities.

Natural pattern – The order of entities created and governed by the law of nature.

Neon shell – The spacetime sphere containing the electron octet of $2s^2 2p^6$.

Newtonian time – Absolute, true and mathematical time, of itself, and from its own nature, flows equably, without relation to anything external; Uniform circular motion of time counted by a clock in contrast to sinusoidal time of spherical quantity in the spherical view.

Orbital hybridization – A traditional but wrong hypothesis that when carbon is forming an organic compound, its 2s and 2p orbitals may mix together to produce orbitals of a different property. There are three possible types of orbital hybridization, namely sp^3, sp^2, and sp hybridization.

Orbital synchronization – The way electronic orbitals of various kinds such as $2p_x$, $2p_y$, $2p_z$, and 2s within a neon shell transform by rotatory operations with their spherical quantities proceeding at the same phase at any moment.

Orthogonal – In atomic spacetime, spherical quantities Φ_0 and Φ_1 are orthogonal if they are at an exact dimension interval or have a whole differential or/and integral relationship. A spherical quantity may transform into its orthogonal spherical quantity. This is different from conventional orthogonality in Euclidean space where orthogonal physical quantities are independent and inconvertible.

Ovum – The haploid circular complex function of the space component such as ψ.

Partial rotatory operation, partial rotation – The partial effect of a rotatory operation relative to an observer when the rotatory operation does not proceed swiftly enough from the viewpoint of the observer.

Path of vector calculus – A line segment, a semicircular arc, a hemispherical surface, or a solid sphere along which a vector calculus is carried out in one, two, three, or four space dimensions respectively. This is the geometric path of an electron in Euclidean space. There are paths for vector differential calculus as well as paths for vector integral calculus.

Periodic table of physical quantities – A periodic table listing all physical quantities according to their space and time dimensions in parallel with Mendeleev's periodic table of chemical elements (see Table 10.5).

Physical laws, the laws of physics – The rules and properties of one or more physical quantities; or the relationships that govern two or more physical quantities.

Physical quantity – A quantity that assumes a fixed dimension and a continuously varying magnitude.

Polor – A spherical quantity in the first spherical quadrant.

Process mathematics Dynamic calculus of spherical quantities that expresses the development of events, such as electronic motion and biological activity.

Quaternity axes – Four dimensional factors *1*, *u*, u^2, and u^3 that serve as spacetime axes in the quaternity coordinate system.

Quaternity coordinates A schematic diagram that adopts four dimensional factors as four axes (*1*, *u*, u^2, and u^3) to illustrate the intricate dimensional relationships between various spherical quantities in dynamic calculus instead of facilitating metric measurements between positions in Cartesian coordinates.

Quaternity equation – A partial differential equation describing spherical quantities of electrons in harmonic oscillations in four dimensions in a neon shell: $\frac{\partial^4 \Phi}{\partial t^4} = u^4 \frac{\partial^4 \Phi}{\partial l^4}$.

Quaternity periodic table of elements – A periodic table of chemical elements tabulated according to the property of their characteristic electrons in quaternity spacetime under the columns of the scalors, the polors, the metors, the vitors, and the scalors (Table 5.1).

Quaternity perspective – The perspective of observing an entity from outside its activity sphere and at a proper distance and time interval that match the size and oscillating rhythm of the entity.

Quaternity spacetime – Four-dimensional spacetime in a neon shell characterized by the polor, the metor, the vitor, and the scalor.

Quaternity space – The notion of four space dimensions (including points, semicircular arcs, hemispherical surfaces, and spheres) within a spherical layer.

Quaternity, the quaternity theory, the theory of quaternity – A four-dimensional spacetime theory that defines the electronic orbitals within a neon shell by spherical quantities in dynamic calculus. Quaternity also refers to four geometric properties (point, line, surface, and solid) of oscillating electrons within atoms, more complete than particle-wave duality.

Rotatory operation, rotation, counterclockwise rotation – A differential operation with respect to a time dimension along with an integral operation over a space dimension upon a spherical quantity in the meantime.

Rotatory vector – A position vector on the X-Y plane that rotates about the origin O for π/2 to represent the smooth transformation of a spherical quantity from the X dimension to the Y dimension.

Scalability – The feasibility of repeatedly using four spherical quantities (the polor, the metor, the vitor, and the scalor) in sequence to characterize a multilayered sphere such as a krypton atom.

Scalor – A spherical quantity in the fourth spherical quadrant.

Selective electron transfer – Electron transfer through an asymmetric carbon along a selective pathway from a covalent bond to another directed by the chirality of the carbon atom in an organic molecule.

Semicircle – A semicircular arc subtending a diameter that connotes the course of an electron in harmonic oscillation or the course of dynamic calculus.

Set – A collection of objects (or elements) or a container of objects. In this book, a set refers to a part of the spacetime continuum within an atomic sphere.

Sex organ, reproductive organ – The gateway function $\dot{\Psi}$ in spherical quantity $A_1A_2(\cos\Psi\cos\psi - \dot{\Psi}\sin\Psi\cos\psi)$, which is capable of initiating or releasing a haploid sperm circular complex function $C_1(\cos\alpha - \omega\sin\alpha)$ in a male; or the gateway function ψ' in spherical quantity $A_1A_2(\cos\psi\cos\Psi - \psi'\sin\psi\cos\Psi)$, which is capable of initiating or releasing a haploid ovum circular complex function $C_2(\cos\beta + r\sin\beta)$ in a female.

Space – True physical existence that transforms in a certain dimension sinusoidally as a chord subtended by the arc of uniform circular motion in the dimension diagram.

Spacetime – The combinations of complementary space and time dimensions in a spherical layer.

Spacetime continuity – The property of atomic spacetime where space and time components of electrons are transforming smoothly and continuously from one state to another without any interruption.

Spacetime continuum – A continuous spacetime sphere as a whole relatively independent of its environment such as a neon atom.

Sperm – The haploid circular complex function of the time component such as Ψ.

Sphere – Traditionally, a sphere is a three-dimensional surface, all points of which are equidistant from a fixed point, but in atomic spacetime, a sphere always refers to a solid spherical object, not a surface. A helium sphere is one-dimensional in atomic space, and a neon sphere is four-dimensional in atomic space.

Spherical layer – An outer sphere minus a concentric inner sphere or the medium between two concentric spherical surfaces. Each spherical layer comprises two atomic shells such as neon and argon shells except spherical layer 1, which comprises two 1s electrons in a helium shell (see Table 5.2). However, in case of not misunderstanding, the neon shell is usually called a spherical layer, too.

Spherical quadrants – The four regions into which a sphere is divided by four spacetime axes (1, u, u^2, and u^3), designated first, second, third, and fourth, counting counterclockwise from the region bounded by the 1 and u axes.

Spherical quantity – A quantity that assumes a fixed magnitude and a continuously varying dimension, complementary to a physical quantity. True physical existence that transforms by dynamic calculus instead of linear algebra. (Please also refer to time and space definitions).

Spherical view – The scientific approach of adopting spherical quantities in dynamic calculus to characterize electrons within atoms; the point of view that entities transform by dynamic calculus of spherical quantities.

Symmetry of space and time – The equivalence of time and space components of a pair of electrons of similar orbital type. For example, within a helium shell, both Ω_0 and Ω_2 electrons are symmetric because the time component of one electron is equivalent to the space component of the other electron, and the space component of one electron is equivalent to the time component of the other electron. Within a neon shell, both 2s electrons are symmetric, so are both $2p_x$ electrons, etc.

Time – True physical existence that transforms in a certain dimension sinusoidally as a chord subtended by the arc of uniform circular Newtonian time in the dimension diagram.

Uniform circular motion – The kinematic movement of an object in a circle at a constant speed. The projection of uniform circular motion onto a one-dimensional profile is harmonic oscillation.

Virtual photon – The hypothetical photon through which electrons interact with each other within an atomic sphere. It physically implements the synchronization of electrons within the atom by dynamic calculus.

Vitor – A spherical quantity in the third spherical quadrant.

Wave function – A physical quantity satisfies a certain wave equation for describing the motion of a particle; or a spherical quantity expressed in sine and/or cosine functions.

Yin and yang theory – A fundamental theory underlying traditional Chinese medicine. Yin and yang are two complementary, interdependent, and opposite properties or entities within a greater whole. Yin refers to feminine, passive, and negative principle in nature; and yang refers to masculine, active, and positive principle in nature. They are constantly transforming into each other.

Zygote – The diploid cell produced by the fusion of haploid gametic nuclei. Mathematically it may be expressed by $B_1 \cos \Psi \cdot B_2 \cos \psi$, which encapsulates the circular complex functions of time and space, i.e., a sperm Ψ and an ovum ψ.

Appendix III. The Law of Nature in Mathematics

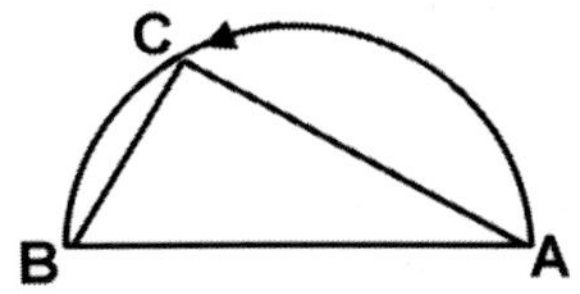

We inherit the ancient Chinese worldview that the earth is a square land under the domed sky and the Chinese philosophy of being smooth and harmonious outside while sticking to a rigid principle inside, but give them a fresh interpretation under a dimension diagram. As a point C traces the arc uniformly, the displacements of AC and BC are always measured by chords. Diameter AB is a dimension that C traverses via its corresponding semicircular arc. At any moment, AC combines with BC to represent the dimension. The law of nature is spherical quantities in dynamic calculus, which invariably describes an entity tracing the arc of the semicircle, but manifesting in diverse geometric forms. Here we list some of them that have been shown in the text. Please read the contexts and see how they are fundamentally equivalent as the paradigms of dynamic calculus. Good arguments are all compatible. The law of nature must be simple to allow diversity and single to achieve unity. In this way, it rules the physical world robustly and unifies the biological kingdoms harmonically. If there is anything that we have achieved with the invention of dynamic calculus in this monograph, then it must be the grand unification of the following various equations in the spherical view.

Harmonic oscillation equations:

$$\frac{d^2\Psi}{dt^2} = -\omega^2\Psi , \tag{1.1}$$

$$\frac{\partial^2\psi}{\partial l^2} = -\frac{1}{r^2}\psi . \tag{1.4}$$

Faraday's law, one of Maxwell's equations:

$$\nabla \times \mathbf{E} = -\frac{\partial \mathbf{B}}{\partial t} , \tag{1.21}$$

$$-\frac{\partial \Omega_0}{\partial t} = \frac{\partial \Omega_2}{\partial l} , \tag{1.19}$$

$$-\int \Omega_0 dl = \int \Omega_2 dt . \tag{1.20}$$

The Duality equation vs. Schrödinger's equation and Maxwell's equations in free space:

$$\frac{\partial^2 \Omega}{\partial t^2} = v^2 \frac{\partial^2 \Omega}{\partial l^2}, \tag{1.13}$$

$$i\hbar \frac{\partial \xi}{\partial t} = -\frac{\hbar^2}{2m} \frac{\partial^2 \xi}{\partial x^2}, \tag{1.38}$$

$$\nabla \times \mathbf{E} = -\frac{\partial \mathbf{B}}{\partial t}; \ \nabla \times \mathbf{B} = \frac{1}{c^2} \frac{\partial \mathbf{E}}{\partial t}. \qquad (10.3, 10.5)$$

Dynamic calculus of circular functions:

$$\frac{d}{d\alpha} \cos\alpha = \cos(\alpha + \frac{\pi}{2}), \tag{1.54}$$

$$\frac{d}{d\beta} \sin\beta = \sin(\beta + \frac{\pi}{2}); \tag{1.55}$$

$$\int \cos\alpha \ d\alpha = \cos(\alpha - \frac{\pi}{2}), \tag{1.56}$$

$$\int \sin\beta \ d\beta = \sin(\beta - \frac{\pi}{2}). \tag{1.57}$$

The symmetry of space and time:

$$-\frac{\partial \Omega_0}{\partial t} = \frac{\partial \Omega_2}{\partial l}; \ -\frac{\partial \Omega_2}{\partial t} = \frac{\partial \Omega_0}{\partial l} \qquad (1.19, 1.63)$$

$$\begin{cases} \Phi_0 = A_1 A_2 (\cos\Psi \cos\psi - \dot{\Psi} \sin\Psi \cos\psi); \\ \Phi_4 = A_1 A_2 u^4 (\cos\psi \cos\Psi - \dot{\psi} \sin\psi \cos\Psi) \end{cases} \qquad (2.18, 2.25)$$

Density distribution of a time dimension of an electron in Euclidean space:

$$P = -\ln(R / R_0). \tag{1.75}$$

Rotatory operation (rotation is marked by superscript ^{o}):

$$\Phi^0 \equiv \int \left(-\frac{\partial \Phi}{\partial t} \right) dl, \tag{2.3}$$

$$\frac{\partial \Phi^0}{\partial l} = -\frac{\partial \Phi}{\partial t}, \tag{2.4}$$

$$\varphi_{00}^{-0} = -\frac{dA_2 \cos\psi}{dl} \int A_1 \cos\Psi dt. \tag{5.1}$$

Integral chain rule:

$$\int A_2 \cos\psi \; dl = -\frac{1}{\psi'} \int A_2 \cos\psi \; d\psi$$

$$= -\frac{1}{\psi'} (A_2 \sin\psi) \tag{2.13, 2.16}$$

$$= A_2 \sin\psi \cdot C_2 \int \cos\beta dl$$

The quaternity equation and three-dimensional Schrödinger's equation:

$$\frac{\partial^4 \Phi}{\partial t^4} = u^4 \frac{\partial^4 \Phi}{\partial l^4}, \tag{2.7}$$

$$i\hbar \frac{\partial \Phi}{\partial t} = -\frac{\hbar^2}{2m} \nabla^2 \Phi + V\Phi, \tag{2.41}$$

$$-\frac{1}{\omega^2 \Psi} \frac{d^2 \Psi}{dt^2} = \frac{1}{\psi'^4 \theta} \left(\frac{d\theta}{dl} + \frac{d^2\theta}{dl^2} + \frac{d^3\theta}{dl^3} \right) + \frac{1}{\psi'^4 \theta} \frac{d^4\theta}{dl^4}. \tag{4.51}$$

Cyclic rotations of an electron octet (rotations are marked by arrows):

$$\Phi_0 \mapsto \Phi_1 \mapsto \Phi_2 \mapsto \Phi_3 \mapsto \Phi_4 \mapsto \Phi_5 \mapsto \Phi_6 \mapsto \Phi_7 \mapsto \Phi_0. \tag{2.29}$$

Dynamic calculus of a spherical quantity in trigonometry:

$$-\frac{d\Psi}{dt} = -C_1 \cos\alpha, \tag{2.30}$$

$$-\frac{\partial \Phi_0}{\partial t} = -A_1 \cos\Psi. \tag{2.33}$$

Dimension encapsulation mechanism:

$$\Phi_{00} = A_1 A_2 \cos\Psi \cos\psi. \tag{2.12}$$

Differential chain rule:

$$\begin{aligned}
-\frac{\partial \Phi_0}{\partial t} &= -A_1 \cos\Psi, \Psi \mapsto \Psi + \frac{\pi}{2} \\
&= -A_1 \dot{\Psi} \sin \Psi \\
&= -A_1 \sin \Psi \cdot (-C_1 \cos\alpha).
\end{aligned} \tag{2.34}$$

The relationship between probability density and flux:

$$\frac{\partial P(x,t)}{\partial t} = -\frac{\partial J(x,t)}{\partial x}. \tag{2.40}$$

Spherical quantities in vector calculus:

$$\begin{cases}
\int \Phi_0 dl_3 = \int_R \Phi_{0x} dx, \\
\int \Phi_1 dl_2 = \int_C \Phi_{1x} dx + \Phi_{1y} dy, \\
\int \Phi_2 dl_1 = \iint_H \Phi_{2x} dydz + \Phi_{2y} dzdx + \Phi_{2z} dxdy, \\
\int \Phi_3 dl_0 = \iiint_V \left(\frac{\partial \Phi_{3x}}{\partial x} + \frac{\partial \Phi_{3y}}{\partial y} + \frac{\partial \Phi_{3z}}{\partial z}\right) dxdydz.
\end{cases} \tag{3.38}$$

$$\begin{cases}
-\frac{\partial \Phi_5}{\partial l_3} = \frac{\partial \Phi_5}{\partial x}, \\
-\frac{\partial \Phi_6}{\partial l_2} = \frac{\partial \Phi_{6y}}{\partial x} - \frac{\partial \Phi_{6x}}{\partial y}, \\
-\frac{\partial \Phi_7}{\partial l_1} = \left(\frac{\partial \Phi_{7z}}{\partial y} - \frac{\partial \Phi_{7y}}{\partial z}\right)\mathrm{i} + \left(\frac{\partial \Phi_{7x}}{\partial z} - \frac{\partial \Phi_{7z}}{\partial x}\right)\mathrm{j} + \left(\frac{\partial \Phi_{7y}}{\partial x} - \frac{\partial \Phi_{7x}}{\partial y}\right)\mathrm{k}, \\
-\frac{\partial \Phi_4}{\partial l_0} = \frac{\partial \Phi_4}{\partial x}\mathrm{i} + \frac{\partial \Phi_4}{\partial y}\mathrm{j} + \frac{\partial \Phi_4}{\partial z}\mathrm{k}.
\end{cases} \tag{3.39}$$

Orbital synchronization:

Green's theorem: $u\int \Phi_0 dl_3 = \int \Phi_1 dl_2$, (3.32)

Stokes' theorem: $u\int \Phi_1 dl_2 = \int \Phi_2 dl_1$, (3.34)

Gauss's theorem: $u\int \Phi_2 dl_1 = \int \Phi_3 dl_0$, (3.36)

The general Stokes' theorem: $\int_{\partial M} \Phi = \int_M d\Phi$. (3.40)

Partial clockwise rotation effect (The special relativity):

$$A_p = P\sqrt{1-\left(\frac{d}{c}\right)^2}\ . \tag{4.6}$$

Dimension factors as operators:

$$\omega\Omega = -\frac{\partial\Omega}{\partial t}, \tag{5.13}$$

$$r\Omega = \int \Omega dl\ , \tag{5.15}$$

$$\eta_i = q^i \eta_0, i = 1,2,\ldots,7\ . \tag{5.105}$$

Electronic rope:

$$\Omega_0{}^0 = \Omega_2\ ;\ \Omega_2{}^0 = \Omega_0, \tag{5.30}$$

$$-\frac{\partial p}{\partial t} = \frac{\partial E}{\partial l}\ . \tag{5.31}$$

Circular complex functions as exponential functions:

$$\Psi = C_1 e^{-\omega\alpha} = C_1(\cos\alpha - \omega\sin\alpha), \tag{5.48}$$

$$\psi = C_2 e^{r\beta} = C_2(\cos\beta + r\sin\beta), \tag{5.49}$$

$$\Theta = A_1 e^{\frac{1}{\Psi}\Psi}, \theta = A_2 e^{-\frac{1}{\psi'}\psi}\ . \tag{5.54}$$

Orbital complement in set theory:

$$\Omega_0 \cup \overline{\Omega}_0 = \Omega,\ \Omega_0 \cap \overline{\Omega}_0 = \emptyset, \tag{5.77}$$

$$\Omega_2 \cup \overline{\Omega}_2 = \Omega,\ \Omega_2 \cap \overline{\Omega}_2 = \emptyset. \tag{5.78}$$

Stepwise *LC* oscillatory circuits of DNA :

$$L\frac{d^2 I}{dt^2} + R\frac{dI}{dt} + \frac{I}{C} = 0\ . \tag{7.6}$$

Human sex identity:

$$Man = \Phi_0 = A_1A_2(\cos\Psi\cos\psi - \dot{\Psi}\sin\Psi\cos\psi), \tag{8.2}$$

$$Woman = \varphi_0 = A_1A_2(\cos\psi\cos\Psi - \psi'\sin\psi\cos\Psi). \tag{8.3}$$

The probability density of biological variables:

$$f(x) = A\sin(ax+b) + C\,. \tag{8.5}$$

The Poisson distribution corollary:

$$\frac{\partial N}{\partial x} = -N - \frac{\partial M}{\partial t}\,. \tag{8.9}$$

The eccentricity of an ellipse:

$$\cos\theta = \sin\delta\,. \tag{9.1}$$

The Pythagorean theorem:

$$(C_1\cos\alpha)^2 + \left|(-C_1\omega\sin\alpha)^2\right| = C_1^2, \tag{9.2}$$

$$(C_2\cos\beta)^2 + \left|(C_2 r\sin\beta)^2\right| = C_2^2, \tag{9.3}$$

$$\cos\alpha = \sin\beta, \tag{9.4}$$

$$\alpha + \beta = \pi/2.$$

The centripetal force upon electron cloud volume at a fixed arc (Coulomb's law):

$$F = L\frac{C_1C_2}{r^2}\cos\frac{2L}{r} = L\frac{C_1C_2}{r^2}\cos 2\beta\,. \tag{9.12}$$

Appendix IV. Special Meanings of Symbols

$\mapsto$ A counterclockwise or clockwise rotatory operation. A counterclockwise rotatory operation or a rotation is a differential operation with respect to a time dimension along with an integral operation over a space dimension: $\int -\frac{\partial}{\partial t}dl$. A clockwise rotation is a differential operation with respect to a space dimension along with an integral operation over a time dimension: $\int -\frac{\partial}{\partial l}dt$.

∇^2 The Laplacian operator.

$\nabla\times$ The *curl* operator.

-0 The superscript denotes a clockwise rotatory operation. For example, $\varphi^{-0} = \int -\frac{\partial\varphi}{\partial l}dt$.

0 The superscript denotes a rotatory operation. For example, $\Phi^0 = \int -\frac{\partial\Phi}{\partial t}dl$.

$\otimes$ The subscript denotes a gateways combination operator.

A A dimension constant.

a The Bohr radius

A_1 A dimension constant of the time component of 2s2p or 3s3p electrons; a dimension constant of the time component of an organism.

A_2 A dimension constant of the space component of 2s2p or 3s3p electrons; a dimension constant of the space component of an organism.

A_3 A dimension constant of the time component of 4s4p or 5s5p electrons.

A_4 A dimension constant of the space component of 4s4p or 5s5p electrons.

A_5 A dimension constant of the time component of 6s6p or 7s7p electrons.

A_6 A dimension constant of the space component of 6s6p or 7s7p electrons.

B Magnetic field strength.

C Capacitance.

C A dimension constant.

c The speed of light.

C_1 A dimension constant of the time component of 1s electrons; a dimension constant of a sperm.

C_2 A dimension constant of the space component of 1s electrons; a dimension constant of an ovum.

D A constant or Dimension constant.

d The speed of a partial clockwise rotating entity.

D_i A dimension constant.

E	Electric field strength.
E	Energy or work.
E_1	Energy for a 1s electron.
E_2	Energy for a 2s2p electron.
F	Force.
f	Frequency.
h	Planck's constant.
$\hbar$	Rationalized Planck's constant.
I	Electric current.
i	The imaginary unit of time.
J	Current density.
J	Flux or probability current.
j	The imaginary unit of space.
j	When placed before a dimension or unit, it shifts the orientation of that dimension to its orthogonal direction.
k	Boltzmann's constant.
L	Arc length.
L	Inductance.
l	A space dimension.
l_0	A space dimension in the fourth spherical quadrant for a 2s orbital within a neon shell; a space dimension in radial direction within a four-dimensional spatial sphere.
l_1	A space dimension in the third spherical quadrant for a $2p_z$ orbital within a neon shell; a space dimension corresponding to two parallel planes within a four-dimensional spatial sphere.
l_2	A space dimension in the second spherical quadrant for a $2p_y$ orbital within a neon shell; a space dimension corresponding to two parallel lines within a four-dimensional spatial sphere.
l_3	A space dimension in the first spherical quadrant for a $2p_x$ orbital within a neon shell; a space dimension corresponding to two symmetric points within a four-dimensional spatial sphere.
m	Meter, the SI unit of space, signifying a space dimension.
m	Reciprocal velocity for 5s5p electrons; a clockwise rotation operator.
n	Reciprocal velocity for 3s3p electrons; a clockwise rotation operator.
p	Momentum.
p_2	Momentum for a 2s2p electron.
P	Probability density.
P	Density distribution of a time dimension of an electron in Euclidean space.
p	Velocity for 6s6p electrons; a rotation operator.
Q	Electric charge.
Q	Heat.
q	Reciprocal velocity for 7s7p electrons; a clockwise rotation operator.
r	The orbital radius for 1s electrons in space; an integral operator over space.
R	The radius of a sphere.
R	Resistance.
R_0	The maximum radius of a sphere.

S Entropy.

s Second, the SI unit of time, signifying a time dimension.

T Temperature.

t A time dimension.

t_0 A time dimension in the first spherical quadrant for a $2p_x$ orbital within a neon shell.

t_1 A time dimension in the second spherical quadrant for a $2p_y$ orbital within a neon shell.

t_2 A time dimension in the third spherical quadrant for a $2p_z$ orbital within a neon shell.

t_3 A time dimension in the fourth spherical quadrant for a 2s orbital within a neon shell.

U The energy density of a 1s electron.

u A rotatory vector.

u Velocity for 2s2p electrons; a rotation operator.

V Electromotive force.

V Spatial volume of a sphere.

v Velocity for 1s electrons; a rotation operator.

W The thermodynamic probability or disorder parameter.

w Velocity for 4s4p electrons; a rotation operator.

α A radian angle for wave functions of 1s electrons in time.

$[\alpha]_D$ The magnitude and direction of rotation of plane-polarized light by an enantiomer.

β A radian angle for wave functions of 1s electrons in space.

Γ Spherical quantities for 4s4p electrons; a circular complex function of time components of 4s4p electrons; a radian angle for wave functions of 6s6p or 7s7p electrons in time.

γ Spherical quantities for 5s5p electrons; a circular complex function of space components of 5s5p electrons; a radian angle for wave functions of 6s6p or 7s7p electrons in space.

Γ_i A spherical quantity for a 4s4p electron..

γ_i A spherical quantity for a 5s5p electron.

$\overline{\Gamma}_i$ A spherical quantity for a $5d^{4s4p}$ or $4f^{4p}$ electron.

δ A radian angle measuring the property of an entity in simple harmonic oscillation deviating from kinematic movement toward matter state transformation.

ε_0 Electrical permittivity of free space.

ζ A nucleotide in the space strand of a DNA double helix.

Z A nucleotide in the time strand of a DNA double helix.

H Spherical quantities for 6s6p electrons.

η Spherical quantities for 7s7p electrons.

H_i A spherical quantity for a 6s6p electron.

η_i A spherical quantity for a 7s7p electron.

θ A circular complex function of space components of 2s2p electrons; a radian angle for wave functions of 4s4p or 5s5p electrons in space.

ϑ A rotating angle in a partial clockwise rotation.

ϑ	The angle between the positive Y-axis and the unit tangent vector in direction of increasing arc length.
Θ	A circular complex function of time components of 2s2p electrons; a radian angle for wave functions of 4s4p or 5s5p electrons in time.
$\hat{\theta}$	A circular complex function of space components of 3s3p electrons.
$\hat{\Theta}$	A circular complex function of time components of 3s3p electrons.
λ	Wavelength.
μ	Expectation of a normally distributed variable.
μ_0	Magnetic permeability of free space.
ξ	A wave function for the motion of a particle in Schrödinger's equations.
ρ	Charge density.
σ	Standard deviation of a normally distributed variable.
Φ	Magnetic flux.
Φ	Spherical quantities for 2s2p electrons.
φ	Spherical quantities for 3s3p electrons.
Φ_i	A spherical quantity for a 2s2p electron.
φ_i	A spherical quantity for a 3s3p electron.
ϕ_T	The tangential component of Φ_1 .
$\overline{\Phi}_i$	A spherical quantity for a $3d^{2s2p}$ electron.
$\overline{\Phi}_i$	A spherical quantity for a $5d^{2s}$, $4f^{4s}$ or $4f^{2p}$ electron.
ψ	A circular complex function of space components of 1s electrons; a radian angle for wave functions of 2s2p or 3s3p electrons in space; a biological haploid ovum.
Ψ	A circular complex function of time components of 1s electrons; a radian angle for wave functions of 2s2p or 3s3p electrons in time; a biological haploid sperm.
ψ'	A gateway function of a spherical quantity; a female gonad.
$\dot{\Psi}$	An angular velocity; a gateway function of a spherical quantity; a male gonad.
ω	The angular velocity for 1s electrons; a differential operator with respect to time.
ϖ	The angular velocity for an *LC* circuit.
Ω	Spherical quantities for 1s electrons.
Ω_i	A spherical quantity for a 1s electron.
$\overline{\Omega}_i$	A spherical quantity for a $3d^{1s}$ electron.

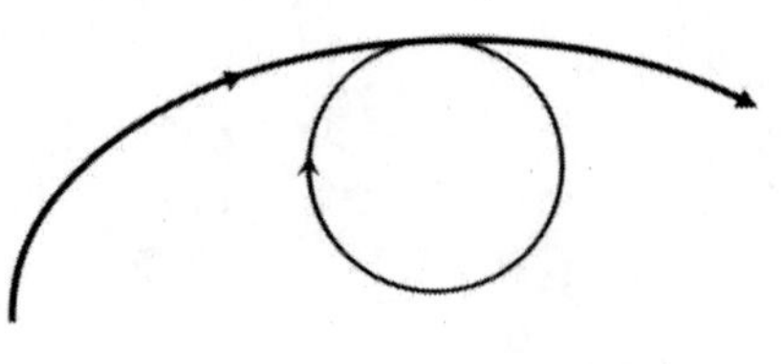

Motif 4. Taking a roundabout route, adding a new dimension, attachment, pregnancy, tangency, join, and confluence.

References

[1] Xu, K. Novel spacetime concept and dimension curling up mechanism in a neon shell, *Chinese Contemporary Thoughts Treasury*, 2005, 638-650.

[2] Xu, K. *Novel spacetime concept and dimension curling up mechanism in a neon shell*, http://arxiv.org/abs/physics/0511020. (http://xxx.lanl.gov/abs/physics/0511020).

[3] Xu, K. Electronic orbitals within a helium atom characterized by spherical quantities in dynamic calculus, *Physics Essays*, 2011, 24(4), 531-540.

[4] Xu, K. Geometrical and physical interpretations of electronic harmonic oscillations in four space dimensions, http://arxiv.org/abs/physics/0512089.

[5] Xu, K. Quaternity spacetime with dimensions curled up in 2s2p electronic orbitals, *Scientific Research Monthly,* 2007, 4, 54-64.

[6] Xu, K. A novel spacetime concept for describing electronic motion within a helium atom, *Scientific Research Monthly,* 2006, 2, 1-7.

[7] Xu, K. A novel spacetime concept for describing electronic motion within a helium atom, http://arxiv.org/abs/0705.4331.

[8] Xu, K., Ma, L., Li, C. The partitioning methods of four space dimensions in a sphere, China invention patent, publication date: Jan. 9, 2018. (in Chinese).

[9] Einstein, A.; Podolsky, B.; and Rosen, N. Can quantum-mechanical description of physical reality be considered complete? *Physical Review*, 1935, 47, 777-780.

[10] d'Espagnat, B. The quantum theory and reality, *Scientific American*, 1979, 241(5), 158-175.

[11] Lula, R. A. High manganese austenitic steels: past, present and future, *Conference Proceedings of high manganese high nitrogen austenitic steels*, edited by R. A. Lula. ASM International, 1987, 1-80.

[12] Xu, K. *The law of nature: Spherical quantity in dynamic calculus.* Nova Science Publishers, Inc. New York. ISBN 978-1-61668-345-0, April 2011, 1-261.

[13] Xu, K. *The atomic structure and law*. Nova Science Publishers, Inc. New York. ISBN 978-1-63463-420-5, 2015, 1-291.

[14] Xu, K. Discovery of four space dimensions in a sphere for electronic orbitals in a neon shell, *Journal of Physics: Conference Series*, 2021 (1739), 012028.

[15] Brewster, J. H. A useful model of optical activity I. Open chain compounds, *Journal of American Chemical Society*, 1959, 81, 5475-5483.

[16] Brewster, J. H. The optical activity of saturated cyclic compounds, *Journal of American Chemical Society*, 1959, 81, 5483-5493.

[17] Zhu, H-J; Ren, J.; Pittman Jr., C. U. Matrix model to predict specific optical rotations of acyclic chiral molecules, *Tetrahedron*, 2007, 63, 2292-2314.

[18] Parr, G. P.; Szentplay, L. V; Liu, S. Electrophilicity index, *Journal of American Chemical Society*, 1999, 121, 1922-1924.

[19] Bada, J. L.; Luyendyk, B. P.; Maynard, J. B. Marine sediments: Dating by racemization of amino acids, *Science*, 1970, 170, 730–732.

[20] Bailey, J. Chirality and the origin of life, *Acta Astronautica*, 2000, 46, 627-631.

[21] Garay, A. S. On the role of molecular chirality in biological electronic transport and luminescence, *Life Sciences*, 1971, 10 (24), 1393-1398.

[22] Xu, K. Anisotropic 2s2p orbitals as simple descriptors of the chirality of carbon centres, *Australian Journal of Chemistry*, 2016, 69, 775 -784.

[23] Xu, K. Chiral carbon center by anisotropic 2p-orbitals for selective electron transfer in biomolecules, *Scientific Research Monthly,* 2007 (5) 6-11, 25.

[24] Xu, K. The theory and application of anisotropic geometries of 2p electronic orbitals within a neon shell, *Physics Essays,* 2015, 28 (1), 33-42.

[25] Snart, R. S. Photoelectric effects of DNA. *Biopolymers*, 1968, 6, 293-297.

[26] Klein, H.; Wilke, R.; Pelargus, C.; Rott, K.; Pühler, A.; Reiss, G.; Ros, R.; Anselmetti, D. Absence of intrinsic electric conductivity in single dsDNA molecules. *Journal of Biotechnology*, 2004, 112, 91-95.

[27] Porath, D.; Bezryadin, A.; de Vries, S.; Dekker, C. Direct measurement of electrical transport through DNA molecules. *Nature,* 2000, 403, 635-638.

[28] Okahata, Y.; Kobayashi, T.; Tanaka, K.; Shimomura, M. Anisotropic electric conductivity in an aligned DNA cast film. *Journal of American Chemical Society*, 1998, 120, 6165-6166.

[29] Kasumov, A. Y.; Kociak, M.; Gueron, S.; Reulet, B.; Volkov, V. T.; Klinov, D. V.; Bouchiat, H. Proximity-induced superconductivity in DNA. *Science,* 2001, 291, 280-282.

[30] Hall, D. B.; Holmlin, R. E.; Barton, J. K. Oxidative DNA damage through long-range electron transfer. *Nature*, 1996, 382, 731-735.

[31] Fukui, K.; Tanaka, K.; Fujitsuka, M.; Watanabe, A.; Ito, O. Distance dependence of electron transfer in acridine-intercalated DNA. *Journal of photochemistry and photobiology B: Biology*. 1999, 50, 18-27.

[32] Wan, C.; Fiebig, T.; Schiemann, O.; Barton, J. K.; Zewail, A. H. Femtosecond direct observation of charge transfer between bases in DNA. *Proceedings of the National Acadamy of Sciences USA* 97, 14052-14055 (2000).

[33] Giese, B. Long-distance charge transport in DNA: The hopping mechanism. *Accounts of Chemical Research*, 2000, 33, 631-636.

[34] Giese, B.; Amaudrut, J.; Köhler, A-K.; Spormann, M.; Wessely, S. Direct observation of hole transfer through DNA by hopping between adenine bases and by tunneling. *Nature*, 2001, 412, 318-320.

[35] Nunez, M. E.; Hall, D. B.; Barton, J. K. Long-range oxidative damage to DNA: effects of distance and sequence. *Chemical. Biology*, 1999, 6, 85-97.

[36] Bixon, M.; Giese, B.; Wessely, S.; Langenbacher, T.; Michel-Beyerle, M. E.; Jortner, J. Long-range charge hopping in DNA. *Proceedings of the National Acadamy of Sciences USA*, 1999, 96, 11713-11716.

[37] Bixon, M.; Jortner, J. Long-range and very long-range charge transport in DNA. *Chemical Physics*, 2002, 281, 393-408.

[38] Schuster, G. B. Long-range charge transfer in DNA: transient structural distortions control the distance dependence. *Accounts of Chemical Research*, 2000, 33, 253-260.

[39] Xu, K. Stepwise oscillatory circuits of a DNA molecule, *Journal of Biological Physics*, 2009, 35, 223-230.

[40] Marcus, R. A. Electron transfer reactions in chemistry theory and experiment. *Journal of Electroanalytical Chemistry* 1997, 438, 251-259.

[41] Giese, B. Electron transfer in DNA. *Current Opinion in Chemical Biology*. 2002, 6(5), 612-618.

[42] Murphy, C. J.; Arkin, M. R.; Jenkins, Y.; Ghatlia, N. D.; Bossman, S.; Turro, N. J.; Barton, J. K. Long-range photoinduced electron-transfer through a DNA helix. *Science,* 1993, 262, 1025–1029.

[43] Lewis, F. D.; Liu, X.; Liu, J.; Miller, S. E.; Hayes, R. T.; Wasielewski, M. R. Direct measurement of hole transport dynamics in DNA. *Nature*, 2000, 406, 51-53.

[44] Ladik, J.; Ye, Y-J.; Shen, L. The a.c. conductivity of aperiodic DNA revisited. *Solid state communications,* 2004, 131, 207-210.

[45] Zhang, W.; Ulloa, S. E. Structural and dynamical disorder and charge transport in DNA. *Microelectronics Journal* 2004, 35, 23-26.

[46] Nogues, C.; Cohen, S. R.; Daube, S.; Apter, N.; Naaman, R. Sequence dependence of charge transport properties of DNA. *Journal of Physical Chemistry* B, 2006, 110, 8910-8913.

[47] Wetmore, S. D.; Boyd, R. J.; Eriksson, L. A. Electron affinities and ionization potentials of nucleotide bases. *Chemical Physics Letters*, 2000, 322, 129-135.

[48] Rajski, S. R.; Jackson, B. A.; Barton, J. K. DNA repair: models for damage and mismatch recognition. *Mutation Research*, 2000, 447, 49-72.

[49] Barnett, R. N; Cleveland, C. L.; Joy, A.; Landman, U.; Schuster G. B. Charge migration in DNA: Ion-gated transport. *Science* 2001, 294, 567-574.

[50] Xu, K. DNA circuit structure and charge transfer mechanism, paper ID 70110, *Oral presentation at The 6th Internatioanl Conference on Bioinformatics and Biomedical Engineering* (iCBBE 2012), Shanghai, China, May 17-20, 2012.

[51] Xu, K. DNA circuit system and charge transfer mechanism, *Engineering*, 2013, 5, 381-385, http://dx.doi.org/10.4236/eng.2013.510B077.

[52] Lovelock, J. E. *The Ages of Gaia: A Biography of our Living Earth*, Oxford University Press, Oxford, 1988, 1-80.

[53] Sagan, L. On the origin of mitosing cells, *Journal of Theoretical Biology,* 1967, 14 (3), 255–274.

[54] Darwin, C. *On the origin of species*, Oxford world's classics, Oxford University Press, 1996, 1-360.

[55] Xu, K. The unification of Pythagorean theorem for electronic orbitals with Kepler's law for planetary orbits, *Scientific Research Monthly,* 2007 (3) 13-19, 25.

[56] Xu, K. The unification of Pythagorean theorem for electronic orbitals with Kepler's law for planetary orbits, http://arxiv.org/abs/0705.4508.

[57] Xu, K. Non-constant-sum rule of electronic transformation dictated by the Pythagorean theorem, *Physics Essays*, 2015, 28(4): 522-528.

[58] Xu, K. *Discovering the spacetime towards grand unification: The theory of quaternity*. Xiamen University Press, ISBN 7-5615-2346-7/0.135, March 2005, 1-136.

[59] Xu, K. The periodic table of physical quantities by space and time dimensions, *Physics Essays*, 2016, 29(4), 546-552.

[60] van Wijk, R.; van der Greef, J.; van Wijk, E. Human ultraweak photon emission and the yin yang concept of Chinese medicine, *Journal of acupuncture and meridian studies*, 2010, 3(4), 221−231.

[61] Browne, C. Taiji variations: yin and yang in multiple dimensions, *Computers & Graphics*, 2007, 31, 142-146.

[62] Legge, D. Yin and yang surfaces: an evolutionary perspective, *Journal of Acupuncture and Meridian Studies*, 2014, 7(6), 281-290.

[63] Zhang, T.; Yan, L.; Ma, S.; He, J. Human biological rhythm in traditional Chinese medicine, *Journal of Traditional Chinese Medical Sciences* 2016, 3, 206-211.

[64] Ma, Z.; Jia, C.; Guo, J.; Gu, H.; Miao, Y. Features analysis of five-element theory and its basal effects on construction of visceral manifestation theory, *Journal of traditional Chinese medicine*, 2014, 34(1), 115-121.

Index

A

B

C

D

E

S

T